# First European Biennial Workshop on Nuclear Physics

Megève (France), March 25-29, 1991

# First European Biennial Workshop on Nuclear Physics

Megève (France), March 25-29, 1991

Edited by

**D. Guinet and J. R. Pizzi**

*Institut de Physique Nucléaire de Lyon*
*CNRS – IN2P3*
*Université Claude Bernard – ISM*
*France*

**World Scientific**
*Singapore • New Jersey • London • Hong Kong*

*Published by*

World Scientific Publishing Co. Pte. Ltd.

P O Box 128, Farrer Road, Singapore 9128

*USA office:* Suite 1B, 1060 Main Street, River Edge, NJ 07661

*UK office:* 73 Lynton Mead, Totteridge, London N20 8DH

**FIRST EUROPEAN BIENNIAL WORKSHOP ON NUCLEAR PHYSICS**



ISBN 981-02-0839-1

Printed in Singapore by Utopia Press.

*"Dulcibus quædam otiis plures labori"*

"Devote time to fun but more time to work"

*"Consacrez du temps au plaisir, mais encore plus au labeur"*

# First European Biennial Workshop on Nuclear Physics

## Megève (France) March 25–29, 1991

---

### ORGANIZING COMMITTEE

| | |
|---|---|
| D. GUINET | (Lyon) |
| R. KAMERMANS | (Utrecht) |
| H. NIFNECKER | (Grenoble) |
| J. R. PIZZI | (Lyon) |
| B. REMAUD | (Nantes) |
| W. TRAUTMANN | (Darmstadt) |

# CONTENTS

Opening Speech xiii

## I. 4π Detectors in Heavy-Ion Domain

Production and Decay of Hot Nuclei
(Neutron Ball — French–German Collaboration) 3
*U. Jahnke et al.*

Hot Nuclei with High Spin States 23
*J. Galin*

The 4π Facility at SIS/ESR First Data Taking 35
*F. Jundt*

First Results Obtained with the 4π Amphora Detection Array 52
*D. Drain*

The Rise and Fall of Multi-Fragment Emission 72
*C. A. Ogilvie*

## II. Theoretical Aspects of Nuclear Collisions

Dense Matter from Chiral and Scaling Lagrangians 85
*M. Rho*

Chaos and Order in Excitation Functions of Heavy-Ion Collisions 95
*S. Yu. Kun*

Symmetry Adapted Non-Spurious Wavefunctions
for Multi-Cluster Systems 105
*J. Katriel & A. Novoselsky*

Theoretical Investigation of Transverse Collective Motion
in Heavy-Ion Reactions Below 100 MeV per Nucleon 115
*F. Sebille et al.*

Fermionic Molecular Dynamics for Ground States and
Collisions of Nuclei 125
*H. Feldmeier & K. Bieler*

## III. Nuclear Structure and Exotic Nuclei

Exotic Beams 139
*R. Bimbot*

Decay Studies of Nuclei Far From Stability: From MeV/u to Gev/u Heavy-Ion Beams 167
*E. Roeckl*

Superdeformation in the Mass 190 Region 182
*C. W. Beausang*

$\pi$ and K Meson Production Below the Nucleon–Nucleon Threshold 200
*G. Ericsson & B. Jakobsson*

Analogy Between Ground-State Bands in Neutron-Rich A $\simeq$ 100 Isotopes and Super Deformed Bands in Heavy Nuclei 214
*G. Lhersonneau et al.*

Exotic Nuclei — The Nuclear Structure Origin of Isotopic Anomalies in the Allende Meteorite 218
*K.-L. Kratz et al.*

Spin-Flip Transition at N = 20, The $^{33}$Al $\rightarrow$ $^{33}$Si Beta Decay 227
*M. Bounajma et al.*

Isotopic A- and Isobaric Z-Distributions in Charged Particle Induced Fission 236
*P. P. Jauho*

## IV. Various Aspects of Heavy-Ion Reactions

Limiting Excitation Energy in Heavy-Ion Fusion Reactions 245
*W. Bohne et al.*

Non-Equilibrium Emission and the Decomposition of the Cross Section in Fusion-Like Reactions 263
*H. K. W. Leegte et al.*

Dynamics and Thermalization in Argon Induced Collisions Around 30 MeV/Nucleon 274
*M. F. Rivet et al.*

Proton and Subthreshold Pion Production Correlated with Light Fragments: A Statistical Process? 292
*J. L. Laville et al.*

Projectile Breakup and Nuclear Astrophysics 301
*H. Utsunomiya*

Systematics of the Power Law Parameter and the Role of Angular Momentum in the Emission of Intermediate Mass Fragments 309
*W. Trautmann et al.*

**V. Ultra Relativistic Heavy-Ion Beams**

Dilepton Production with Proton and Ion Beams at the CERN-SPS 317
*D. Contardo*

A Dilepton Spectrometer Project for SIS 330
*L. Meritet*

**VI. Future Equipment in Heavy-Ion Domain**

Agor — A Superconducting Cyclotron for Light and Heavy Ions Plans for Experimental Facilities and Physics Program 343
*S. Galès*

Eurogram: A New Gamma Spectrometer for a New Dimension of Nuclear Structure Studies 365
*F. A. Beck*

INDRA: A New $4\pi$ Detector for Charged Particles and Nuclear at the GANIL Facility 378
*E. Plagnol et al.*

# OPENING SPEECH

Welcome to Megève.

It is a pleasure for me to open the "First European Biennial Workshop on Nuclear Physics".

Why this Workshop?

Before answering, let me recall some past events. In 1966 a few physicists from l'Institut de Physique Nucléaire de Lyon decided to organize a meeting on Nuclear Physics. The first one was in 1967, which gathered nuclear physicists from various French laboratories. The aim was to discuss both theoretical and experimental subjects to explain recent results.

With the advent of the heavy-ion accelerators like SARA, GANIL or SATURNE, and recently SIS, and also in the near future the VIVITRON or AGOR, nuclear physics is now becoming international and more specially European for us. Indeed European collaborations have settled down. For example, for the GANIL or SATURNE experiments half the physicists come from European countries and some come from USA or Japan. And it is sure now that the future of nuclear physics must be planned on an European scale.

So it appeared in 1989 during the 10th meeting of Aussois that this European opening should be done. It is the reason why we are here today for the "First European Biennial Workshop on Nuclear Physics". The recent start-up of SIS at GSI in Darmstadt, the various $4\pi$ detector projects such as Eurogram and INDRA, and the many recent results in the heavy-ion domain suggested the topics of this first workshop.

I am confident of fruitful discussion among the physicists coming from the various collaborations; the speakers' quality does not let subsist any doubt on that. And I hope that we will succeed in reviewing this physics ranging from a few MeV up to GeV per nucleon.

To conclude I would like to thank the CEA and l'IN2P3 for their financial support which helped us to organize this workshop.

I wish you a good week working and skiing in the station of Megève, which is certainly the oldest French ski station.

Thank you.

J. R. PIZZI
Director of l'Institut de Physique Nucléaire de Lyon

## LIST OF PARTICIPANTS

| | | |
|---|---|---|
| ASTIER A. | I.P.N. - Lyon (France) | ASTIER@FRCPN11 |
| BECK F.A. | C.R.N. - Strasbourg (France) | BECK@FRCPN11 |
| BENRACHI S. | I.P.N. - Lyon (France) | |
| BIMBOT R. | I.P.N. - Orsay (France) | |
| BOHNE W. | H.M.I. - Berlin (Germany) | BOHNE@VAX.HMI.DBP.DE |
| BEAUSANG C.W. | Dept. of Physics - Liverpool (England) | cwb@uk.ac.liverpool.physics.gec-b |
| BOUNAJMA M. | C.R.N. - Strasbourg (France) | |
| CONTARDO D. | I.P.N. - Lyon (France) | CONTARDO@FRCPN11 |
| CAMPI X. | I.P.N. - Orsay (France) | CAMPI@FRIPN51 |
| CHBIHI A. | Ganil - Caen (France) | CHBIHI@FRCPN11 |
| CHEON T. | Dept. of Physics - Tokyo (Japan) | CHEON@TKYVAX.HEPNET |
| DURAND M. | I.S.N. - Grenoble (France) | DURAND@FRCPN11 |
| DRAIN D. | I.P.N. - Lyon (France) | DRAIN@FRCPN1 |
| DUCLOUX A. | I.P.N. - Lyon (France) | |
| DOGNY S. | I.P.N. - Orsay (France) | |
| FELDMEIER H. | G.S.I. - Darmstadt (Germany) | TP36@DDAGSI3 |
| GALES S. | I.P.N. - Orsay (France) | GALES@FRIPN51 |
| GALIN J. | Ganil - Caen (France) | GUERREAU@FRCPN11 |
| GIORNI A. | I.S.N. - Grenoble (France) | GIORNI@FRCPN11 |
| GUAITA P. | L.N.S. - Saclay (France) | |
| GUINET D. | I.P.N. - Lyon (France) | GUINET@FRCPN11 |
| HU X.C. | I.P.N. - Lyon (France) | |
| HABS D. | M.P.I. - Heidelberg (Germany) | HABS@DHDMPI5 |
| JUNDT F. | C.R.N. - Strasbourg (France) | |
| JAUHO P. | Dept. of Physics - Jyvaskyla (Finland) | JAUHO@JYFL.JYU.FI |
| JAKOBSSON B. | Div. of Cosmic - Lund (Swed) | KOSUBO@SELDC52 |
| JAHNKE U. | H.M.I. - Berlin (Germany) | |
| KATRIEL J. | Dept. of Chemistry - Haïfa (Israël) | CHR09KT@TECHNION |
| KUN S. | M.P.I. - Heidelberg (Germany) | KUN@DHDMPI50 |
| KUNDE G. | G.S.I. - Darmsadt (Germany) | GJK@DDAGSI5 |
| KRATZ K. | I.F.K. - Minz (Germany) | KLKRATZ@DMZNAT51 |
| LAVILLE M. | L.P.C. - Caen (France) | LAVILLE@FRCPN11 |

| | | |
|---|---|---|
| LLERES A. | I.S.N. - Grenoble (France) | LLRES@FRCPN11 |
| LHENRY I. | I.P.N. - Orsay (France) | LHENRY@FRIPN51 |
| LONGEQUEUE J.P. | I.N2.P3 - Paris (France) | |
| LHERSONNEAU G. | I.F.K. - Mainz (Germany) | LHERS@DMZNAT51 |
| MERITET L. | L.P.C. - Clermont-Ferrand (France) | MERITET@FRCPN11 |
| MARTELLI F. | Novarra (Italy) | |
| OGILVIE C. | G.S.I. - Darmstadt (Germany) | OGILVIE@DDAGSI5 |
| PIZZI J.R. | I.P.N. - Lyon (France) | |
| PLAGNOL E. | Ganil - Caen (France) | PLAGNOL@FRCPN11 |
| RHO M. | C.L.N. - Saclay (France) | RHO@POSEIDON.CEA.FR |
| RIVET M.F. | I.P.N. - Orsay (France) | RIVET@FRCPN11 |
| RAMI F. | C.R.N. - Strasbourg (France) | RAMI@FRCPN11 |
| ROECKL E. | G.S.I. - Darmstadt (Germany) | KC24@DDAGSI3 |
| SOUZA S. | L.N.S. - Saclay (France) | SERGIO@FRCPN11 |
| SCHUTZ Y. | Ganil - Caen (France) | SCHUTZ@FRCPN11 |
| SEBILLE F. | L.P.N. - Nantes (France) | SEBILLE@FRCPN11 |
| SELTZ R. | C.R.N. - Strasbourg (France) | |
| SANN H. | G.S.I. - Darmstadt (Germany) | SANN@DDAGSI5 |
| SIMON R. | G.S.I. - Darmstadt (Germany) | SIMON@DDAGSI5 |
| SANDOVAL A. | G.S.I. - Darmstadt (Germany) | SANDOVAL@CERNVM |
| SCONPARIN E. | Torino (Italy) | |
| TRAUTMANN W. | G.S.I. - Darmstadt (Germany) | TRAUT@DDGSI5 |
| UTSONOMIYA H. | I.P.N. - Orsay (France) | UTSUNOMI@FRIPN51 |
| WALTER G. | C.R.N. - Strasbourg (France) | WALTER@FRCPN11 |
| WESSELS J. | G.S.I. Darmstadt (Germany) | WESSELS@DDAGSI5 |
| WIENOLD T. | M.P.I. Heidelberg (Germany) | WIENOLD@DDAGSI5 |
| WILSCHUT H. | K.V.I. - Groningen (Netherland) | WILSHUT@HGRRUG51 |

# I. $4\pi$ Detectors in Heavy-Ion Domain

# PRODUCTION AND DECAY OF HOT NUCLEI

( Neutron Ball - French- German Collaboration )

U. JAHNKE[1] , S. BRESSON[2] , J.L. CHARVET[3] , J. GALIN[2] , B. GATTY[4] , D. GUERREAU[2] , D. JACQUET[4] , B. LOTT[5] , M. MORJEAN[2] , E. PIASECKI[6] , J. POUTHAS[2] , E. SCHWINN[1] , A. SOKOLOV[2] .

[1] Hahn-Meitner-Institut Berlin, Glienicker Str. 100, D 1000 Berlin 39, Germany
[2] GANIL, B.P. 5027, 14021 Caen Cedex, France
[3] CE Bruyères-le-Châtel, B.P. 12, 91680 Bruyères-le-Châtel, France
[4] IPN, B.P. 1, Orsay Cedex, France
[5] CRN Strasbourg, B.P. 20 CRO, 67037 Strasbourg Cedex, France
[6] Inst. Exp. Physics, Warsaw University, Hoza 69, PL-00-681 Warsaw, Poland

**Abstract:** A report will be given of the most important results obtained over the last three years with high-efficient 4π- neutron detectors at Ganil. The various ways to create very hot nuclei is the central issue, as well as the observation of their decay pattern. The multiplicity of evaporated neutrons, recorded eventwise in the scintillator tank, allows to deduce the excitation energy and also to select different decay modes of the hot composite nucleus. Ar- projectiles have been used over a wide range of incident energy from $E_{inc}$ = 10 to 77 MeV/u, and also more massive beams of Kr ( at 32 MeV/u ) and Pb ( at 29 MeV/u ), to induce reactions on Au-, Th- and U- targets.
With Ar- ions the energy dissipation becomes rather ineffective above 30 MeV/u : only about 25% of the available energy, or 700 MeV, is thermalized on the average in central collisions at $E_{inc}$ = 77 MeV/u. Conversely, 1000 MeV or more are accessible with Kr- and Pb- ions already at $E_{inc}$= 30 MeV/u. The decay modes change drastically from conventional fission to a complete disassembly into small fragments at very high excitation.

## I Introduction : 4π - Neutron Multiplicity Detectors

It is 3 1/2 and 2 1/2 years ago by now, respectively, since the completion of the neutron tank detectors at HMI and at Ganil. These two detectors ( and also similar detectors developed in Rochester- and Texas A & M- University ) are the members of a new generation of neutron scintillator tanks specifically dedicated to heavy-ion accelerator experiments. The very old principle of neutron detection (Re 54, Di 55), fig. 1, by slowing them down in a vast volume of scintillator liquid by elastic collisions with H-atoms and having them finally captured by cadmium or gadolinium atoms which are added in small quantities ( 0.3 to 0.4 % by weight ) to the liquid, proves to be well adapted ( Ja 82 ) to the new requirements: a large number of neutrons ( 20, 40, 60 or more ) to be counted from each single nuclear reaction. It is worthwhile to note that, in this old principle the essential inconvenience of neutron detection, namely the missing Coulomb interaction with the detector material, has been turned to advantage: the neutrons which are simultaneously released in the reaction can for that reason be stored in the detector volume for a long time up to 50 μsec and are sequentially counted during this time by the fast (n,gamma)-capture process and the low dead-time ( 50 nsec ) registration of the resulting scintillator light flashes in the photomultiplier system.

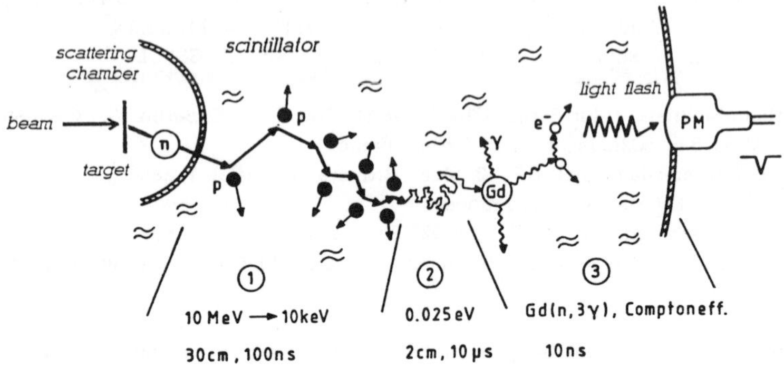


*Fig. 1: Illustration of the neutron detection with Gd- loaded liquid scintillators. Characteristic energies, lengths and times are given for the 3 steps in the process: ( 1 ) thermalization, ( 2 ) diffusion and storage and ( 3 ) radiative capture and registration of the light flash from Compton- scattered electrons by the phototube.*

The HMI - detector, fig. 2, is a sphere with 140 cm in diameter comprising 1500 l of scintillator liquid and has 24 fast 4-inch phototubes on its surface. In its middle it houses a 42 cm in diameter reaction chamber for the targets and the various charged particle detectors. With regard to the higher beam energies and heavier projectiles available at Ganil, as compared to the Vicksi- accelerator in Berlin, the Ganil- detector

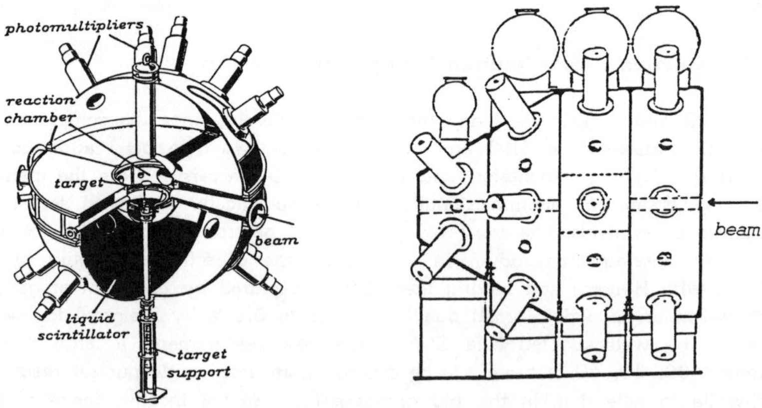


*Fig. 2 : Sketch of the two neutron detectors from HMI ( left ) and Ganil ( right ).*

has cylindrical shape made from 4 separated slices ( Ga 90 ). The scintillator thickness is thus increased in the forward direction. It has 3000 l of liquid, 22 phototubes and also a cylindrical chamber in its middle, fig. 2. Since the end of last year the Ganil-detector, called Orion, is once more being increased in size to 4000 l of scintillator volume, because the existing inner scattering chamber has to be enlarged in order to assimilate larger multidetector systems.

Both neutron detectors, at HMI and Ganil have a very high detection efficiency, fig. 3, which is normally set to 88 % for neutrons from the $^{252}$ Cf calibration source ( $E_n$= 2MeV ), thereby tolerating 0.3 to 0.7 background events during the 50 μsec counting intervall. For higher neutron energies the efficiency drops slowly, more slowly for the larger Orion - detector, because the neutrons can progressively escape from the liquid before being completely thermalized. At $E_n$ = 20 MeV, which might be considered as upper energy limit for evaporation from a hot ( T=6 MeV ) and slowly ( 2 MeV/u ) moving source, the efficiency has dropped to about 60% for the HMI- tank and to 70% for Orion. For neutron energies of 30 to 50 MeV or higher, where we expect neutrons from preequilibrium processes, the detection probability is reduced to the 30%-level, although, of course, ideally the detectors should be transparent for these neutrons, because only evaporated neutrons should be counted in order to deduce excitation energies.

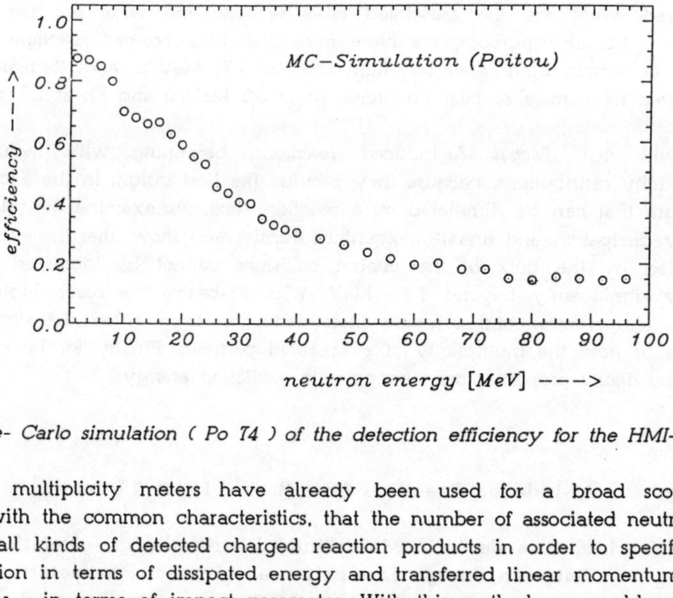


*Fig. 3 : Monte- Carlo simulation ( Po 74 ) of the detection efficiency for the HMI- tank.*

The neutron- multiplicity meters have already been used for a broad scope of experiments with the common characteristics, that the number of associated neutrons is used to tag all kinds of detected charged reaction products in order to specify the creating reaction in terms of dissipated energy and transferred linear momentum or - less stringently - in terms of impact parameter. With this method, we could separate primary and secondary projectile-like fragments ( Ja 83, Mo 91 ), light charged particles from evaporation, preequilibrium or break-up processes ( Fu 85 ) and, as a further example, fission fragments from peripheral and from central reactions ( Ja 86, Ga 88 ) and investigate their proper angular distributions ( Sc 89 ).

In these experiments, we have generally bombarded heavy target nuclei like Au, Th and U in order to strengthen the contribution of neutrons to the deexcitation chain, because we have so far no analogous $4\pi$- detector for charged particles at our disposal in order to count them for each reaction as we do for the neutrons. For these very neutron rich systems and up to high excitation energies of 500 to 750 MeV, however, the neutron contribution amounts to 70% or more, whereas only about 30% of the thermal energy is carried off by other light particles, mostly H- and He-isotopes ( see below, Ar + Th ). Heavier particles are much less abundant in the deexcitation line, and even so, their emission is substantially subsidized by large positive Q-values as in fission. For this reason, the emission of light particles is the dominant cooling mechanism and their multiplicity is largely independent of the decay modes ( evaporation, fission or multifragmentation ) of the hot composite system as well as from its exact mass number ( Fr 90 ). In this respect neutron - or light particle - counting provides a minimum bias access to the thermal excitation energy, at variance, for instance with the folding- angle method to determine the linear-momentum transfer, which relies on the binary fission decay.

The central issue of the recent Ganil- experiments and of our contribution to this workshop is the highest excitation energies: What are the most effective tools ( choice of projectile , target and incident energy ) to reach high excitations; how can they be measured; what are the associated cross sections and what are the disintegration pattern ? For this purpose, we have investigated Ar-induced reactions over a wide range of bombarding energies from 10 to 77 MeV/u and thereafter we have employed more massive beam particles, Kr at 32 MeV/u and Pb at 29 MeV/u.

We will first discuss Ar-induced reactions, beginning with inclusive neutron multiplicity distributions, because they provide the best insight in the excitation energy spectrum that can be stimulated by a reaction. Next, we examine the relation between energy dissipation and linear- momentum transfer and show, that the excitation energy provided by the bulk of the central collisions cannot be increased with incident energy significantly beyond 750 MeV. With Kr-beams we reach higher excitations, higher temperatures and compare their deduction from either the slopes of energy spectra or from the multiplicity of evaporated particles. Finally for Pb + Au, we show how the decay properties can change with excitation energy.

## II Ar-induced Reactions from Einc = 10 to 77 MeV/u

### II a Inclusive Neutron- Multiplicity Distributions - Relation Between Energy Dissipation and Impact Parameter

Inclusive neutron- multiplicity distributions have been gathered almost as a byproduct - because they are so easy to measure and contain at the same time valuable information - in many experiments with the neutron tanks at Vicksi and Ganil. They extend by now over a broad range of projectile- target combinations and incident energies. Fig. 4 shows a small sample of them for Ar-induced reactions.

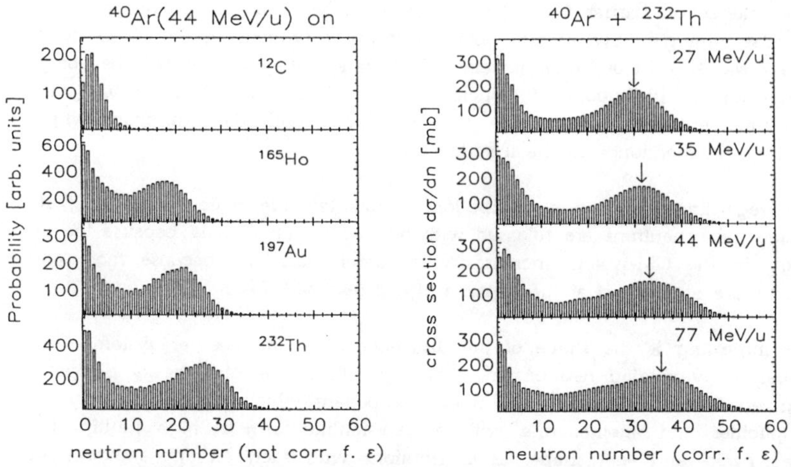


*Fig. 4 : Inclusive neutron- multiplicity distributions for Ar- induced reactions: Raw data taken at Einc = 27 MeV/u for a sample of target nuclei ( left ) and calibrated spectra (corrected for efficiency and in absolute cross section) for Einc= 27 to 77 MeV/U (right).*

We measure these distributions with the pulsed beam, because then the occurrence of a reaction is recognized by a coincidence between the beam pulse signal, indicating the arrival of the beam burst on the target, and the prompt response signal of the neutron detector. ( This prompt signal results from reaction gamma-rays and proton recoil events. The threshold on this signal is set so low, that also very gentle reactions with very few or only one neutron are safely registered. ) The coincidence opens the 50 μsec - neutron gate, during which all ( n, gamma )- capture events, i.e. the number of released neutrons are counted. Then the next cycle starts with a new beam burst on the target. The beam intensity thereby is reduced, in order to keep the probability for initiating two reactions per burst small; but already some 15 min of beam time is sufficient to collect a complete distribution.

The distributions thus exhibit the probability for the occurrence of a reaction, for which the detector has registered the number of emitted neutrons, left side of fig. 4 for 44 MeV/u Ar impinging on C, Ho, Au and Th targets. The probabilities can be converted to absolute cross sections by normalizing to the Rutherford scattering and the number of detected neutrons can be corrected for the known detector efficiency. Both has been done for the neutron number spectra on the right side of fig. 4 : Ar + Th for incident energies from 27 to 77 MeV/u. The integral of the plotted differential cross sections gives total reaction cross sections in agreement ( within the error bars of 10 to 20 % ) with the systematics ( Wi 80 ).

These distributions now characterize the energy- dissipation capability of the reaction, a global quantity which is otherwise hardly accessible. They resemble to the total kinetic energy- loss spectra which are observable for strongly damped reactions close to the interaction barrier ( Hu 76 ) or, at the other end of the energy scale, to the transverse- energy flow spectra obtained with $^{16}$ O- ions in the 10 to 200 GeV/u range ( Na 88 ). In our intermediate energy range, they can at best be compared to folding- angle distributions of coincident fission fragments which in turn reflect the linear- momentum transfer profile of a reaction - with the restrictions which have already been mentioned in the introduction.

When regarding the neutron distributions on the left side of fig. 4, we see that on the average more neutrons are released with heavier targets. This is, because the available energy in the CM-system increases with target mass and because more and more neutrons are evaporated at the expense of charged particles.

More intriguing is the shape of the distributions for the heavier systems, where the number of evaporated neutrons more closely reflects the total energy dissipation. It is composed from two contributions : one component which decreases slowly from low multiplicities and another one with a gaussianlike form at high multiplicity - the horse's back- and -tail shape, as distributions with such an appearance have been named at BNL- and Cern- energies.
This shape reflects the geometry of the collision. It comes about ( Ja 88 ), because various impact parameters contribute differently to the dσ /dn -spectrum. Touching collisions give rise to gamma- deexcitation ( zero neutron ) only. With decreasing impact parameter the energy damping and consequently the neutron number increases roughly linearly. The partial cross sections thereby decrease, due to the diminuation of the geometrical weight of the smaller impact parameters or lower partial waves. However, from a certain impact parameter on, the damping becomes maximal ( with broad fluctuations certainly ) under the influence of the mean field and hardly increases any more for still closer collisions. All these "central" collisions therefore accumulate to the gaussianlike cross section at high multiplicity in the inclusive spectra.

This simple geometrical interpretation of the shape of the inclusive multiplicity spectra is strongly supported by the observation, that this shape is largely invariant with incident energy ( cf. fig. 4 ) and with mass asymmetry in the entrance channel ( cf. figs. 10 and 14 ). The roughly equal weight of the peripheral and the central component in the spectra indicates, that the central collisions with full dissipation begin at an impact parameter b, which corresponds to about half-overlap between projectile and target nucleus, e.g. at $b = R_{target}$ . For Ar + Th this is true throughout the range incident energy from 27 to 77 MeV/u. The central- collision cross section amounts to about half the total reaction cross section, which varies from 4.7 to 5.3 barn in this range.

The above interpretation, when substantiated with a reaction model, could provide a more quantitative relation between neutron multiplicity and impact parameter.

## II b Correspondence Between Energy Dissipation and Linear Momentum Transfer

The next topic is, to relate the neutron multiplicity to fission-fragment angular correlation studies which have been pursued before for Ar + Th, in particular in the range of 31 to 44 MeV/u ( Co 85 ). We have made cross measurements between both quantities at 10 ( Sc 89 ), 27, 35, 44 and 77 MeV/u.

For the present purpose, however, we pick up one example only: at 27 MeV/u. The correlated fission fragments have been observed on both sides of the beam axis with position- sensitive solid state detectors. Their folding- angle spectrum, fig. 5, exhibits the prominent double-humped structure from peripheral and central reactions which has been observed often before ( Jc 85 ). At this incident energy, the average momentum transfer for the fusionlike peak near 110 degrees still amounts to 80%, as can be inferred from the fractional momentum scale below the figure. On top of the folding- angle distribution, we have plotted the associated average neutron- and charged-particle (sum of H- and He-isotopes) multiplicity. The neutron number grows strongly with increasing momentum transfer from 8 neutrons up to about 32 and then saturates slowly near 34 neutrons, when approaching the fusionlike maximum. The charged particles follow the tendency of the neutron number, though on a much weaker scale. They level off with 5 particles , i.e. 15% in number or 20% in reduction of excitation energy when compared to the neutral component.

The gain in particle evaporation with increasing momentum transfer is, of course, what is expected from the simple massive- transfer picture: energy and momentum dissipation is mediated by particle transfer from the projectile to the target nucleus. They are thus correlated quantities.

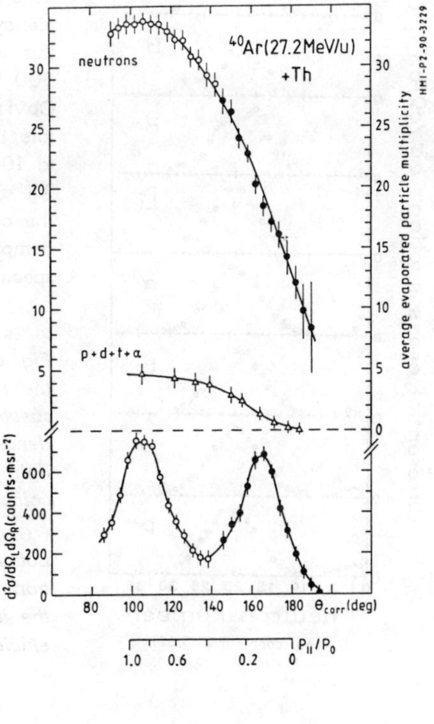


*Fig. 5 : Folding- angle distribution for Ar + Th at Einc= 27.2 MeV/u and the associated average multiplicity of evaporated neutrons ( corrected for efficiency ) and of evaporated H- and He- isotopes. The scale below the figure indicates the linear- momentum transfer $p_{II}/p_0$ . Lines are to guide the eye.*

This correlation is also reflected by the similarity of the inclusive neutron- multiplicity spectra and the folding- angle distributions. However, the similarity is restricted to the gross features only, i.e. the double-humped structure. For a closer comparison one would have to unfold the influence of the fission process ( the decreasing fission probability for very low momentum transfer, the varying angular anisotropy of fragment emission etc. ) from the folding angle distribution, in order to regenerate the undisturbed linear-momentum transfer spectrum. This hardly solvable task is presumably also the reason, why until now the analysis of angular- correlation data with reaction models has not been too successful ( Jc 90 ).
We will pick up this subject once more in the next paragraph.

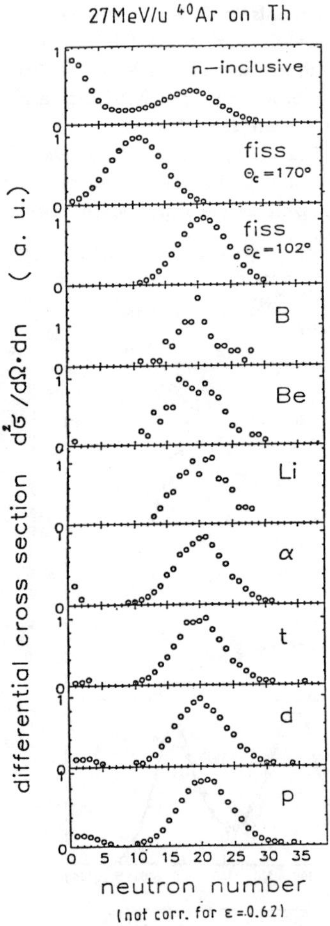


The multiplicity of charged particles shown in fig. 5, has actually not been observed directly as a function of folding angle, as has been done before ( Jc 85 ). Instead, we recorded their emission at backward angles together with the associated neutron multiplicity. Fig. 6 shows the exclusive neutron- number spectra in coincidence with Z= 1 to 5- particles. All these spectra are bell-shaped. They start out with about 10 neutrons and cumulate at 20 detected neutrons ( or 32.5 neutrons when corrected for the tank efficiency, which has been exceptionally low in this early experiment, because we used a small 500 l neutron detector from Bruyères-le-Châtel ). Obviously, no charged particles are released before the system is heated up enough, to emit up to 10 neutrons ( or up to 16, after correction ). All charged particles originate from the bulk of the central collisions, as can be inferred from the comparison with the inclusive neutron- number spectrum on top of figure 6.

*Fig. 6 : Neutron- multiplicity spectra recorded for the reaction Ar ( 27.2 MeV/u ) + Th : Inclusive spectrum and distributions observed in coincidence with backward ( $\Theta_{lab}$ = 163 ° ) emitted light particles ranging from p to Boron, and in coincidence with two correlated fission fragments ( $\Theta_{fold}$ = 102 and 170 °, see fig. 5 ). The contributions in the p, d, t and α- spectra at low neutron numbers arise from the carbon backing of the target. Neutron numbers are not corrected for efficiency.*

From the exclusive spectra, we deduce absolute multiplicities dM(p,d,t... B)/dn by correcting for the solid angle of the detector and the Jacobian, assuming isotropic emission in the frame of the source and by dividing by the inclusive n-spectrum ( Ji 89 ). These absolute multiplicities have then been incorporated in fig. 6 as a function of the associated neutron number.

Also, once more in order to illustrate the broadening involved in both experimental observables, neutron number and folding angle, we have added to fig. 6 the neutron distributions recorded on top of the two maxima in folding angle, at 102 degrees for the central and at 170 degrees for the peripheral component.

## II c Evolution of Excitation Energy and Cross Section for Central Collisions from $E_{inc}$ = 10 to 77 MeV/u

In the previous section we have discussed the close correspondence between energy dissipation and momentum transfer, or rather its experimental observables, the neutron multiplicity and the fission folding angle, on basis of the cross correlation experiment at 27 MeV/u, as an example. We will now compare the results of the two inclusive observations over the whole energy range from Einc = 27 to 77 MeV/u.

In fig. 7, we have plotted to the right the inclusive correlation- angle data, mostly from the Saclay-Darmstadt collaboration ( Co 85 ) and only supplemented at 27 and 77 MeV/u by our own measurements. The distributions have been somewhat arbitrarily normalized, to give the same yield in the maximum at large folding angles, in view of the problems connected with an absolute calibration. The left hand side of the figure exhibits the neutron- number distributions from fig. 4.



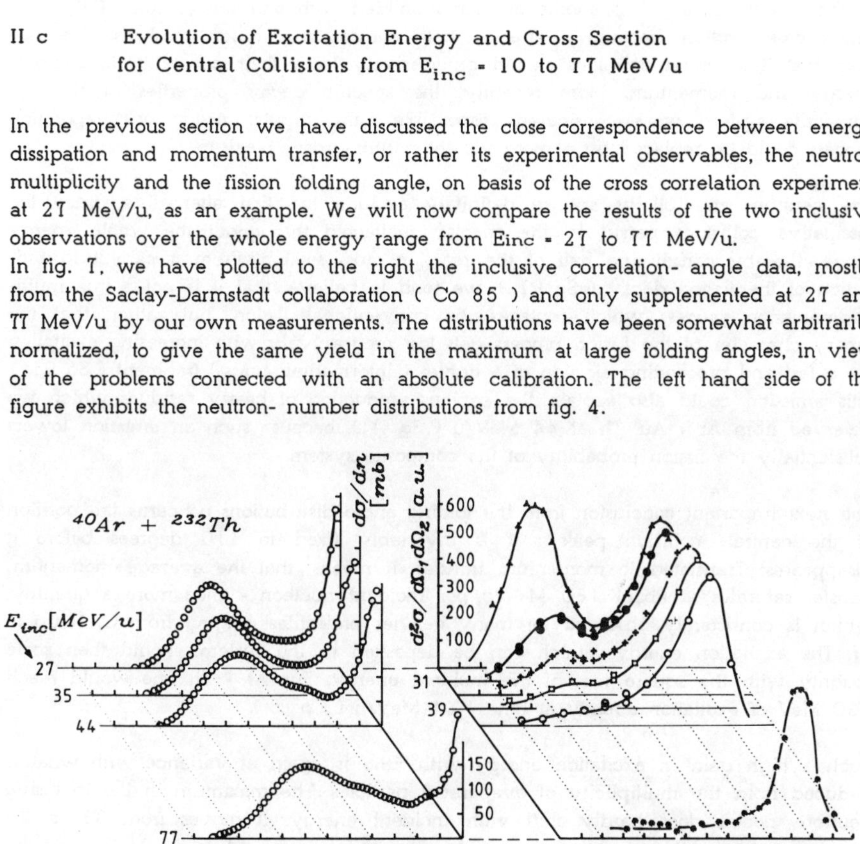


*Fig. 7 : Comparison of neutron- multiplicity spectra and folding- angle distributions for Ar + Th over the whole range of incident energy from 27 to 77 MeV/u.*

The most striking feature of the angular correlations is the disappearence of the central collision bump within Einc = 31 to 44 MeV/u. This sudden decrease has been a puzzle and also a challenge ever since its first observation about 6 years ago ( for a review see S. Leray, Le 86 ). Meanwhile, it became clear, that it is a more general property of heavy- ion reactions at intermediate energy, since it occurs also with lighter ( like N, O ) and with much more massive projectiles ( Kr ). Likewise it shows up with light low-fissile systems, where the central reaction peak disappears from the recoil- velocity spectrum of the heavy evaporation residues, once more for bombarding energies in excess of 35 MeV/u.

In the first instance, the phenomenon was identified with a disappearence of the total fusion cross section, because it affected both conventional fusion products. The idea was, that the central collisions should progressively loose their capability to dissipate energy and momentum. More recently, the specific decay properties of the hot composite system were considered responsible : the regular fission or evaporation process could be replaced by a more violent multifragment breakup.

The neutron- multiplicity spectra definitely exclude the first alternative, since the dissipative collisions persist in the spectra unaffected throughout the whole energy range, thereby constituting half of the reaction cross section. From a careful investigation of the fission decay ( Sc 91 ), we tend to believe, that it is not a true multifragmentation process which replaces the conventional fission, but rather, that the binary character of the fission process gets lost progressively with increasing excitation by a fast and preceeding ejection of a lighter ( intermediate-mass-) fragment ( So 91 ). This emission could also explain the copious production of heavy residues which we observed from Ar + Au, Th at 44 MeV/u ( Sc 91 ), because such an emission lowers substantially the fission probability of the composite system.

The next important conclusion from the folding angle distributions concerns the position of the central- collision peak : it is invariably fixed to 110 degrees before it disappeares. Translated to momentum transfer, it means, that the average momentum transfer saturates at about 180 MeV/c per incident nucleon - once more a quantity, which is consistently observed for many heavier projectiles ranging from C, N up to Kr. The excitation energy, which can be deposited in the system, should then scale roughly with the square root of the incident energy. For Ar + Th one would reach 930 MeV of excitation already at Einc = 44 MeV/u ( Co 85 ).

Such a high gain in excitation energy with Einc is again at variance with what is deduced from the multiplicity of evaporated particles. The maximum in the inclusive neutron spectra does hardly shift with incident energy. It moves from 33 to 35 neutrons, when nearly doubling the bombarding energy from 44 to 77 MeV/u. At the same time, the inactivity of the neutron emission is not counterbalanced by an enhanced evaporation of charged particles.

In fig. 8, we summarize our findings concerning the growth of light- particle evaporation with Einc . The neutral as well as the charged components augment only very slowly above Einc = 30 MeV/u. From these multiplicities, we deduce the total excitation energies which can be stimulated with the bulk of the central reactions in an iterative procedure. Step by step, we allow an energy of ( $B_n + 2T$ ) and ( $B_{ch} + 2T + V_c$ ) to be removed per neutron or charged particle, respectively, $B_n$ and $B_{ch}$ being the binding energies, $V_c$ the Coulomb barrier and $T = ( 10/A \times E_{ex} )^{1/2}$ the initially unknown temperature of the system.

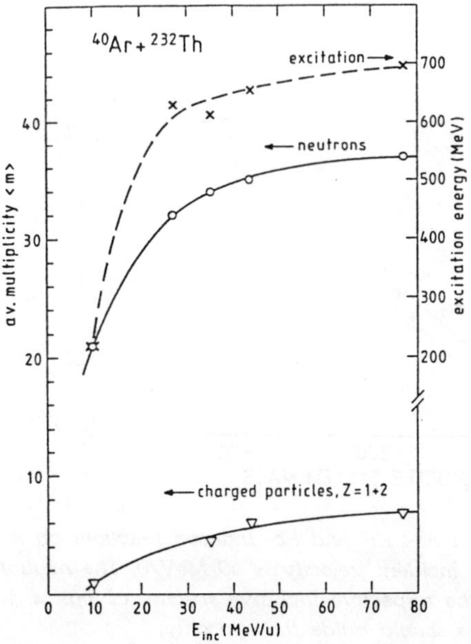


The excitation energy deduced in this manner follows the trend of the multiplicities, fig. 8 : after a strong rise till about $E_{inc}$ = 30 MeV/u, it gradually saturates near 700 MeV of excitation ( Ji 89 ). From this behavior it then seems rather unlikely that much higher excitation energies can be realized with Ar- projectiles, although it would indeed be worthwile, to replenish the present systematics towards higher beam energy.

*Fig. 8 : Evolution with incident energy of the average multiplicity of neutrons ( corrected for efficiency ) and light charged particles ( sum of Z= 1 and 2 ), released from central collisions of Ar+ Th. From these multiplicities we estimate the deposited excitation energy ( with an uncertainty of ± 100 MeV ), right hand scale. The 10 MeV/u- data are from ( Sc 89 ), lines are to guide the eye.*

It should be added, that in a recent theoretical reconsideration ( Fr 90 ) of our data a similar saturation has been found, although not quite as pronounced as originally suggested ( Ji 89 ).
Also, it is important, to note that our consideration concerns the bulk of the central-collision cross section only. By no means, can it be excluded that in the tails of the distributions, i.e. with much weaker probability, higher excitations may be reached, in contrary, it appears quite conceivable ( sect. IV ).

## III Systematics of Neutron Multiplicity from Central Reactions with Ar, Kr and Pb - Projectiles

Once we have accepted that nuclear excitations accessible with Ar- projectiles are limited to about 700 MeV or 3 MeV/nucleon in heavy systems, it seems logical to try heavier beam particles, because the totally transferred momentum could be larger, even if, on the per- nucleon basis, it might be restricted to 180 MeV/c.

For a quick overview, we compare the inclusive neutron- number spectra for reactions induced with Ar-, Kr- and Pb- beams on Au- and Th-, U- targets at the same incident velocity of about 30 MeV/u, thereby implicitly assuming, that the neutrons are in all reactions representative for the total evaporation. More specifically, we infer the neutron multiplicity from the centroids of the well pronounced central- collision bumps in these spectra, correct it for the detector efficiency which varies between 65 and 80% in the different experiments, and display it in fig. 9 versus the total mass of the collision system.

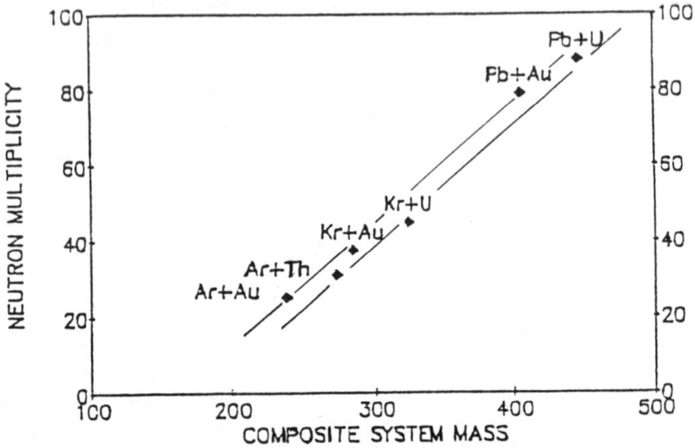


*Fig 9 : Systematics of neutron production in Ar-, Kr- and Pb- induced reactions on Au and Th, U- target nuclei at about the same incident velocity of 30 MeV/u. The neutron multiplicity is taken from the maxima in the respective inclusive spectra, cf. figs. 4, 10 and 14, and is corrected for efficiency. Lines should guide the eye only.*

The most obvious feature from this comparison is, that the neutron multiplicity increases strongly for the heavier projectiles. This increase is much faster, than that of the total number of nucleons in the system: for Ar + Th, as an example, the emitted neutrons represent 12% of all nucleons, whereas for Pb + Au (with the same N/Z proportion) this ratio is 19%, and this gain would even be more important, if it would be based on the number of nucleons in the hot composite system, which has actually been formed.

For the systems formed with Kr- ions, we deduce excitation energies around 900 MeV ( produced by the bulk of the central collisions, see next section ), 200 MeV more, than for the ones formed with Ar-ions. For Pb + Au, U we still have to postpone the determination of excitation energy, because the charged- particle evaporation has not yet been evaluated. However, since the neutron emission has roughly doubled as compared to Kr + Au, Th, whereas the total mass increased by 40% only, one might once more expect considerably higher excitation energies and higher temperatures for Pb + Au, U.

Also from fig. 9. we find systematically larger neutron numbers for the heavier targets Th and U, than for Au. This effect can be attributed to the lower available energy in the CM- system and to a slightly stronger contribution of charged particles in the deexcitation chain in the case of Au.

The most important message from this comparison is, that indeed, with heavier projectiles one can stimulate considerably higher thermal excitations than with lighter ones. Although Einc = 30 MeV/u has been chosen here for the comparison for the practical reason, that this was the highest available beam energy for Pb at Ganil, it may also be a suitable choice in view of the fact, that at a higher beam energy one may encounter similar saturation effects, as already observed with Ar- projectiles.

## IV Kr + Au at 32 MeV/u : Maximum Excitation Energies Deduced from Multiplicity and Energy Spectra of Evaporated Particles

The system $^{84}$ Kr + $^{197}$ Au has been studied at 32 MeV/u, i.e. at a moderate energy, where a strong momentum transfer of about 80% of pinc can still be expected. We focus here on the analysis of energy spectra of evaporated alpha-particles from this experiment ( Cr 91 ).

For orientation, we show in fig.10 the inclusive neutron number distribution as well as the one in coincidence with He-particles, detected with solid- state telescopes at 159 degrees. Both are raw spectra, i.e. they are not yet corrected for background, neutron detection efficiency and are not normalized to give absolute cross sections.
We recognize the familiar two-component shape in the former and the dumb-bell form in the latter distribution. The exclusive distribution is centered at n=28, verifying the origin of the alpha-particles from central collisions. From the two spectra we deduce the differential multiplicities dM(Z=2) /dn for the emission towards 109 and 159 degrees, shown in the lower part of fig.10. They still have to be corrected for the motion of the emitting source and integrated over angle to arrive at the total multiplicity, which will be addressed to later on.

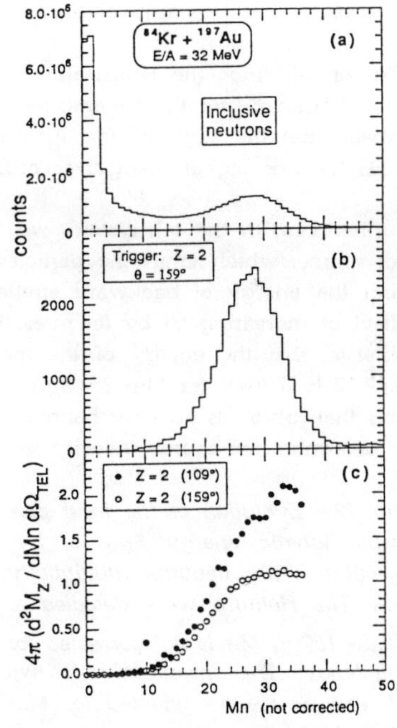


*Fig. 10 : $^{84}$Kr + $^{197}$Au at 32 MeV/u : ( a ) Inclusive neutron- number pattern, ( b ) in coincidence with backward ( 159 ° ) emitted He- particles and ( c ) differential Z= 2- multiplicity, measured at 109 and 159 °. Neutron number without efficiency correction.*

The next fig. 11 exhibits the energy spectrum of He- isotopes, observed at 159 degrees with a nearly perfect exponential slope, which extends over 3 orders of magnitude.

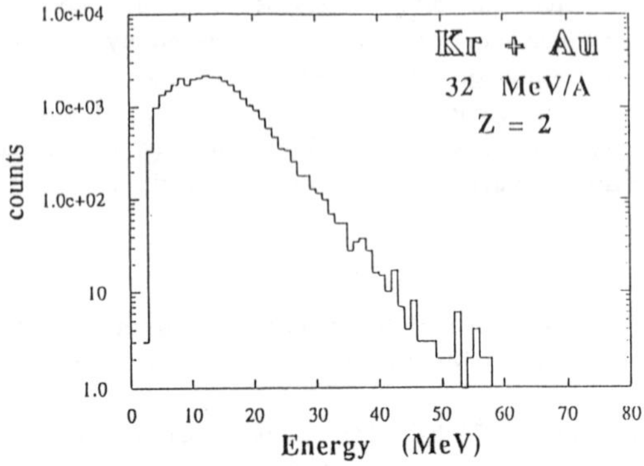


*Fig. 11 : Laboratory energy spectrum of He- particles from $^{84}Kr$ ( 32 MeV/u ) + $^{197}Au$, recorded at $\Theta_{lab}$ = 159 degrees.*

What do we expect, when we decompose the He-energy spectrum according to the number of simultaneously observed neutrons ?

First of all, since the temperatur T of the source and its size ( and thereby the Coulomb barrier $B_c$ for the He-emission ) increase with neutron multiplicity $M_n$, we would expect, that the slope of the spectra decreases and their most probable energy $E_{mp}$ shifts towards higher energy, according to $E_{mp} = B_c + T$ ( assuming surface emission ).

Furthermore, in the lab-system, we have to consider that the recoil velocity $v_s$ of the hot source, which emits the particles also increases with $M_n$, thereby, however, reducing the energy of backward emitted particles. With a projectile as heavy as Kr, the effect of increasing $v_s$ by far overwhelms the one evoked by T or $B_c$. As a result, we observe, that the energy of the maximum, $E_{mp}$, strongly decreases from 17 MeV for $M_n$= 15 to 4 MeV for $M_n$= 35; fig.12. This then gives us an easy handle to determine the source velocity $v_s$.

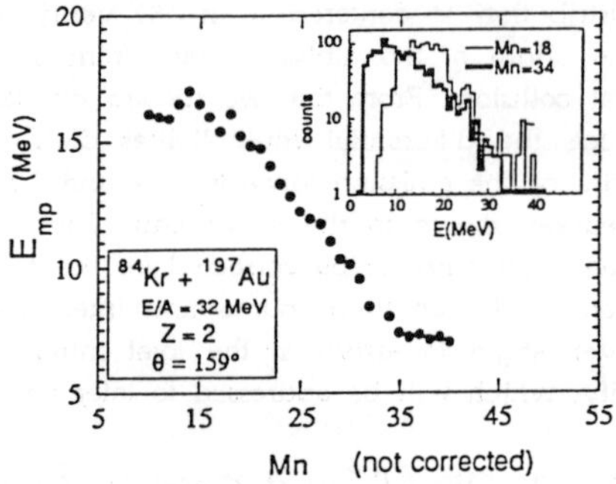


*Fig. 12 : Evolution of the most probable kinetic energy $E_{mp}$ of He-ejectiles with neutron multiplicity $M_n$. The Heliums were detected at $\Theta_{lab}$= 159 °; $M_n$ is not corrected for efficiency. The insert exhibits two He-energy spectra selected for $M_n$= 18 and 34 as examples.*

In practice, we decompose the He-spectrum according to 5 bins in $M_n$ : A ( $M_n$= 11 to 15 ), B ( $M_n$= 16 to 20 ) . . E ( $M_n$= 31 to 35 ); the weighted average $\langle M_n \rangle$ for the five bins is given in table 1, after correction for the efficiency. For the same windows, we determine $v_s$ from the shift of $E_{mp}$ and once we know $v_s$, we can also evaluate absolute multiplicities $M_Z$ for Z= 1 and 2-particles ( table 1 ). From $M_n$ and $M_Z$ in turn, we calculate the excitation energy $E_{ex}$ ( section IIc ) and therefrom the temperature T with the simple relation $E_{ex} = A/10 \times T^2$ . The last step then serves as a cross check for these temperatures T from particle multiplicity: the five energy spectra are transformed into the reference system of the emitting source and apparent temperatures $T_{fit}$ are extracted from Maxwellian fits, fig. 13.

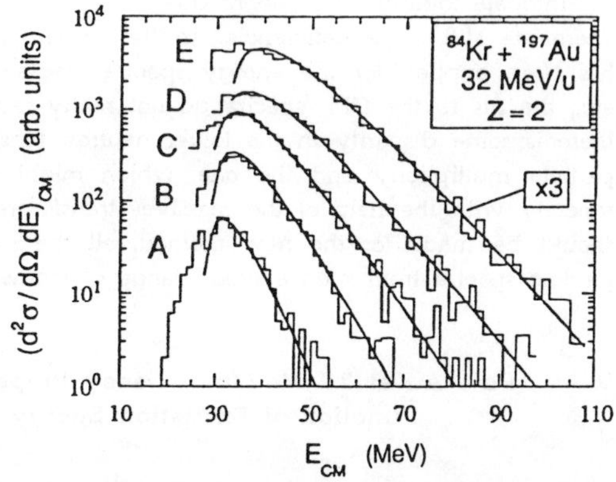


*Fig. 13 : Energy spectra of He-particles from Kr + Au, selected according to five bins in $M_n$ ( A to E, table 1) and transformed in the reference system of the emitting source. The lines are Maxwellian fits for the extraction of the slope parameter $T_{fit}$.*

From the entries in table 1, we see that the central collisions, which create the light charged particles, actually comprise a broad band of momentum transfers ranging from 40 to 85% of $p_{inc}$ with the maximum at 75% in bin D. The excitation energy varies accordingly, the most probable being 860 MeV in bin D. However, for still higher neutron multiplicity, bin E beyond the maximum in the inclusive neutron distribution, even 1000 MeV, that is 4 MeV/u or a temperature in excess of 6 MeV is reached. Therewith, the associated cross section in bin E is far from being negligible, it still amounts to about 10% of the total reaction cross section.

From the last two columns of table 1, it appears, that the particle multiplicity provides temperatures, which are consistent with the ones from the slope of the energy spectra.

*Table 1 : Parameters for the reaction $^{84}$Kr ( 32 MeV/u ) + $^{197}$Au, as deduced from neutron multiplicity and charged- particle evaporation, for five bins in $M_n$, see text.*

| n- windows | $\langle M_n \rangle$ (corrected) | $v_s / v_{FMT}$ | $M_{Z=1}$ (θ=159°) | $M_{Z=2}$(θ=159°) / $M_{Z=2}$(θ=109°) | $E^*$ (MeV) | T (MeV) | $T_{fit}$ (MeV) |
|---|---|---|---|---|---|---|---|
| A: 11 - 15 | 18.3 ± 0.9 | 0.45 | 0.5 ± 0.1 | 0.4 ± 0.1 / 0.4 ± 0.1 | 300 ± 40 | 3.5 ± 0.5 | 3.3 |
| B: 16 - 20 | 24.6 ± 1.2 | 0.60 | 1.7 ± 0.2 | 1.5 ± 0.2 / 1.2 ± 0.2 | 480 ± 60 | 4.5 ± 0.6 | 4.3 |
| C: 21 - 25 | 30.1 ± 1.5 | 0.70 | 3.9 ± 0.6 | 2.5 ± 0.4 / 2.4 ± 0.3 | 670 ± 90 | 5.2 ± 0.7 | 5.1 |
| D: 26 - 30 | 36.2 ± 1.8 | 0.81 | 5.6 ± 0.8 | 3.5 ± 0.5 / 3.6 ± 0.3 | 860 ± 130 | 5.8 ± 0.8 | 6.3 |
| E: 31 - 35 | 40.5 ± 2.0 | 0.90 | 7.3 ± 1.1 | 4.0 ± 0.6 / 4.1 ± 0.4 | 1030 ± 150 | 6.2 ± 0.8 | 7.0 |

From this experiment, we have learned so far, that higher excitation energy can be deposited in the composite system with Kr- ions than with the lighter Ar-ions. At 32 MeV/u of incident energy, the bulk of the central collisions realizes 860 MeV and 1000 MeV are excited still with a sizeable cross section - higher energies than might have been supposed before. Also, with the heavier projectile, the charged- particle evaporation, which has to be observed anyhow for the extraction of the excitation energy, provides an easy access to the linear- momentum transfer, thus superseding the intricate folding- angle technique.
There are still some deficiencies in the present analysis : the simplest parametrisation has been applied for the energy spectra, neglecting angular momentum, deformation etc.; the fits to the CM- spectra do not really reproduce the low energy part and also there is some disparity in the lower neutron bins between the excitation energy from particle multiplicity and the one, which might be inferred from the deduced recoil velocity with the help of the massive- transfer recipe. In this respect, some allowence should be made for the new method, all the more, as it seems difficult, to analyse particle spectra from such a broad range of momentum transfers without a proper filter.

## V Pb + Au at 29 MeV/u : Decay Properties as a Function of Excitation Energy

We have already addressed in sect. III to the energy dissipation in the system Pb + Au at $E_{inc}$ = 29 MeV/u, as far, as it is sizeable at the present stage of the analysis.
In the inclusive neutron- number spectrum, fig. 14, we find the central reactions centered at $M_n$= 50, or approximately at $M_n$= 78, when taking into account an estimated neutron- detection probability.

It is not too difficult, to imagine that a good part of the totally available energy of 7.2 MeV/u must have been dissipated, in order to create this enormous emission, which comprises nearly one third of all participating neutrons.

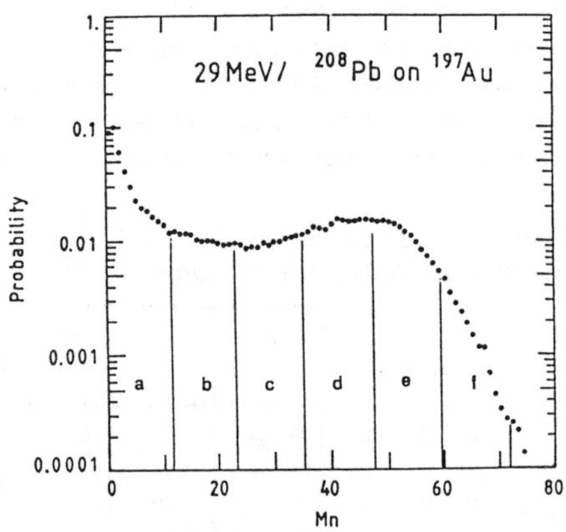


*Fig. 14 : Inclusive neutron- number pattern for $^{208}$Pb (29 MeV/u) + $^{197}$Au ( raw data, no efficiency correction ). The windows labelled a to f are used as a filter for the selection of decay products, figure 14.*

From a compagnion investigation ( Qu 91 ) of the spectral and angular distribution of the neutron emission with the time-of-flight method, it appears, that, even after very dissipative collisions with kinetic- energy losses up to 1.3 GeV, the neutrons are mostly

evaporated sequentially from two heavily excited target- and projectile-like residues. This study will also provide the average neutron kinetic energy for a reliable correction of our detection efficiency.
From the Pb + Au experiment, we choose here ( Pi 91 ) only a sequence of images, fig. 15, showing the abundancy of fragments from the Pb-projectile in a charge vs. energy matrix for 6 consecutive neutron bins, which cover the whole range in the inclusive neutron multiplicity distribution, fig. 14. These pictures demonstrate, how the violence of the reaction manifests itself in the decay properties, i.e. in size, frequency and energy of the remnants, or conversely, they emphasize the advantages of the neutron multiplicity as a filter on the impact parameter.

*Fig. 15 : Pictorial sequence of decay products from $^{208}$Pb ( 29 MeV/u ) + $^{197}$Au for increasing violence, or increasing neutron multiplicity of the reaction. The fragments are displayed in the E vs. δE- plane. The intensity scale ( N ) changes from linear to logarithmic after the third Mn- bin, so as to accentuate the nearly exponential slopes of the distributions. The Mn- bins are also indicated in the inclusive spectrum, figure 14.*

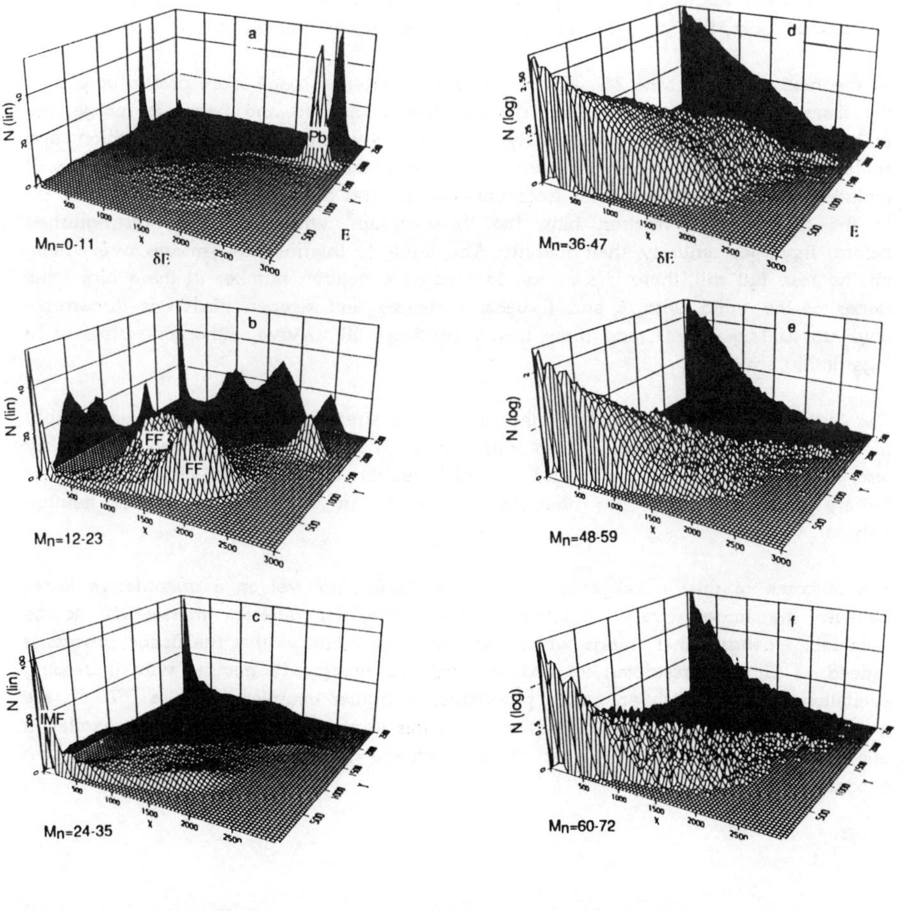

The charged fragments were detected in a forward dE × E-telescope, covering the angular range from the grazing angle near 6 out to 20 degrees. The 200 μm (dE)- and the 500 μm (E)-detectors therein, had a surface of 24 × 24 $mm^2$ and were made from 6 strips each, which were oriented horizontally and vertically, respectively, for tracing the particle impact. The depth of the dE- detector actually imposes a threshold of about 16 MeV/u on the detection of heavier fragments with atomic numbers above 30. In this way, and also by the kinematics of the reaction, our detector registers nearly exclusively debris from the projectile, the others being emitted mostly towards larger angles.

In the lowest bin $M_n$= 0- 11, fig. 15, the detector sees a strong spike of elastically or near- elastically scattered Pb-ions and some tiny indications from sequential fission of the Pb- ion in the middle of the plane.
In the next bin, $M_n$= 12- 23, the Pb- peak spreads in Z and E due to increasing inelasticity, and the fission fragments now dominate the scene. They have a double-humped structure from the two kinematical solutions : the fragments are emitted forward and backward from the faster moving Pb-like source. The latter ones appear somewhat more abundant in the picture, due to the enhanced kinematical focussing for this branch. Also, the yield of very light ions has strongly increased.

In the next window, $M_n$= 24- 35, the scenery experiences once more drastic changes : the fission- and the projectile-like- groups have broadened and begin to merge, the projectile fragments being completely detached now from the elastic position and shifted towards fission. More important, however, is the stretching of the light group. It includes now intermediate- mass fragments and bridges the whole gap to fission.
In the higher three neutron bins, the three groups, which could be distinguished before, have lost entirely their identity. The light- to intermediate- masses overwhelm all the rest. But still, there is an evolution with the neutron number in these bins : the slopes of the yield with Z and E become steeper and steeper. Fission is discernible only up to $M_n$= 36- 47, and some heavy residues still survive, although visible on a logarithmic scale only.

The highest bin, $M_n$=60- 72, if not the one before already, exhibits a total disassembly of the Pb-nucleus into small pieces with low Z. The Au-nucleus by then presumably also falls completely in pieces and the total neutron excess in the system N-Z= 79 is largely evacuated as single neutrons, which we in turn observe in the neutron detector.

The outcome of this investigation so far is - though not yet on a quantitative level, because the largely varying kinematical focussing for different masses has to be unfolded, amongst other things which bias the observation - that the fission process is indeed ( Gr 86 ) restricted to medium excitation energy. Its decline with increasing excitation proceeds with an ample production of lighter complex particles ( Tr 89, see also the contribution of C.A. Ogilvie et al. to this workshop ). At the highest excitation energy everything falls in pieces, mass or charge yields approach the well known exponential slopes.

( For a forthcoming discussion of this experiment, see the contribution of J. Galin ).

## VI Summary

In a series of experiments with Ar-, Kr- and Pb induced reactions on Au- and Th, U-target nuclei, we have investigated the effect of projectile-size and -energy on the amount of energy, that can be converted into intrinsic excitation or heat in the composite system, and we have observed its decay properties as a function of excitation energy.

From the inclusive neutron- multiplicity distributions and their geometrical interpretation, it follows, that the central or dissipative collisions exhaust about one half of the total reaction cross section, nearly independent of the bombarding energy and of the particular choice of the projectile-target combination.

In particular, for Ar + Th, the excitation energy, which can be stimulated by the bulk of the central reactions ( i.e. with a large cross section of about 2.5 barn ) saturates gradually for incident energies in excess of 30 MeV/u and approaches slowly 700 MeV at $E_{inc}$= 77 MeV/u.

The reaction Kr + Au creates higher excitation energy : 850 MeV for the most probable central reactions and even 1000 MeV for the most dissipative collisions (with a share of still 10% in the total cross section ). The composite system thereby attains a temperature of 6.2 MeV.

It is thus not a property of the intermediate complex, but rather a dynamical hindrance in the entrance channel, which generates the saturation effect observed with the lighter Ar-projectile at higher velocity.

The main issue in the Pb + Au experiment so far, is the decay characteristics of the extremly heated Pb-like fragment. Its fission decay is restricted to medium excitations, whereas the highest excitations lead to a complete disassembly of the nucleus.

As far as our basic instrument, the neutron multiplicity meter, is concerned, this series of experiments has demonstrated its indispensability for these kind of investigations. Its sensitivity to the reaction parameters, i.e. excitation energy, linear- momentum transfer and impact parameter, could still be improved by complementing it by an analogues device for light charged particles. This would allow to register eventwise practically the total deexcitation energy and is all the more important at very high excitations, because up there, the neutron number looses some sensivity ( Lo 91 ) in favour of the then relatively stronger increasing charged particle emission.

It is a pleasure to thank our colleagues B. Cramer, H. Doubre, J. Fréhaut, G. Ingold, D.X. Jiang, C. Magnago, Y. Patin, Y. Pranal and J.L. Uzureau, who contributed to the earlier stages of the programme, as well as R. Bougault, J. Colin, E. Crema, A. Genoux-Lubain, D. Horn, J. Jastrzebski, A. Kordyasz, C. Le Brun, J.F. Lecolley, M. Louvel, L. Pienkowski, B. Quednau, F. Saint-Laurent, W.U. Schröder, W. Skulski, J. Töke and X.M. Wang, who participated more recently.

References :


Co 85 M. Conjeaud et al. , Phys. Lett. B 159 (1985) 244

Cr 91 E. Crema et al. , Phys. Lett. B 258 (1991) 266

Di 56 B.C. Diven, H.C. Martin, R.F. Taschek, J. Terrell, Phys. Rev. 101 (1956) 1012

Fr 90 W.A. Friedman , University of Wisconsin-Madison preprint Mad/NT/90-08, (1990)

Fu 85 H. Fuchs et al. , Phys. Rev. C31 (1985) 465

Ga 88 J. Galin et al. , Z. Phys. A - Atomic Nuclei 331 (1988) 63

Ga 90 J. Galin et al. , in Proc. 2nd IN2P3-Riken Symposion, Obernay, ed. B. Heusch and M. Ishihara ( World Scientific, 1990 )

Gr 86 D.H.E. Gross, Zhang Xiao-ze, Xu Shu-yan, Phys. Rev. Lett. 56 (1986) 1544

Hu 76 J.R. Huizenga, J.R. Birkelund and W.U. Schröder, Phys. Rev. Lett. 37 (1976) 885

Ja 82 U. Jahnke et al. , in Proc. Symp. on Detectors in Heavy-Ion Reac., Berlin 1982, ed. W. von Oertzen, Lecture Notes in Physics 178 (1983) 179, Springer Verlag

Ja 83 U. Jahnke et al. , Phys. Rev. Lett. 50 (1983) 1246

Ja 86 U. Jahnke et al. , Phys. Rev. Lett. 57 (1986) 190

Ja 88 U. Jahnke et al. , in Proc. 20th Summer School on Nucl. Phys. Mikolajki, Poland ed. Z. Wilhelmi, G. Szeflinska, Adam Hilger, Bristol and Philadelphia, 1989

Jc 85 D. Jacquet et al. , Phys. Rev. C32 (1985) 1594

Jc 90 D. Jacquet et al. , Nucl. Phys. A511 (1990) 195

Ji 89 D.X. Jiang et al. , Nucl. Phys. A503 (1989) 560

Le 86 S. Leray, Journ. de Phys. Colloque C4 (1986) 275

Lo 91 B. Lott et al. , to be published

Mo 91 M. Morjean et al. , Nucl. Phys. A524 (1991) 179

Na 88 S. Nagamiya , Nucl. Phys. A488 (1988) 3c

Pi 91 E. Piasecki et al. , Phys. Rev. Lett. 66 (1991) 1291

Po 74 J. Poitou, C. Sigarbieux , Nucl. Instr. Meth. 114 (1974) 113

Qu 91 B.M. Quednau et al. , in Winterworkshop on Nucl. Dynamics, Key West, 1991

Re 54 F. Reines, C.L. Covan, F.B. Harrison, D.S. Carter, Rev. Sci. Instr. 25 (1954) 1061

Sc 89 E. Schwinn et al. , Nucl. Phys. A502 (1989) 551c

Sc 91 E. Schwinn et al. , to be published

So 91 A. Sokolov et al. , to be published

Tr 89 R. Trockel et al. , Phys. Rev. C39 (1989) 729

Wi 80 W.W. Wilcke et al. , Atomic Data and Nucl. Data Tables Vol. 25 (1980) Nos. 5/6

# HOT NUCLEI with HIGH SPIN STATES in collisions between heavy nuclei*.

J.Galin#
GANIL BP 5027 14021 Caen-Cedex

## Introduction

It is now well established that nuclei with temperatures exceeding 4-5 MeV can be formed in violent collisions between heavy nuclei at bombarding energies of a few tens of MeV/u[1)]. But, at such moderate energies, in order to be violent a collision needs not to be central and, as a consequence, large angular momenta are involved, leading to the formation of hot nuclei with high spin states. In addition, and due to the weak overlap between projectile and target nuclei in such collisions, one deals with initially uncompressed nuclei, in contrast with what happens for more central collisions. In this contribution, we would like to first show how it is possible to select hot nuclei in peripheral collisions, infer their temperature from exclusive neutron multiplicity measurements and, then, demonstrate, by utilizing fission properties, that such nuclei have pretty large spin states.

## Peripheral collisions studied in 32 MeV/u Kr+Au reactions: evidence for the formation of hot nuclei.

Projectile-like fragments have been detected by a standard telescope made of Si detectors. This telescope sits at 7° i.e. close to the grazing angle and two types of events can be distinguished on an identification matrix (fig.1). Those with a velocity close to the beam velocity, extending from Z= 37 down to Z= 2, represent one wing in the matrix. In addition, there is a second wing made of events with similar Z's but with velocities strongly reduced as compared to the beam velocity. Due to the detection threshold, only part of the latter could be investigated (fig.1). Moreover, below Z=20 or so, the two contributions start overlapping, leading to broad energy spectra. A more detailed description of the two types of events is provided in the scatter plot of fig.2 where the number of accompanying detected neutrons is presented as a function of the kinetic energy of the measured fragments. For most of the latter (38>Z>20), one can identify two blobs of data points that could be attributed to two different trajectories, corresponding to well distinct impact parameters. On the one hand, both the high fragment velocity and the relatively low number of neutrons sign a collision at high impact parameter with a trajectory dominated by the coulomb field between projectile and target. On the other hand, the events with a low kinetic energy, correlated with a high

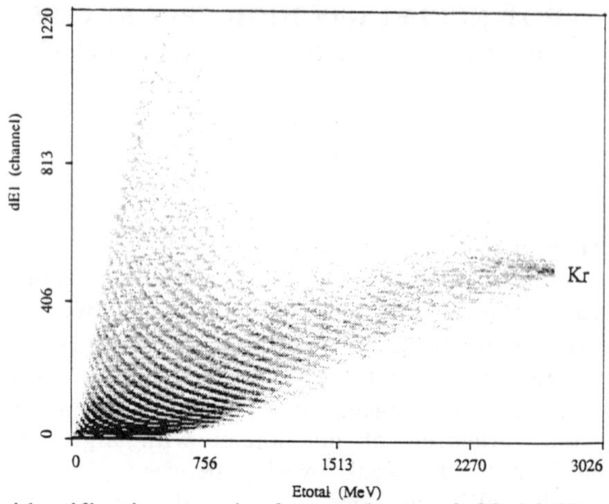


Fig.1 ΔE-E identification matrix for products of 32 MeV/u Kr+Au at 7°

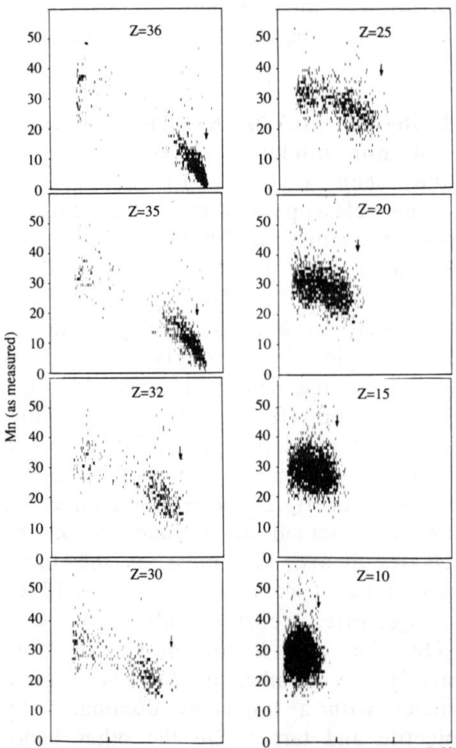


Fig.2 Scatter plot of events of fig.1 identified in Z and characterized by their kinetic energy and associated number of detected neutrons.

neutron multiplicity whatever the Z of the ejectile, could arise from less peripheral collisions and correspond to trajectories crossing the beam direction, because of the strong attraction of the nuclear field. This interpretation appears to be more likely than the hypothesis of binary fission after large momentum transfer and this is for two reasons. First, the kinetic energy of these nuclei exceeds the one expected for binary fission, even when considering the extreme case of full momentum transfer, and second, above Z=37 this second component vanishes whereas even heavier fission fragments should still be present.

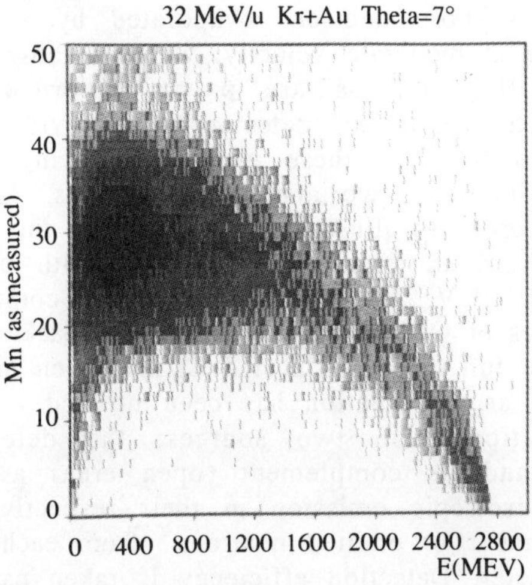


Fig.3 Same as for fig.2 when summed over events of all Z's.

The evolution from "positive" to "negative" trajectories as a function of impact parameter has been observed in dynamic calculations of the Landau-Vlasov type, for Ar+U at a similar beam velocity, by D. Jacquet et al[2]. When gathering all experimental data obtained at 7° (fig.3) on a single scatter plot, one is struck by the strong correlation between kinetic energy of the ejectile and neutron multiplicity. As it will be shown later on, most detected neutrons can be assigned to target-like evaporation, therefore a large number of neutrons means the achievement of high temperatures in the target-like nucleus, thus indicating strong correlation between the size (or/and the kinetic energy) of the detected ejectile and the temperature reached by its partner. Such a correlation has also been observed in Ar induced reactions at similar bombarding velocity by M. Morjean et al[3] and could be essentially understood, at bombarding energies lower than 30 MeV/u, in the frame of a massive transfer reaction: the missing part of the projectile is

essentially captured by the partner nucleus and the kinetic energy of the transferred nucleons degraded into heat. Here, more than the reaction mechanism itself, we are interested in assessing the amount of thermalized energy in the partners and for this purpose we have utilized the data collected by the 4π-neutron detector.
As presented elsewhere by J.Galin et al[4] and in this meeting by Jahnke[5], the ORION detector is a liquid scintillator detector, loaded with Gadolinium, made of four independent sectors, covering 4π, whose main virtue is a high detection efficiency (>75%) for neutrons of less than 10 MeV. In contrast, the detection efficiency decreases steadily for high energy neutrons. Such a detector is thus well suited to register those neutrons evaporated by a target-like hot nucleus recoiling at moderate velocity, whereas those issued from a rapid projectile-like nucleus are poorly recorded. The second advantage of such a sectorized detector is to provide information on the spatial distribution of these neutrons, making it possible to unfold the data between two major contributions. An example has been chosen in order to illustrate how sensitive this procedure can be. Events with Z=28 have been selected with average kinetic energies of 1600 MeV. The distribution of the coincident neutrons in the four sectors of ORION (A, B, C, D from backward to forward) as well as in the full detector (labeled E) are given in fig.4 by open dots. A Monte Carlo simulation has been utilized to determine the neutron distribution from two sources: the detected fragment (hatched area) and its complement (open area) assuming binary kinematics and isotropic emission in their respective rest frames. The absolute number of emitted neutrons from each source is left as a free parameter. Detection efficiency is taken care of for each emitted neutron, characterized by its velocity vector as obtained from the output of the Monte-Carlo evaporation simulation code in such a way that experimental data, as measured, can be directly compared with the simulation output. It can be seen that, as expected, most of the registered neutrons (95% in the present case) are emitted by the target-like nucleus. A severe constraint on the number of neutrons evaporated by the target-like nucleus is provided by the backward measurements. Thirty evaporated neutrons are needed to reproduce the data. As for the neutron emission from the projectile-like nucleus, it cannot be entirely fitted by considering the complementary part measured forward. There is some room left for an extra contribution in the most forward sector (D) and that could be due to preequilibrium emission, disregarded in the present simulation. Therefore, the uncertainty on the actual number of projectile-like neutrons remains large and this makes difficult an estimate of the corresponding thermal energy (3 have been considered in the fit of

fig.4). Moreover for such a medium mass nucleus, charged particle emission cannot be ignored.

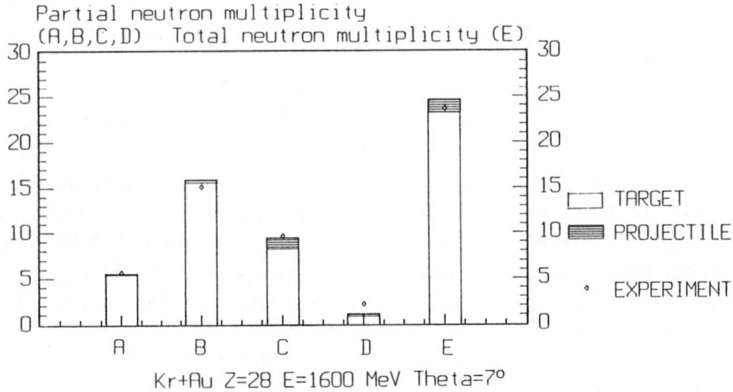


Fig.4 Number of neutrons detected from backward (A) to forward (D) and in the whole ORION detector (E). Experimental data are given by the open dots. Emission simulated from projectile and target-like nuclei are given by histograms. For detail see text.

On the contrary, there is no such an ambiguity for the target-like emission. In order to infer an excitation energy from the deduced multiplicity of 30 evaporated neutrons, evaporation codes have been used, leading to $E^*$=700 MeV or T=6 MeV. There is thus strong evidence that pretty hot nuclei can be formed, even in rather peripheral collisions. Indeed, for the collisions considered before it must be stressed that a nucleus of Z equal to 28 has survived with about 85% of the beam velocity. This implies that projectile and target have overlapped very little, as it was illustrated by M.F. Rivet[6] in Landau-Vlasov calculations performed on a different system but at a similar beam velocity .

**Peripheral collisions studied in the 29 MeV/u Pb+Au collisions: evidence for the formation of nuclei with high spin states.**

When an excited, spinning nucleus decays by particle emission or by fission, these particles (or fission fragments) are focused in a plane perpendicular to the spin vector and this effect is all the more important that the spin value is large. One can thus exploit out-of-plane measurements to infer the strengh of the spin alignment and the spin modulus. When such measurements are performed in peripheral collisions the reaction plane is defined by means of the beam axis and the scattering direction of either the projectile-like or target-like nucleus. Emitted particles or fission fragments are distributed preferentially in this plane (see for instance R. Babinet et al [7]), P. Dyer et al [8]), D.V. Harrach et al [9]), J.C. Steckmeyer et al [10]). In all previous experiments the violence of the collision and reaction plane were selected by detecting a given projectile-like fragment. In the following, we would like to show that a selection on impact parameter (or violence of the collision) can be performed using the neutron multiplicity as a filter.

The studied system is 29 MeV/u Pb+Au (S.Bresson et al[11]), E.Piasecki et al[12]) and the data analysis being not completed yet, we will restrict ourselves to semi-quantitative arguments. The experimental set-up was made of a large area (24*24 $mm^2$) telescope built from two silicon strip-detectors with thicknesses of 200 μm and 500 μm for ΔE and E respectively. The projectile-like fission fragments could be identified in Z without any ambiguity and their position in theta and phi was determined from the fired strips in ΔE and E, mounted orthogonal to each other.

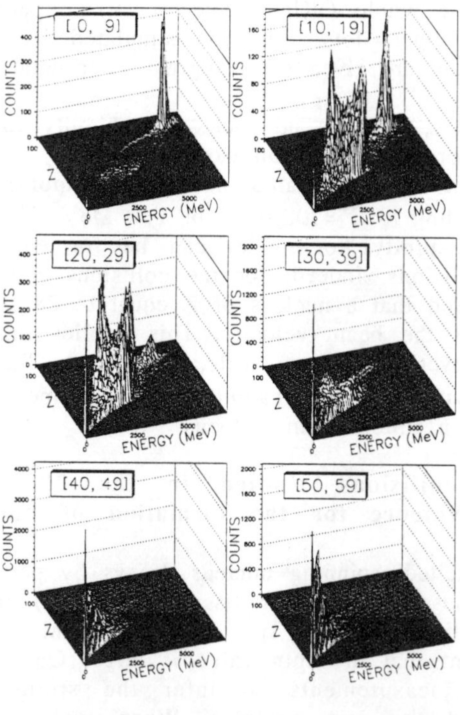


Fig.5 Distribution of reaction products detected between 6° and 20° for the 29 MeV/u Pb+Au system, as a function of their Z and kinetic energy for several neutron multiplicity gates (in brackets)[11]).

As shown in fig.5[11]), fission fragments of the projectile-like ejectile, defined as fragments with a velocity close to the beam velocity and a Z of about one half of the projectile Z, extend in a domain of collisions characterized by intermediate neutron multiplicities. For the most peripheral collisions (low neutron multiplicities) the projectile-like nucleus does not receive enough spin or/and excitation energy to undergo binary fission; instead it ends up in an evaporation residue. At the other extreme i.e. for the most violent collisions (high neutron multiplicities) binary fission might either be followed by a sequential fission of the primary formed fragments or give way to a one-step multifragmentation process. Whatever the decay process might be, mostly fragments of Z=30 and below (the so-called intermediate mass fragments) are then detected. In the following we will mainly focus on binary fission-like events and, accordingly, events with measured neutron multiplicities between 5 and 34 will be selected. The number of detected neutrons should be roughly increased by 50% to correct for the detector efficiency and get the actual number of emitted neutrons. Indeed, a Monte Carlo simulation shows that, on the average, about 64% of the neutrons emitted by both target-like and projectile-like nuclei are registered. The actual neutron gates, after efficiency correction, thus extend from about 7 up to 51 neutrons (i.e. from about 3 to 25 neutrons for either the projectile-like or target-like nucleus of the nearly symmetric system).

The fission fragment data have been sorted as a function of Z for several gates in neutron multiplicity. Their invariant cross sections are given in fig.6[11]) as a function of the two components in velocity, parallel and perpendicular to the beam direction. Since only the Z's and kinetic energies have ben measured, masses were

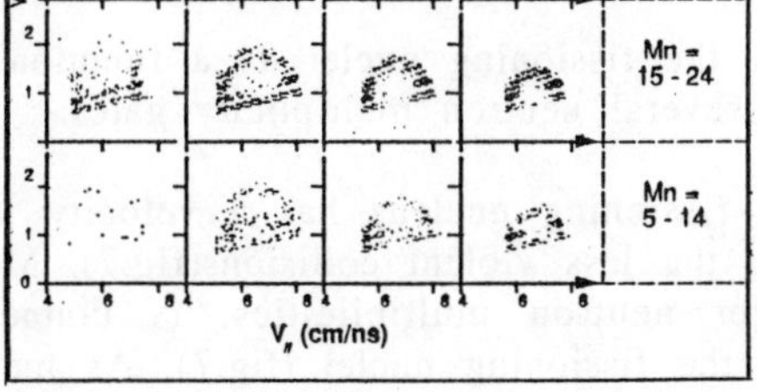


Fig.6 Galilean invariant cross sections of fission fragments versus $v_{par}$ and $v_{per}$ for several Z's and neutron multiplicity gates.

assigned as following the bottom of the ß stability valley, in order to determine the velocity. As it could be discussed later on, this assumption may have some influence on the derivation of several quantities.
The regular pattern observed in fig.6 calls for several comments about the general features. First, nice rings show up which are experimentally troncated at small angles, simply by lack of measurements, and at low energy because of the finite thickness of the ΔE detector. Second, it is obvious that the rings are not centered in the beam direction but rather at 5-6°, but, despite of this, the pictures are not blurred. This can be interpreted as due to a strong selectivity in impact parameter and deflection angle thanks to the neutron multiplicity filter. The quasi-projectile deflected by the Coulomb field of the target nucleus undergoes fission and the fission fragments sit, in a velocity diagram, on a circle centered at the tip of the velocity vector of the parent nucleus. By fitting the locus of the data points with circles one can precisely determine the characteristics of the fissioning nucleus (velocity in modulus and angle) and of the fission fragments (velocity).

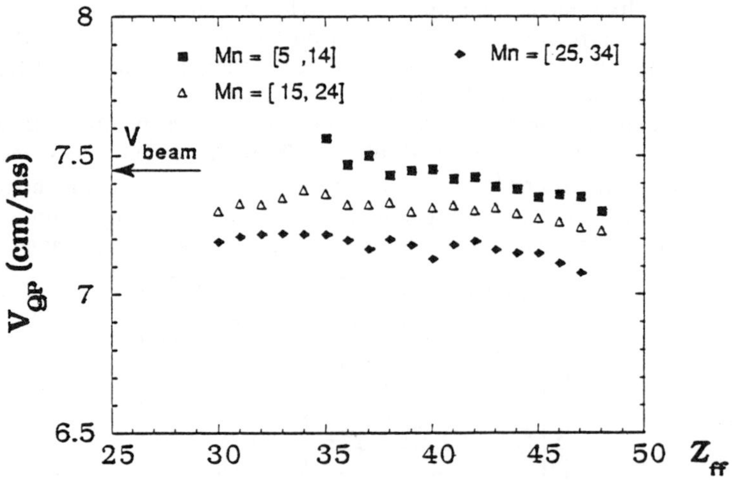


Fig.7 Velocity of the fissioning nuclei as a function of the fission fragment Z's for several neutron multiplicity gates.

As expected, the fissioning nucleus has a velocity identical to the beam velocity for the less violent collisions(fig.7). More dissipation, selected by larger neutron multiplicities, is connected with a slowing down of the fissioning nuclei (fig.7). As mentioned earlier, the fact that the velocities were not directly measured but inferred from E and Z may slightly modify the velocities given in fig.7, but qualitatively the evolution with neutron multiplicity is the expected one. As for the deflection angle of the fissioning nucleus, its precise determination suffers the fact that only fraction of the circle has been actually measured. All primary scattering angles are determined as being slightly below the grazing angle but an evolution with neutron multiplicity could not be derived unambiguously. The intrinsic velocity of the fragments follows what is expected from the Viola systematics and momentum conservation laws (fig.8). The slow evolution towards lower velocity in the more dissipative collisions is not yet understood. The Z distribution centered at Z= 41 (half the projectile Z) for the less dissipative collisions is shifted by a few units for hotter fissioning nuclei as expected if charge particle evaporation competes with neutron emission either before or after fission (fig.9).

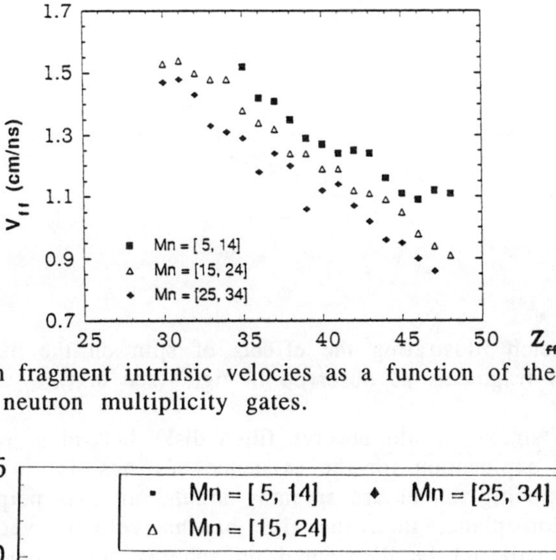


Fig.8 Fission fragment intrinsic velocies as a function of their Z's and for several neutron multiplicity gates.

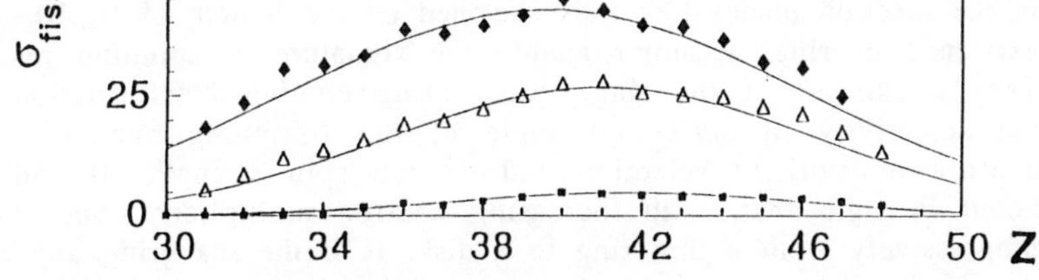


Fig.9 Integrated cross sections of the fission fragments as a function of their Z for several neutron multiplicity gates.

The most intriguing and interesting feature from these data is connected with the spin of the fissioning nuclei. The data reveal that pretty large amounts of spin are present. The finger-print is the existence of "rings" in the invariant cross section picture of fig.6. Why are there "rings" of events and not "disks" (or filled "rings")? In contrast with fusion-like reactions[13], the velocity vector of the fissioning nucleus is not unique but describes a cone of revolution around the beam axis. Therefore one would expect the tips of the velocity vectors of the fission fragments to be located on spheres that strongly overlap, as suggested on top of fig.10.

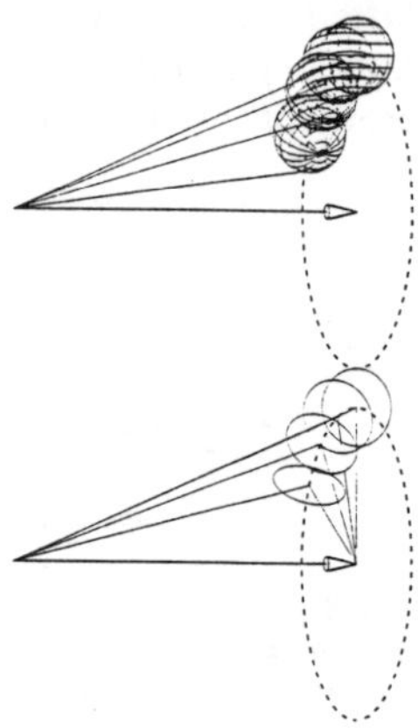

Fig.10 Sketch illustrating the effects of spin on the distribution of the fission fragments as observed in fig.6 (see text).

If it were so, we would observe filled disks instead of open rings in fig.6. The explanation for the existence of rings lies in spin effects. If the fissioning nuclei are spinning around an axis perpendicular to the reaction plane, then the tips of the velocity vectors of the fragments will not be distributed on spheres but essentially focused in the reaction planes like it is sketched on the bottom of fig.10. The existence of rings is unmistakably the signature of spinning nuclei. The "thickness" of the ring must bring combined information on the selectivity in deflection angle of the fissioning nucleus after neutron multiplicity selection and on the spin strength. It can be noted in fig.6 that, with increasing neutron multiplicity, one shifts progressively from a thin ring to a disk. Is it the scattering angle of the fissioning nucleus which gets poorly determined or the spin value which decreases? Evaporation prior to fission is expected to broaden the primary angular location of the fissioning nucleus and to diminish the spin value of the nucleus when undergoing fission.

Both effects should contribute to the filling of the ring. Monte-Carlo simulations are clearly needed to estimate all these effects. Moreover, in a forthcoming experiment, we will try to get more constraints in detecting, in addition to the present observables (one fission fragment and the neutron multiplicity), both the scattered target-like nucleus (to define the reaction plane) and the complementary fission fragment.

To summarize, in the first part of this contribution we have shown that pretty hot nuclei could be obtained in peripheral collisions of Kr+Au. The collisions considered in the chosen example give rise to a nucleus of Z=28 with a kinetic energy of 1600 MeV (i.e. a velocity close to 27 MeV/u to be compared with the 32 MeV/u of the beam). The excitation energy deposited in the non-detected target like-nucleus, deduced from the neutron multiplicity measurements, amounts to 700 MeV ( T= 6 MeV). In the second part of the contribution, one used the well known properties of fission, and particularly its sensitivity to spin, to show in a qualitative way that pretty high spin values are into play. A more quantitative analysis together with additional measurements are still needed in order to infer precise figures of spin. It can be noted that for the 29 MeV/u Pb+Au reaction $l_{max}$ amounts to 1700 ħ. If we assume that the sticking or rolling conditions can be fulfilled for initial angular momenta of about 2/3 $l_{max}$, then a projectile-like (and its target partner) could acquire an intrinsic spin of about 160ħ. The behavior of a Pb-like nucleus brought in such an exotic state (T=6 MeV and J=160ħ) is certainly worth to be studied in detail. It is also worth recalling that, when obtained in peripheral collisions, the hot nuclei thus formed do not suffer much initial compression at variance with what happens in more central collisions. There is thus an interesting field to be explored of hot, high spin but uncompressed nuclei[14].

* Experiments performed at the GANIL facility

# The data presented in this contribution have been obtained in collaboration with a large number of people listed below. Special tribute should be paid to E. Crema and S. Bresson who have performed the data reduction of the Kr and Pb parts respectively.
**Kr experiment**: E.Crema, S.Bresson, H.Doubre, J.Galin, B.Gatty, D.Guerreau, D.Jacquet, U.Jahnke, B.Lott, M.Morjean, E.Piasecki, J.Pouthas, F.Saint-Laurent, E.Schwinn, A.Sokolov and X.M.Wang.
**Pb experiment**: S.Bresson, R.Bougault, J.Colin, E.Crema, J.Galin, B.Gatty, A.Genoux-Lubain, D.Guerreau, D.Horn, D.Jacquet, U.Jahnke, J.Jastrzebski, A.Kordyasz, C.Le Brun, J.F.Lecolley, B.Lott, M.Louvel, M.Morjean, C.Paulot, E.Piasecki, L.Pienkowski, J.Pouthas, B.Quednau, W.U.Schröder, E.Schwinn, W.Skulski, J.Töke

References:

1) E.Crema et al. Phys. Lett. in press, and GANIL Preprint P 90-19a
2) D.Jacquet et al. Nucl. Phys. **A511** (1990) 195
3) M.Morjean et al. Phys. Lett. **B 203** (1988) 215 and Nucl. Phys. **A524** (1991) 179
4) J.Galin et al. in Proceedings of the Second IN2P3-RIKEN Meeting on Heavy-Ion collisions, edited by B.Heusch et M.Ishihara (World Scientific, Singapore, 1990), p102
5) U.Jahnke et al., this colloquium
6) M.F.Rivet et al., this colloquium
7) R.Babinet et al. Z. Physik **A 295** (1980) 153
8) P.Dyer et al. Nucl. Phys. **A322** (1979) 205
9) D.V.Harrach et al. Phys. Rev. Lett. **42** (1979) 1728
10) J.C.Steckmeyer et al. Nucl. Phys. **A427** (1984) 357
11) S.Bresson et al. to be published
12) E.Piasecki et al. Phys. Rev. Lett. **66** (1991) 1291
13) Y.Blumenfeld et al. Phys. Rev. Lett. **66** (1991) 576
14) C.Ngô et al. International Workshop XIX Gross Properties of Nuclei and Nuclear Excitations Hirschegg (1991) LNS Preprint/Ph/91-01

# THE 4π FACILITY AT SIS/ESR FIRST DATA TAKING

F. JUNDT
Centre de Recherches Nucléaires and Université L.Pasteur, 67037 Strasbourg (France)
for the 4π collaboration *

*In the last fifteen years of heavy-ion research, the existence of collective effects, strong nuclear stopping and directed flow of nuclear matter, long time predicted, have been evidenced in an enlighting way with the first generation of 4π detectors. Howewer, the existence of nuclear shock waves still has not yet been demonstrated in a direct way, only features compatible with them but which can be explained otherwise have been shown. The present paper presents the status of the 4π facility at SIS/ESR applied to the study of complex fragment production in Au + Au collisions in the 100 to 800 MeV/A range.*

In 1956 S.Z. Belenk'ji and L.D. Landau [1] in their paper "Hydrodynamics theory of multiple production of particles" suggested that nuclear matter as a whole could be studied using hydrodynamics. Thus, if in an energetic violent nuclear collision the projectile is partialy or fully stopped by the target nucleus, the nuclear matter will be strongly compressed and may not be able to escape rapidly enough of the interaction zone. A shock front is then likely to be produced and if it is travelling at a velocity greater than the sound velocity in the nuclear matter a Mach cone will develop as calculated by A.E. Glassgold et al. [2] (Figure 1).

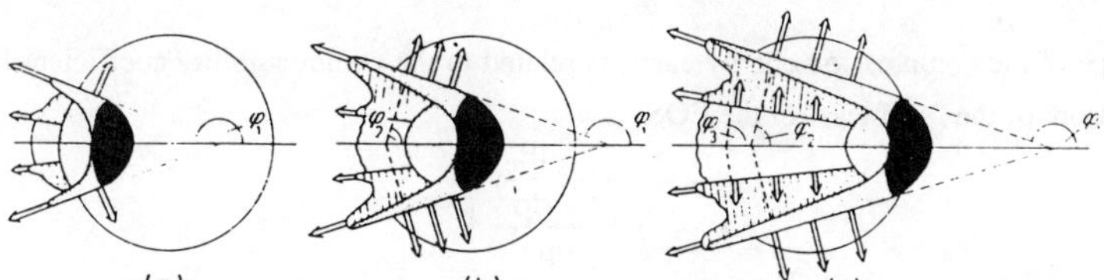


Figure 1

***4π Collaboration** : R. Bock, Z.G. Fan, Z. Fodor, A. Gobbi, K.D. Hildenbrand, S.C. Jeong, M. Krämer, W. Reisdorf, D. Schüll, U. Sodan, K.M. Teh, J. Wessels **(GSI Darmsdadt)** ; N. Hermann, W. Jakel, R. Linke, D. Pelte, R. Schlesier, M. Trzaska, T. Wienold **(Univ. of Heidelberg)** ; T. Blaich **(Univ. of Mainz)** ; C. Cerruti, J.P. Coffin, P. Fintz, G. Guillaume, F. Jundt, A. Houari, C. Kuhn, C.F. Maguire, F. Rami, R. Tezkratt, P. Wagner **(CRN Strasbourg)** ; J.P. Alard, N. Bastid, M. Crouau, P. Dupieux, L. Fraysse, M. Jorio, G. Montarou, P. Morel **(LPC Clermont-Ferrand)** ; M. Bini, P. Maurenzig, A. Olmi, G. Pasquali, G. Poggi, N. Taccetti **(Univ. and INFN Florence)** ; P. Boccaccio **(INFN Legnaro)** ; A. Buta, I. Legrand, D. Moisa, M. Petrovici, V. Simion **(IPNE Bucharest)** ; Z. Basrak, N. Cindro, R. Caplar, S. Höbling **(RBI Zagreb)** ; J. Erö, J. Kechkemeti, P. Konez, Z. Seres **(CRIP Budapest)** ; I.M. Belayev, Y. Grigorian, A. Lebedev, M.A. Vasiliev, A.L. Zhilin **(KIAE and ITEP Moscow)** ; R. Kotte, J. Mösner, W. Neuberyt, D. Wohlfarth **(ZfK Rossendorf)**.

Such collective phenomena in relativistic heavy ion collisions are a way to study nuclear matter in a domain of very high density and excitation and therefore gain information on the nuclear equation of state (E.O.S.) which describes the response of the nuclear medium to different temperatures and densities.

$$U(\rho,T) = E_c(\rho) + E_{th}(\rho,T)$$

Typically, the E.O.S. is a relation between internal energy, nuclear matter density and temperature. Up to now only the properties of equilibrated, cold nuclear matter around the ground state density are known.

At the saturation point:

$\rho_o$ = 0.17 nucleon/fm$^3$

T = 0

$U_o$ = -16 MeV

Relativistic heavy ion collisions will offer the possibility to create highly compressed nuclear matter at high temperature and density, specially in the region where the density is 2 or 3 times the ground state density.

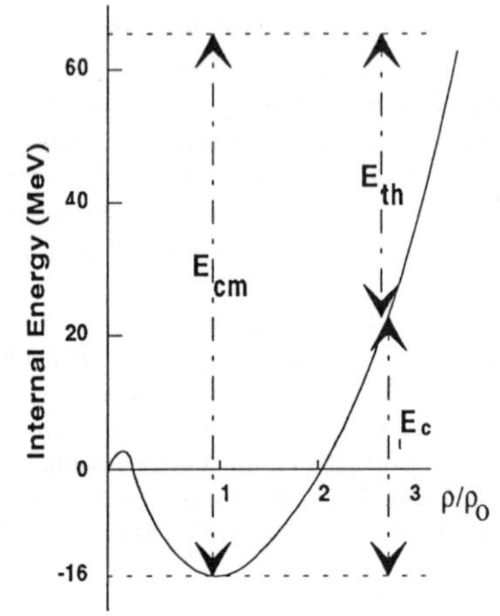


Figure2

The slope of the compression energy curve is related to the compressibility coefficient K and is a measure of the "stiffness" of the EOS:

$$K = 9 \, \frac{d(\rho^2 \frac{dE_c}{d\rho})}{d\rho} \Big|_{\rho=\rho_o}$$

So far, results do not remove the ambigüities in the nuclear compressibility modulus at $\rho > \rho_o$ and the in medium dependence of the effective nucleon - nucleon cross-section $\sigma_{nn}$ which is related to viscosity.

The knowledge of the properties of the E.O.S. would also help in the understanding of many astrophysical problems like:

Supernovae explosions

Stability of neutron stars

Dynamics of the early universe.

Since 1975, a first generation of experiments has investigated on collective effects in energetic heavy ion collisions with the aim to test the hydrodynamical theory in using $4\pi$–geometry detector systems. These are the Streamer Chamber and the Plastic Ball at the LBL and Diogene at Saturne. The first experimental evidence of nuclear stopping and local thermal equilibrium was obtained by the Plastic Ball group in the 400 MeV/A, Au + Au experiment [3].

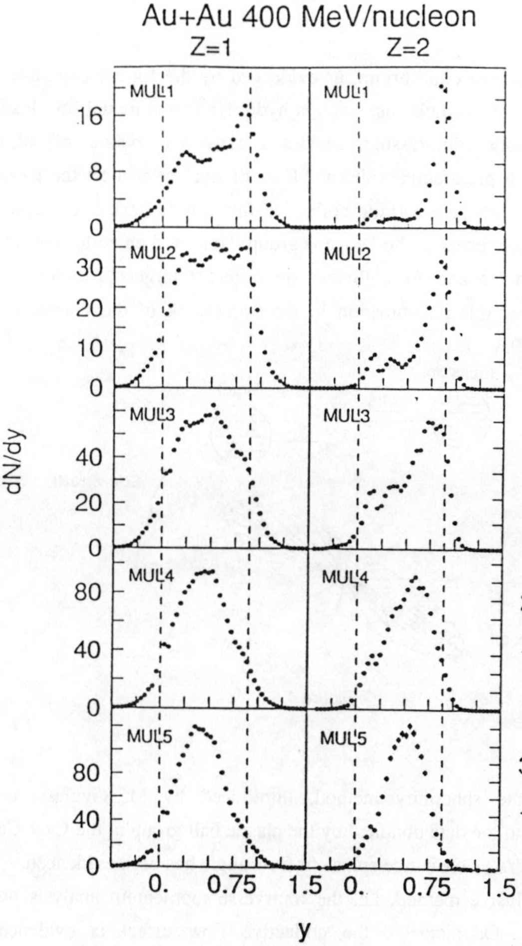


Figure 3

Rapidity distributions for Z=1 and Z=2 particles, from the 400 MeV/A, Au + Au reaction, in five multiplicity bins going from peripheral to central. The dashed lines are at target and projectile rapidities.

These results show that a simple geometrical participant spectator model is applicable: for peripheral collisions, remnants of both the projectile and the target are present. For semi-central and central collisions, almost complete thermalization is reached as indicated by the nearly gaussian shape, almost no remnants of target or projectile are present in the higher multiplicity bins, that means complete stopping is achieved. However, no shock wave is evidenced, perhaps the stopping power is such that the projectile is stopped too quickly for a Mach cone to develop?

Nuclear stopping and thermal equilibrium, as evidenced by the former experiments, do not reveal compressional effects. Simulations using a hydrodynamical model for close to central collisions show that shock compression, besides a transversal bounce-off of target and projectile, should cause a preferential sideward flow of nuclear matter: the so called side-splash. First observations of such a finite collective flow angle have been reported by the Plastic Ball group[4) . more recently, the Diogene group[5) reported on collective flow of pions and light composites in Ne and Ar collisions on different targets. Molitoris et al.[6) and Peilert et al.[7) investigated this phenomenon in the framework of the Quantum Molecular Dynamic model (QMD)[8).

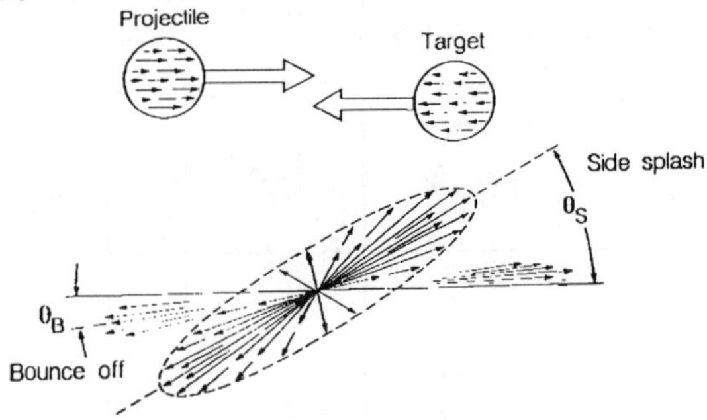


Figure 4

Indeed, by applying the sphericity method, introduced by M. Gyulassi et al.[9) and P. Danielewicz et al[10) , to the data obtained by the plastic ball group in the Ca + Ca, Nb + Nb and Au + Au at 400MeV/A, a clear side splash or flow angle has been evidenced.

By using the most exclusive method, i.e. the transverse momentum analysis proposed by P. Danielewicz and G. Odyniec[11) , the collective flow effect is evidenced by the characteristic S shape curve when plotting the transverse momentum versus the rapidity, the angle of flow being deduced from the slope of that curve at mid-rapidity. Figure 5 represents the mean transverse momentum per nucleon projected onto the reaction plane as a function of the center of mass rapidity at different multiplicity bins for the same experiments as on figure 3.

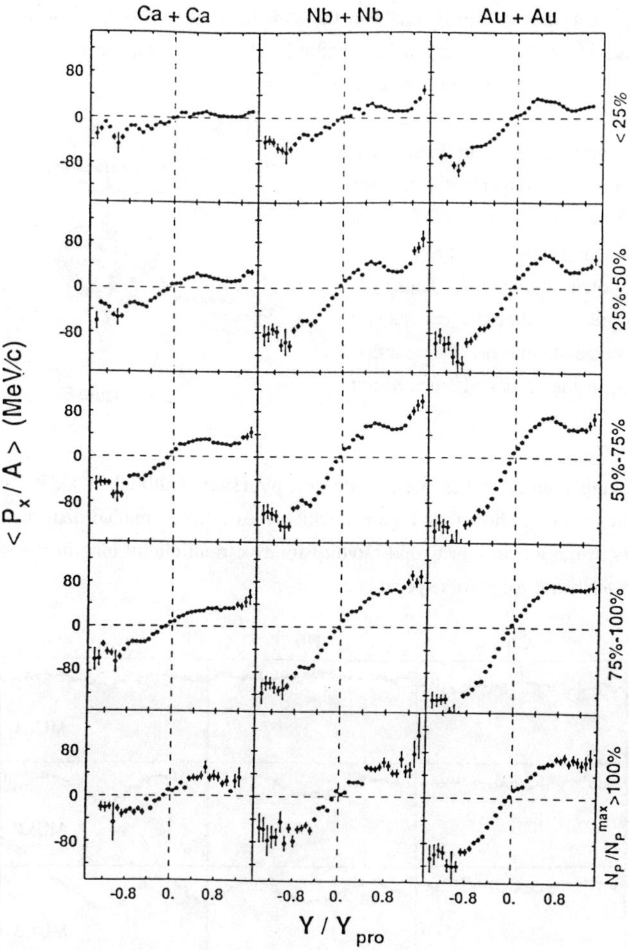


Figure 5
Transverse momentum analysis for five bins of multiplicity.

By applying the hydrodynamical model in the case of very violent head-on collision, the Frankfurt group[12)] predicted that nuclear matter, from the compression zone, could be squeezed-out perpendicularly to the reaction plane.

Indeed, for very central collisions, at almost zero impact parameter, the nuclear system is expected to evolve from a somewhat oblate shape to a prolate one (as in less violent collisions). Thus, the building up pressure should eject nuclear matter out perpendicular to the reaction plane leading to the so called squeeze-out emission.

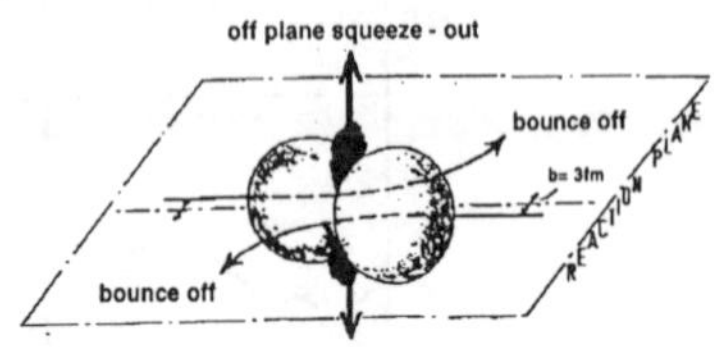


Figure 6

This challenging prediction has been evidenced by H. H. Gutbrod et al.[13)] and also by M. Desmoulins et al. as shown on Figure 7 which shows the azimuthal distributions around the flow axis in the window near mid-rapidity for five multiplicity bins in the 400 MeV/A Ca + Ca, Nb + Nb and Au + Au experiments.

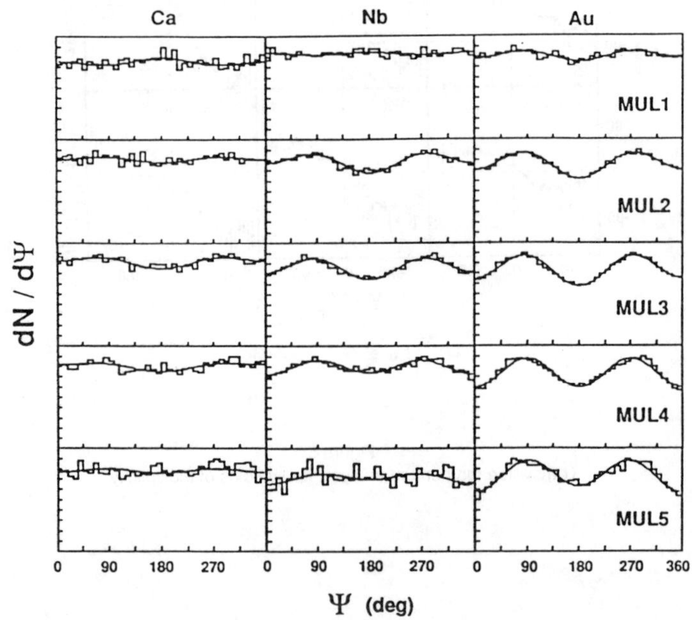


Figure 7

## Motivations for the short range studies with the 4π detector

The first major project of the 4π collaboration[14)] is to investigate the cluster production in Au + Au collisions in the 100 to 800 MeV/A bombarding energy range. The 4π Facility could yield triple differential cross sections in momentum space for fragments identified in atomic number for $Z = 1$ to about 16. Full 4π coverage will be achieved only in Phase II in early 1992. The now completed Phase I has been extensively tested during the past year at incident energies already investigated by the LBL Plastic Ball group.
Comparison of the preliminary results with those of former experiments shows that light particles and clusters (Li for example) have angular distributions peaked well within the angular acceptance of the Phase I setup and that it's different components yield consistent multiplicities in the overlap region. For heavier particles the forward peaking is even more pronounced for kinematical reasons.

The study of the Au + Au system for 100, 150, 250 and 400 MeV/A has begun very recently and the data calibration and reduction is now underway.

The focus is put on:

- ❑ Production of ***Intermediate Mass Fragments*** ( $5 < A < 30$).
- ❑ ***Transverse momentum*** excess and ***rapidity*** in order to determine the ***reaction plane*** and the ***directed flow***.
- ❑ ***Nuclear stopping*** effect and ***transparency***.

which sets a number of requirements on the detector performances.

## The 4π detector for charged particles at SIS

**Existing 4π systems:**

| | | |
|---|---|---|
| Plastic Ball | ΔE - E, electronical | LBL |
| Streamer chambers | magnet, optical readout | LBL, Dubna |
| Diogene | magnet, electronical | Saclay |

**SIS 4π system requirements:**

- ❑ Coverage of a dominant fraction of 4π
- ❑ Sufficient granularity for Au + Au with low multi-hit probability
- ❑ Ability to measure $1 \leq Z \leq 15$ simultaneously
- ❑ Good light particle separation (pions, K's, p, d, t, $^3$He, $^4$He)
- ❑ Accurate flow measurement, axial symetry
- ❑ Good trigger selectivity for central collisions - rare processes
- ❑ Modular setup, posibility to couple with other devices: ($\pi^o$, $\eta$, neutrons, K, photons)

*To be used in conjunction with other devices:*
*TAPS, ALADIN, LAND, solid-state TELESCOPES, MWPC, ....*

## PHASE I SET - UP

This part of the $4\pi$ detector system is operating at GSI since august 1990 and allows the detection of light particles and fragments up to Z=16 in the angular range 1° - 30°. The fragments are analysed by a dE/dx - TOF method, thus giving the velocity and atomic number Z. Figure 8 shows the layout of the SIS/ESR facility at GSI, Darmstadt and figure 9 represents a general view of the Phase I (right hand part) and future Phase II (left) setup in cave B at GSI.

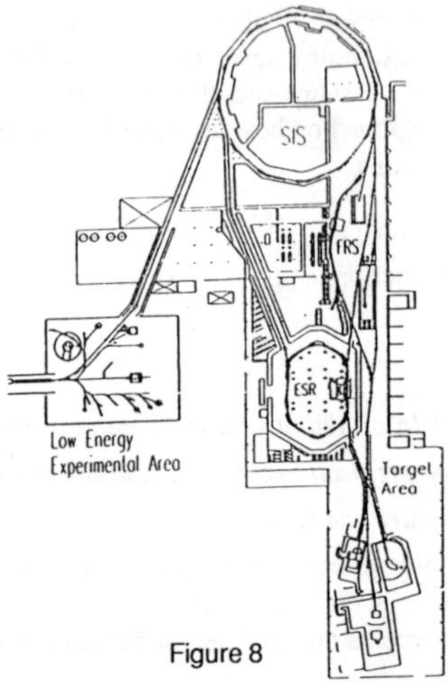


Figure 8

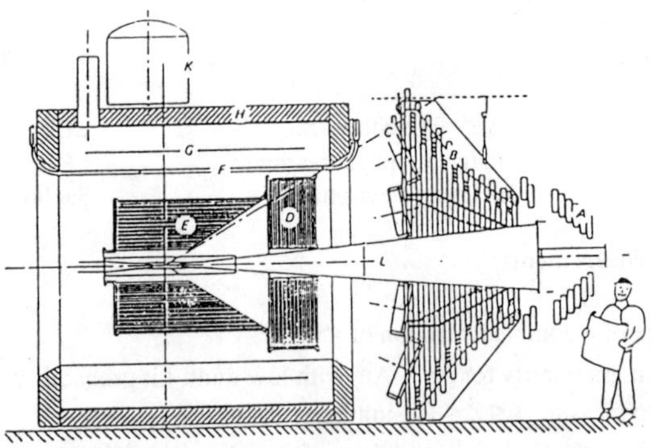


Figure 9

Structure of the $4\pi$ experimental setup at GSI

A = Zero degree detector and Rosace; B = external wall, C = Parabola, D = Helitron;

E= Central Drift Chamber

Light particles (Z = 1 to 6) are detected with the Plastic Wall consisting of two parts:

- an internal wall called the **Zero-Degree Detector** covering the polar angle range 1° to 7.5°. It consists of 7 concentric rings for a total of 252, 20mm thick trapezoïdal plastic scintillator modules coupled to photomultipliers by the mean of short light guides. Fast timing allows precise determination of the time of flight for each detected particle coming from the target.

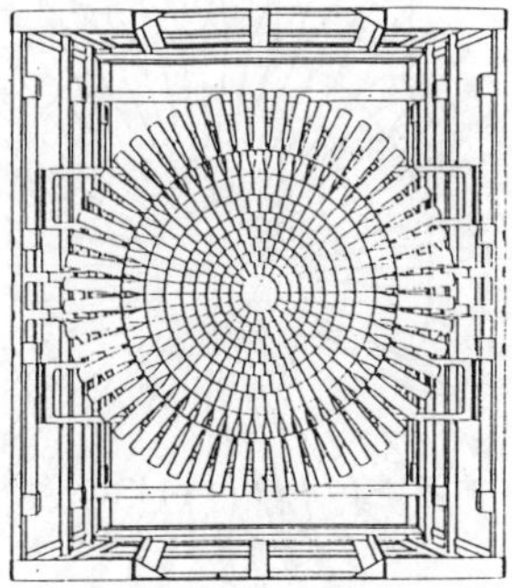

Figure 10

- an external wall called **Plastic Wall** covering the angular range 5° to 30°, composed of 8 sectors of 64 scintillator strips, 25 mm thick plastic, viewed at each end by photomultipliers. Fast timing ($\Delta t \sim 200ps$) is used to measure i) the time of flight of each detected particle ii) the position of the impact along each strip, allowing the localisation of each particle in $\theta$ ($\Delta\theta \sim 0.3°$) and $\varphi$ ($\Delta\varphi \sim 1°$ at $\theta = 30°$; 4° at $\theta = 7°$).

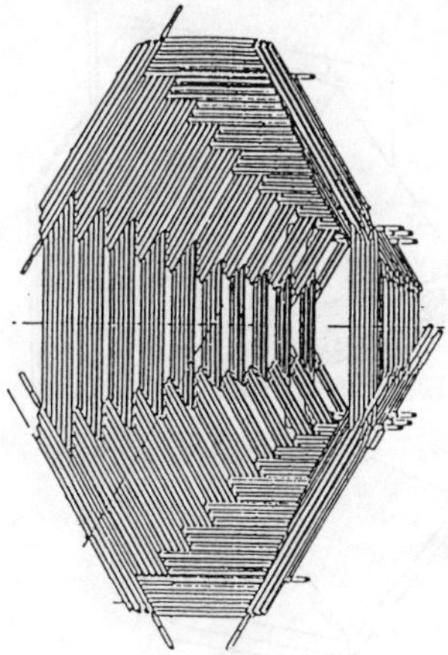

Figure 11

The external Plastic Wall

The heavier fragments are detected with a thin plastic scintillator assembly called the **Rosace** for the 1° to 5° polar angular domain and by an ionization chamber array called the **Parabola** for the 6° to 30° range.

- The **Rosace** consists of 60, 2mm thick plastic scintillator paddles sustending 6° in azimuthal angle each.

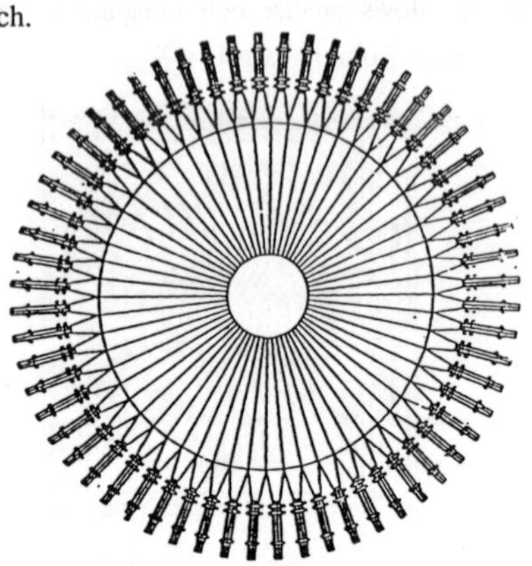

Figure 12
The Rosace, constisting of 60 independant plastic scintillator paddles.

- The **Parabola** is constitued by two rings divided in 8 sectors each. Ring 1 covers the 6° to 17° polar angle domain and ring 2 the 18° to 30° angles . Each of these 16 sectors is a gas filled double ionization chamber with the anode plane beeing subdivided in 8 independant sections.

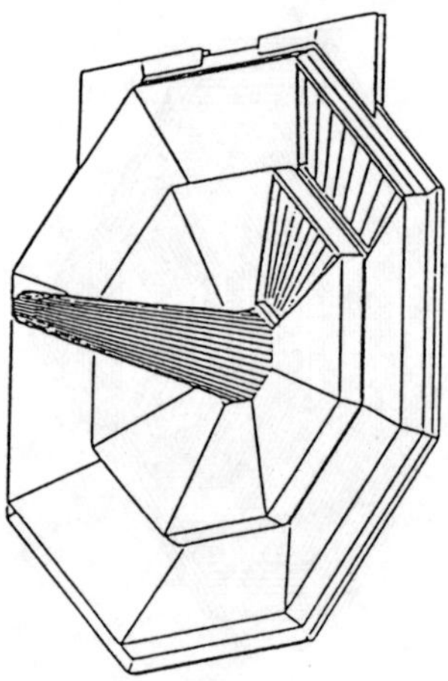

Figure 13
The Parabola, with two rings of ionization chambers (128 individual channels)

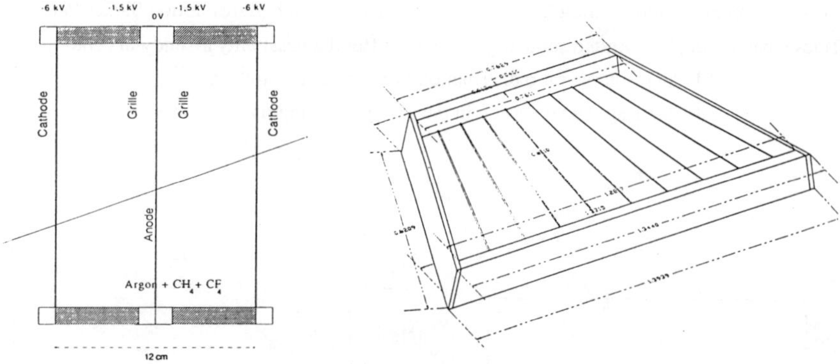


Figure 14

principle of the double ionization chamber and layout of the anode strips in an external ring module

The Ionization Chamber (I.C.) energy loss signals must be correlated with the corresponding velocity parameters derived from the Plastic Wall (P.W.) scintillator signals, this requires tracking algorithms which must take into account:

- the ions stopped in the intermediate air gap between I.C. and P.W.
- the multiple tracks through same anode strip
- the multiple hits in a scintillator strip
- the very different sensitivities of the gas and scintillator.

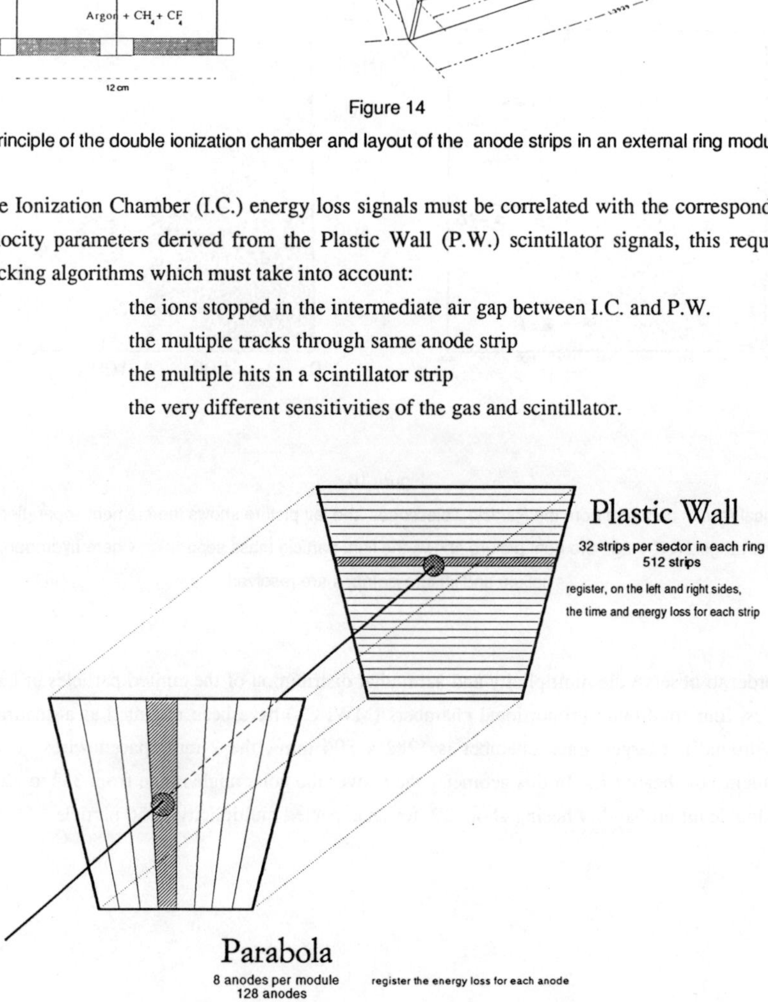


Figure 15

This 4π detector system is complemented by an ensemble of 8 high-resolution Si-CsI Particle Telescopes which, in a limited solid angle domain, offer the possibility to study in detail :

- the isotopic ratio distribution of the reaction products,
- the fragment emission in the squeeze-out region.

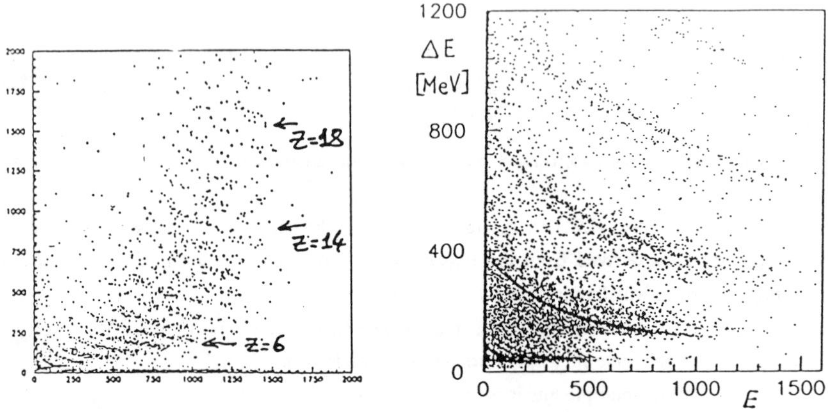


Figure 16

Typical results obtained from the Particle Telescopes, the left picture shows the element separation in the range 2 < Z < 20 , the right picture shows the light particle mass separation where hydrogen, helium and lithium isotopes are resolved.

In order to observe the multiplicity and azimuthal distribution of the emited particles at large angles, four multiwire proportional chambers (MWPC's) have been mounted as a quadratic box around the target. Each chamber is 1282 x 896 $mm^2$, the 2 mm spaced wires beeing parallel to the beam axis. In this geometry they cover the polar angle range from 35° to 128°, the double hit probability beeing about 2% for an expected multiplicity of 50 particles.

## First Results

Figure 17 shows a typical two dimensional spectrum obtained with the Zero degree detector for 200 MeV/A, Au + Au. Figure 18 presents the $\Delta E$ versus velocity spectra registered with the 8 I.C.'s of theParabola's inner ring for the same reaction at 100 MeV/A, the fragment velocity plotted on the horizontal axis is in cm/ns, for linearization purposes $v^*\sqrt{\Delta E}$ is plotted along the vertical axis.

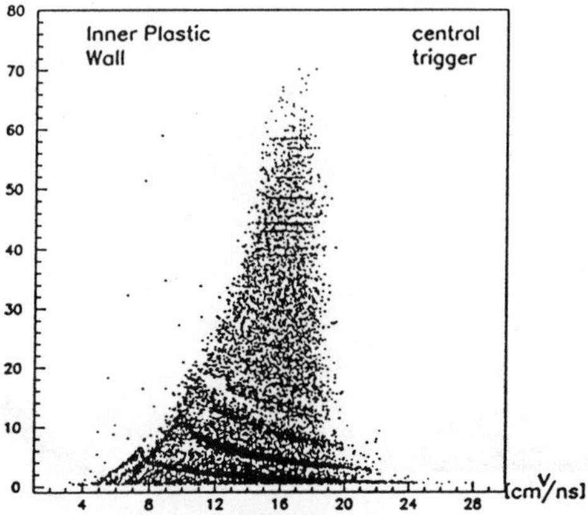


Figure 17
$\Delta E$ versus fragment velocity spectrum from the Zero degree detector

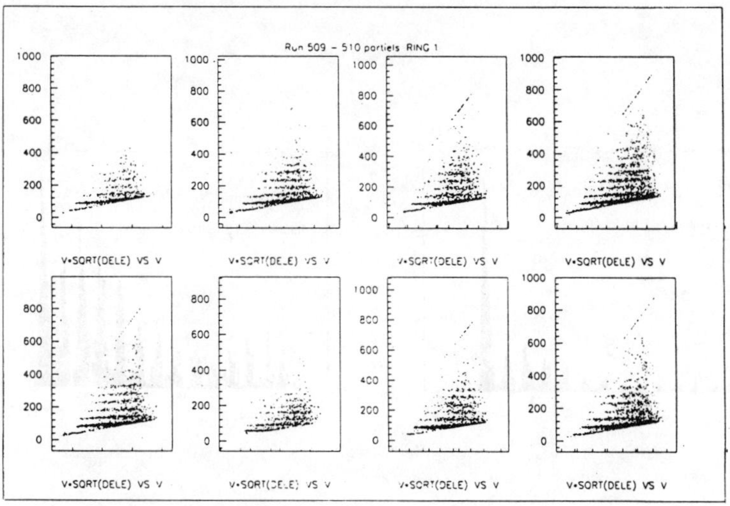


Figure 18
$v^*\sqrt{\Delta E}$ versus fragment velocity spectra for the 8 I.C. of the Parabola inner ring. Each spectrum is the sum of the spectra obtained with the 8 separated anodes.

Figure 19 presents ΔE versus velocity spectra registered with the Parabola (left hand plot) and the outer Plastic Wall (right) for Au + Au at 100 MeV/A, the fragment velocity plotted on the horizontal axis is in cm/ns. The two detectors are complementary for higher Z , lower velocity fragments which are stopped in the plastic scintillator and therefore are registered in the quenching line of the Plastic Wall.

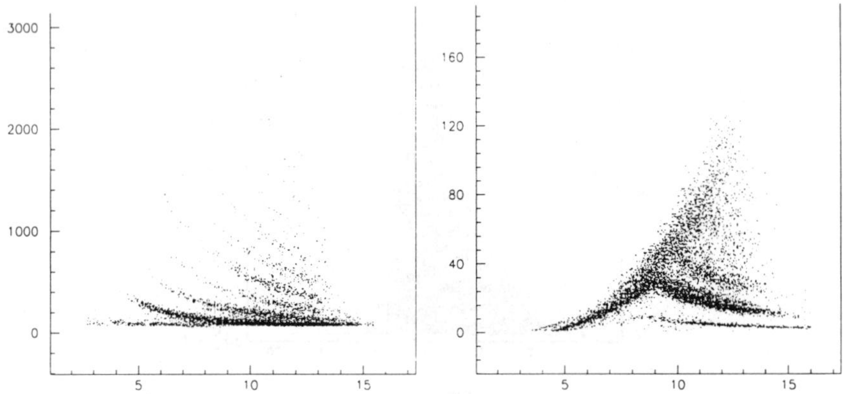

Figure 19

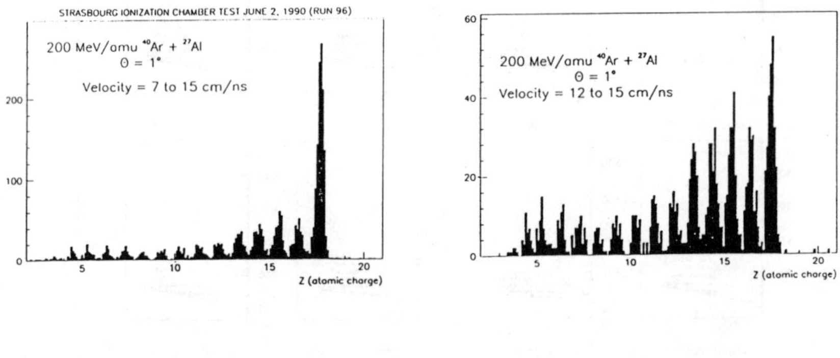


Figure 20

Reduced Z spectra obtained after calibration for the Ar+Al collision at 200 MeV/A for two velocity bins.

We have started the measurement of the excitation function for composite particle production in the Au + Au collisions in last february at energies of 100, 150 ,250 and 400 MeV/A. The data are under analysis, only very preliminary results have been extracted from test runs performed last august at 200 MeV/A

Event triggering was done by using the charged particle multiplicity information of the outer Plastic Wall shown on figure 21, five multiplicity cuts have been choosen.

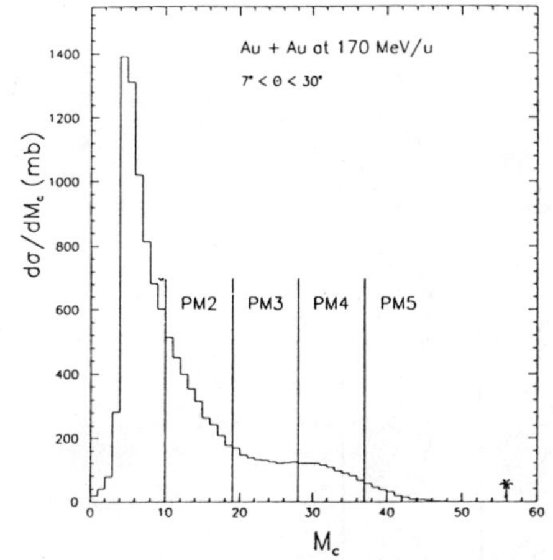


Figure 21: Measured charged particle multiplicity distribution in the Plastic Wall

Due to background contributions only multiplicity bins PM3, PM4 and PM5 can be used reliably at present. The laboratory frame angular distribution of Li are well peaked in the acceptance of the $4\pi$ detector system. The average multiplicity of the integrated fragment yield (for $Z > 2$) increases with centrality of the collision as shown on figure 22.

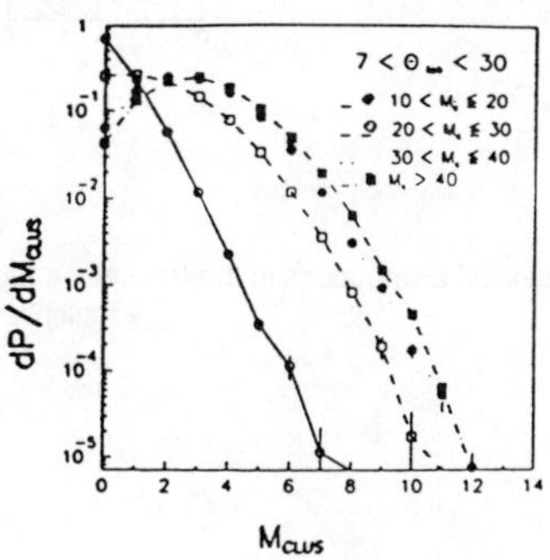


Figure 22
Cluster ($Z > 2$) multiplicity distribution in various multiplicity bins.

Information about the production mechanism can been obtained by analysing the cluster distributions in momentum space, e.g. in a transverse momentum per nucleon (pt/mc) versus rapidity (y) representation. In figure 23 we present such a representation for nuclei with Z = 1,2,3,4 and 5 for the higher multiplicity bins.

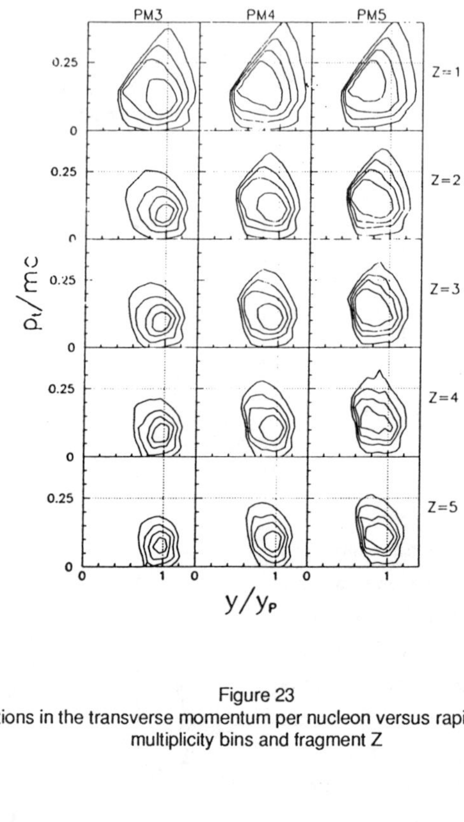


Figure 23
Measured distributions in the transverse momentum per nucleon versus rapidity for various indicated multiplicity bins and fragment Z

When the analysis will be completed, it will be possible to extend the experimental information to the mid-rapidity region and toward higher Z specialy at the lower bombarding energies. A detailed comparison with theoretical models, such as Quantum Molecular Dynamic approach is under way and will hopefully yield a better understanding of the reaction mechanism, nuclear stopping power, directed flow, entropy and equation of state.

# REFERENCES


1) S.Z. Belenk'ji and L.D. Landau, Supplemento al Vol II, serie X, Nuovo Cimento, N°1, 15 (1956)

2) A.E. Glassgold, W. Heckrotte and K.M. Watson, Ann. Phys. 6, 1 (1959)

3) H. Gutbrod, H.K. Kampert, B.W. Kolb, A.M. Kostanzer, H.G. Ritter and H.R. Schmitt, Z. Phys. C (1990)

4) H.A. Gustafsson et al., Phys. Rev. Lett. C52, 1590 (1984)

5) M. Desmoulins et al. Phys. Lett. B241, 476 (1990)

6) J.J. Molitoris et al. MSU Report MSUCL-530 (1985)

7) G. Peilert, H. Stöcker, W. Greiner, A. Rosenhauer, A. Bohnet, J. Aichelin, Phys. Rev. C39, 1402 (1989)

8) J. Aichelin , A. Rosenhauer, G. Peilert, H. Stöcker and W. Greiner., Phys. Rev. C37, 2451 (1988)

9) M. Gyulassi, K.A. Fraenkel and H. Stöcker Phys.Rev. Lett. 110B, 185 (1982)

10) P. Danielewicz and M. Gyulassi Phys. Lett. 129B, 283 (1983)

11) P. Danielewicz, G. Odyniec, Phys. Lett. 129B, 146 (1985)

12) H. Stöcker, L.P. Csernai, G. Graebner, G. Buchwald, H. Kruse, R.Y. Cusson, J.A. Maruhn and W. Greiner Phys. Rev. C25, 1873 (1982)

13) H. Gutbrod et al. Phys. Lett. B216, 267 (1989)

14) J.P. Coffin et al. Workshop on experiments and experimental facilities at SIS/ESR, GSI Darmstadt (1987) and proposal submitted by the Darmstadt - Giessen - Heidelberg - Münster Collaboration (1987).

# FIRST RESULTS OBTAINED WITH THE $4\pi$ AMPHORA DETECTION ARRAY

**D. DRAIN***

***AMPHORA group :**

I.P.N. Lyon (France) , I.S.N. Grenoble (France), H.M.I. Berlin (Germany)

**and associated collaborations :**

GANIL Caen (France), Hope College (U.S.A.), INFN Legnaro (Italy), Stony Brook National Laboratory (U.S.A.) and Texas A & M (U.S.A.)

**Abstract :**

First results obtained with the $4\pi$ multidetector array AMPHORA installed at the SARA facility are presented. Peripheral collisions have been investigated by studying events for which light particles and intermediate mass fragments are detected. Sequential decay of the projectile is observed to be the most likely process. IMF production has been investigated for different reactions at various incident energies. The role of angular momentum in highly excited nuclei has been studied by azimuthal angular correlations between emitted particles.

## INTRODUCTION

In the range of intermediate energies of the order of 25 A·MeV to 40 A·MeV available at the SARA facility, it is of great interest to investigate the evolution of reaction mechanisms. As the incident energy increases, intermediate mass fragments are observed, the origin of which is not well known. The possibility of a saturation of the temperature at a value of the order of 6 MeV for excitation energies higher than 3 A·MeV is suggested [1]. Furthermore the question is raised of the onset of a possible fast multifragmentation [2].

The main difficulty encountered in the study of these mechanisms is that all reaction products need to be detected, within large energy and mass ranges, in order to be able to reconstruct the charge and linear momentum balance of the reaction and to deduce the energy deposited in the system.

The $4\pi$ multidetector array AMPHORA [3] has been installed at the SARA facility a few years ago but first results have only recently been obtained due to the complexity and novelty of the data analysis.

# I. THE $4\pi$ AMPHORA DETECTION ARRAY

## 1) Description and performance :

Figure 1 sketches the 140 detectors of the AMPHORA array. 48 detectors are set on a forward wall (b in figure 1) at a distance of 140 cm away from target whereas 92 detectors are set on a backward "ball" (a in figure 1) and arranged on 7 rings covering angles ranging from 18° to 164° and distances from 44 to 12 cm away from target. The total geometrical coverage is 82% of $4\pi$.

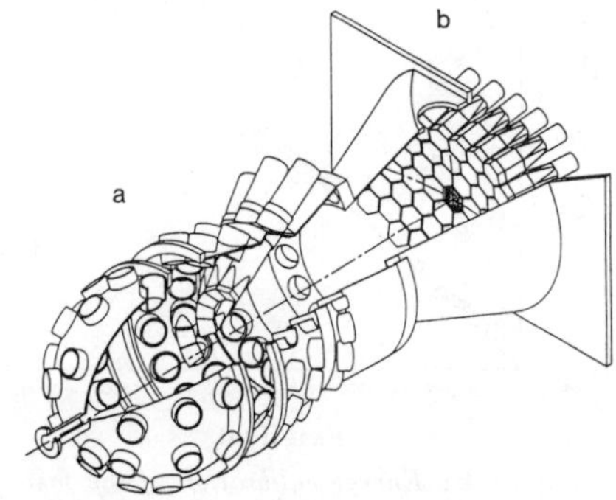


Figure **1** : *Schematic view of the 4 $\pi$ AMPHORA detection array.*

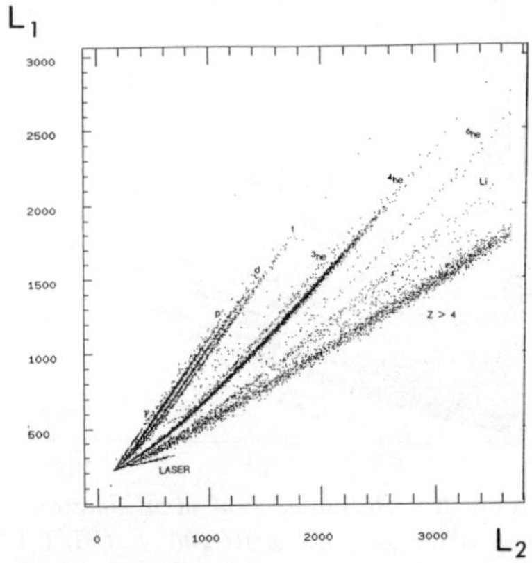


Figure **2** : *Two-dimensional identification map obtained with a CsI(Tl) crystal.*

For all of the experiments reported here, the detector elements were comprised either of CsI(Tl) crystals alone, identifying all light charged particles up to charges Z = 3 (figure 2) and detecting (with no charge identification) intermediate mass fragments (IMF) of charge larger than 4, or of BC400+CsI(Tl) cells, identifying in charge IMF's (figure 3). The inhomogeneity in the BC400 thickness or light efficiency often makes identification uncertain. For this plastic+CsI(Tl) configuration, the identification threshold depends on the thickness of the BC400 scintillator, which is 200 $\mu m$ thick both for the forward wall and for the ring set at 20° and 100 $\mu m$ thick for the rings set between 31° and 47°. The energy threshold is of the order of 14 MeV for $\alpha$'s and 7 A·MeV for heavier ions for a 200 $\mu m$ thick BC400 scintillator.

## 2) Calibrations :

The light response of scintillating materials depends on the detected particle and is not linear as a function of the energy deposited in the material. This makes the energy calibration rather delicate.

*a. Identification :*

Whereas setting identification masks on two-dimensional maps from the CsI(Tl) elements is straightforward, setting masks on plastic-CSI(Tl) detector maps is made difficult by the poor quality of the charge separation often observed. A software routine which sets masks semi-automatically has been developed by P. DÉSESQUELLES [4]. It allows identification of charges with a $\pm 0.5$ charge unit uncertainty for charges lower than 9 and with a $\pm 2$ charge unit uncertainty for higher charges in the worst case.

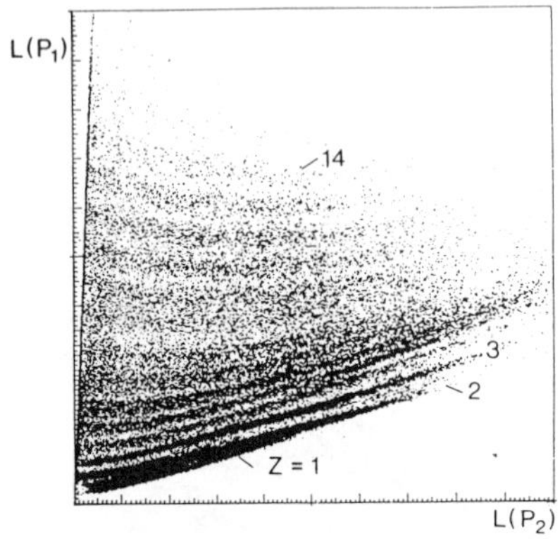


Figure **3** : *Bidimensional identification map obtained with a BC400 + CsI(Tl) phoswich detector.*

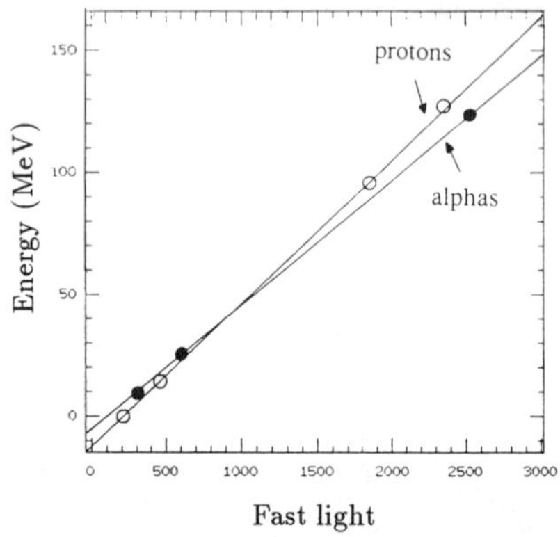


Figure **4** : *Energy calibration of the fast light obtained for protons and α's in a BC400 + CsI(Tl) phoswich detector.*

*b. Energy calibration :*

For CsI(Tl)'s alone, it has been demonstrated [3] that the amount of light integrated within the 400 ns wide fast gate is proportional to energy. Consequently, only a few calibration points (obtained from sources, scattered $\alpha$ particles, punch-through points in the CsI(Tl) or energy losses in a $\Delta E$ Si detector set in front of the CsI(Tl)) are needed in order to get a correct calibration with a $\pm 5\%$ uncertainty (figure 4). Calibrating the BC400+CsI(Tl) modules on the other hand is more delicate as heavy ion beams of various charges and energies are not very easily available. Only the technique of a $\Delta E$ Si detector set in front of the detector to be calibrated is usable in that case but the calibration gets very uncertain for high incident energies since the differential energy loss $dE/dx$ for the Si detector slowly decreases as a function of energy E, introducing a large uncertainty on the determination of the incident energy. An alternative method is to use the set of calibration parameters published by D.W. STRACENER et al. [5].

Our own conclusion on the problem of calibrating scintillators is that the best method would be to measure an angular distribution in a single run performed in an appropriate reaction chamber using a Si telescope and then to normalize the AMPHORA singles spectra to these spectra. This calibration would be less uncertain and more "physical" provided it is admitted that the response is either linear as a function of energy or well known.

**3) AMPHORA as an event filter :**

Although the AMPHORA array covers 82 % of 4 $\pi$, as mentioned above, it still introduces a systematic bias which, together with the finite granularity, distorts the multiplicity spectra. A simulation of the AMPHORA response has been performed by X.C. HU [6] in which physical observables such as source velocities, multiplicities and angular distributions of particles emitted by these sources, and excitation energy carried by the composite system in order to obtain a $4\pi$ multiplicity spectrum. The exact geometry of the AMPHORA cells is then used to deduce the AMPHORA multiplicity spectrum to be compared to the measured one. Figure 5 shows the results of this simulation as compared to experimental results. A significant shift can be seen for high multiplicities of $\alpha$'s and protons. This shift comes on one hand from pile-up events and on the other hand from energy thresholds that are not handled by the simulation. It can be concluded that, not only the measured multiplicity is not the real one but also that events for which the total charge of the system is measured are very unlikely to be observed.

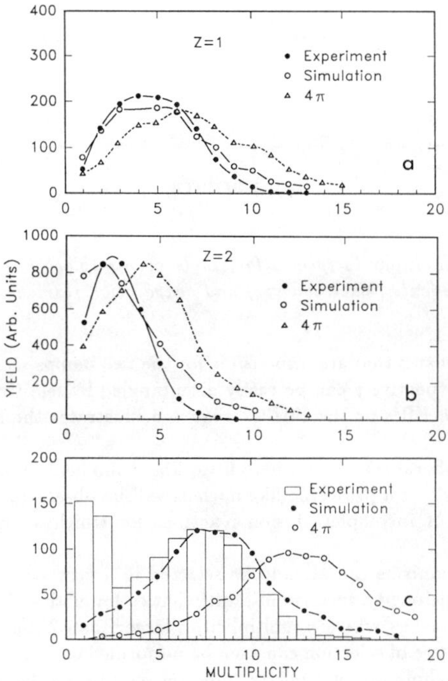


Figure **5** : *Experimental multiplicity spectra for protons (***a***), $\alpha$'s (***b***) and for all charges (***c***) as compared to predictions of reference [6].*

## 4) Selection of various mechanisms :

*a. Selection according to multiplicity :*

Figure 6 shows a typical multiplicity spectrum obtained for light charged particles (regardless of the charge) in the $^{32}S + Ag$ system at 30 A·MeV incident energy [6]. These events were triggered by a set of 4 Si detectors set at 8° in front of the CsI's and that were meant to detect the evaporation residues (ER) as well as the intermediate mass fragments (IMF).

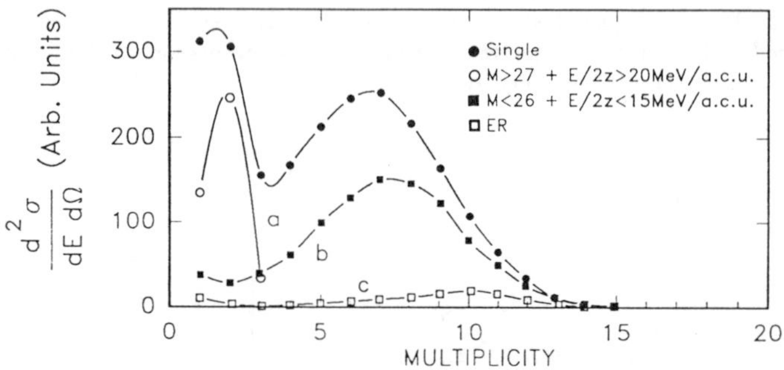


Figure **6** : *Measured multiplicity spectra for single events (black circles), elastic events (open circles), damped events (black squares) and evaporation residues (open squares).*

The various mechanisms that are responsible for the two bumps seen in figure 6 at low and high multiplicity respectively can be easily disentangled by selecting different masses or energies either of the ER's or the IMF's. Figure 6 illustrates the different selections that have been made :

- **a** refers to very peripheral reactions where little linear momentum is transfered to the target and consequently a projectile-like nucleus will be observed,
- **b** refers to complete or incomplete fusion reactions for which damped IMF's will be observed.

One sees that mechanisms can already be selected in a very straightforward way by setting gates on the multiplicity spectrum. Multiplicities lower than 3 can be associated with low momentum transfers whereas multiplicities larger than 7 can be associated with central collisions. This type of selection can even be performed on-line using a fast decision unit which allows a threshold on multiplicity to be set on an event-by-event basis.

*b. Selection using additional triggers :*

Whenever selection of a particle type is needed, we are soon limited by the capability of AMPHORA for detecting all of the reaction products. It is straightforward to set off-line conditions on the IMF's identified at a given angle (provided BC400+CsI(Tl) modules have been mounted at this angle) whereas it is not possible to set any condition on fission products as AMPHORA is not able to identify them. Consequently, although inclusive experiments are interesting to perform in order to get good statistics and to minimize the kinematic bias due to the geometrical position of the trigger, it is very often necessary to use triggers other than standard AMPHORA cells. Trigger detectrors used so far in AMPHORA experiments have been ionisation chambers, with or without position information and used either by themselves for detecting fission products or coupled with CsI's for detecting IMF's. Also Si detectors mounted as $\Delta E$'s in front of CsI crystals have been used in order to detect ER's or PLF's. This type of selection can be supplemented by an off-line threshold on multiplicity or on any other global parameter which is indicative of the reaction mechanism of interest.

We report here on two examples of results obtained in experiments performed with AMPHORA at the SARA facility. These results will hopefully give evidence to the unique capability of a $4\pi$ multidetector array to fully reconstruct events of interest in two different types of reactions, peripheral collisions and central collisions.

## II. STUDY OF PROJECTILE DECAY

In the $^{40}Ca$ + $^{nat}Cu$ reaction [4] at 35 A·MeV incident energy, all the AMPHORA CsI detectors in the polar angular domain $2° \leq \theta \leq 55°$ were equipped with 100 $\mu m$ or 200 $\mu m$ thick plastic foils. With this set-up, reaction products were identified up to charge 20.

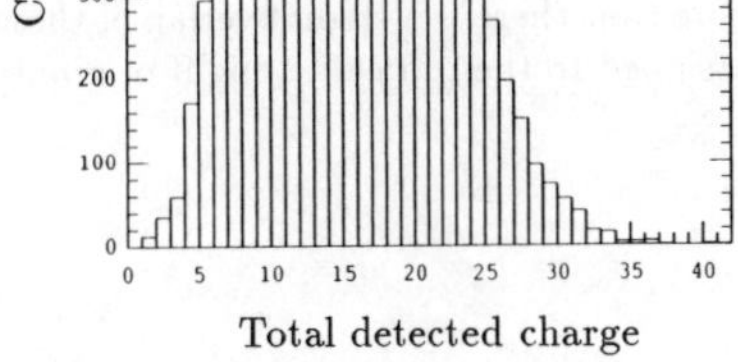


Figure 7 : *Total charge of all reaction products identified in inclusive runs.*

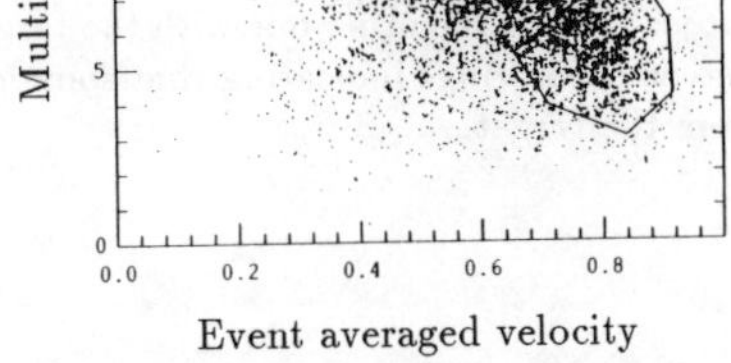


Figure 8 : *Total detected multiplicity as a function of center-of-mass velocity for events of total charge $Z_t \geq 20$. Velocities are normalized to the incident beam velocity $V_b$.*

The distribution of the total charge of the identified products is presented in figure 7. The mean total charge (around $Z_t = 16$) is much lower than the total charge of the system ($Z_t = 49$). The missing charge is due to the incomplete angular coverage of AMPHORA (82% of $4\pi$) and to the detection thresholds which exclude low energy fragments and heavy residues. In a first step of the data analysis, we focused our study on the projectile breakup processes in peripheral collisions. For such reactions, one expects that the geometrical efficiency and threshold effects do not strongly affect the detection of the breakup residues.

### 1) Selection of projectile breakup residues :

The multiplicity of the detected particles is plotted in figure 8 versus their center of mass velocity for events of total charge $Z_t \geq 20$. We observe a strong correlation between these two quantities, which can be directly related to the amount of energy dissipation or to the impact parameter [7]. Very few low multiplicity events are observed since the most peripheral collisions were suppressed by an on-line multiplicity threshold of 4.

In order to select peripheral collisions only events with total charge $Z_t \geq 20$ and included in the contour region plotted in figure 8 were retained. To isolate particles and fragments coming from the breakup of the projectile, the following event-by-event procedure was performed: First, the center of mass velocity of all fragments of charge $Z \geq 3$ was determined. These fragments having, with respect to this center of mass, a velocity lower than a given value $V_z$ were then selected. For each charge, the chosen velocity $V_z$ corresponds to an energy of about twice the Coulomb energy. Then the center of mass velocity of the selected fragments was determined and the total charge associated with all the particles and fragments having, with respect to this new center of mass, a velocity lower than the previously defined $V_z$ values was calculated. Selected products with a total charge equal to 20 were identified as the residues of the projectile breakup. The velocity distribution of the reconstituted projectile has an average velocity of 86 % of the beam velocity $V_b$.

In figure 9, we present the proton velocity distribution ($V_\parallel, V_\perp$) in the reaction plane. The chosen projection reference frame has as origin the projectile center of mass and as parallel axis the direction defined by the projectile and target centers of mass. The target center of mass is derived from momentum conservation. The target mean recoil velocity is 0.08 $V_b$. The spectrum exhibits two emission rings, one of which is associated with the projectile and the other one with the target. We note that there is a slight overlap of these two emission rings indicating that some protons assigned to the projectile might originate from the target.

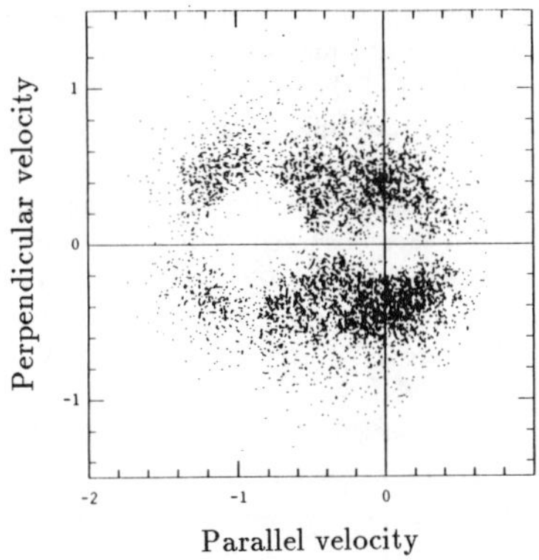


Figure **9** : *Proton velocity distribution (normalized to $V_b$) for events associated with projectile breakup.*

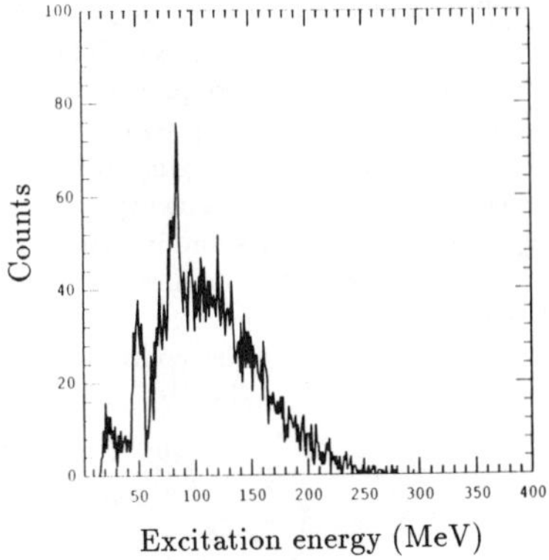


Figure **10** : *Excitation energy spectrum deduced for events associated with projectile breakup.*

**2) Excitation energy, multiplicity and charge distributions :**

The excitation energy of the reconstructed projectile was calculated event-by-event from the measured kinetic energies. The Q-value was determined by assuming the mass of the most stable isotope for fragments of charge $Z \geq 3$. The neutron contribution was estimated by assuming a neutron multiplicity equal to 1.5 times the number of protons and by assuming that each neutron carries away on the average an energy of 12 MeV. The excitation energy distribution is plotted in figure 10. Its mean value is 110 MeV and its width is about 80 MeV.

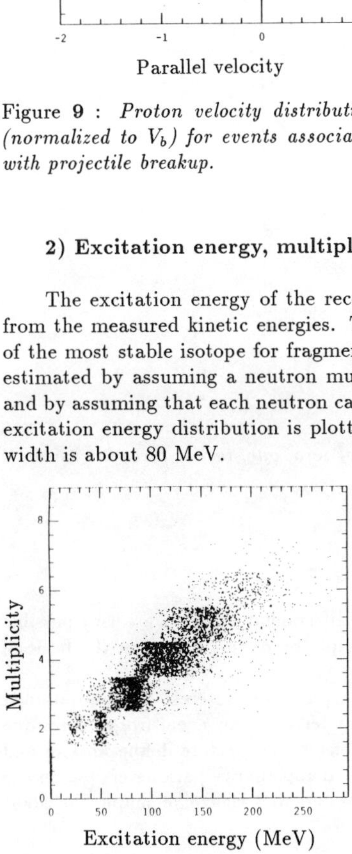


Figure **11** : *Total multiplicity of quasi-projectile fragments as a function of excitation energy.*

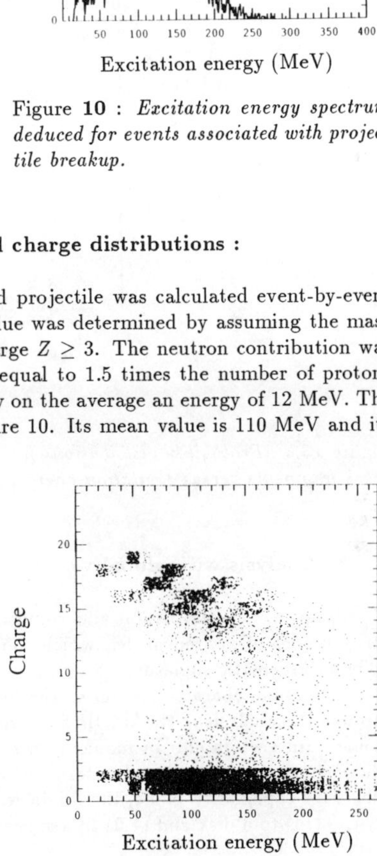


Figure **12** : *Charge of the quasi-projectile fragments as a function of excitation energy.*

The multiplicity and charge distributions versus the excitation energy are shown in figures 11 and 12, respectively. The production yields of intermediate mass fragments, defined as $3 \leq Z \leq 11$, are plotted in figure 13 as a function of the excitation energy.

Figure 11 shows a very strong correlation between the multiplicity and the excitation energy, the multiplicity ranging from 2 to 8 when the energy goes from 10 to 280 MeV. At low energy, we observe one heavy residue and a few light particles (figure 12) ; at high energy, only light particles and intermediate mass fragments are observed. The probabilities for emission of one, two or three IMF increase with the energy up to 160 MeV (figure 13); above 160 MeV, the one IMF emission probability decreases whereas the two and three IMF emission probabilities continuously increase with energy.

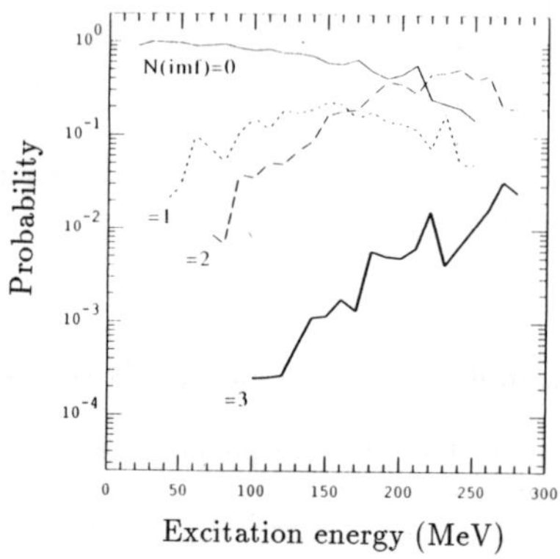


Figure **13** : *Probability distributions for emission of zero, one, two or three intermediate mass fragments versus excitation energy.*

### 3) Analysis with global variables :

Binary sequential fission and simultaneous multifragment breakup are two possible decay modes of excited nuclei, which could in principle be distinguished on the basis of different kinematic signatures [8].

A possible characterization of the breakup mechanisms of interest can be obtained through the analysis of the kinetic flow tensor constructed event-by-event from the residue momentum vectors [9]. In momentum space, the tensor eigenvectors define an ellipsoid whose shape can be analyzed in terms of sphericity and coplanarity parameters [8]. In the sphericity-coplanarity plane, the origin would correspond to a cigar-like shape, the point $(3/4, \sqrt{3}/4)$ to a disk and (1,0) to a sphere.

Figure 14 presents the result of such an analysis applied to the projectile breakup residues. In figure 15 are shown the sphericity-coplanarity plots resulting from the binary sequential fission and multifragmentation calculations performed by B.A. HARMON et al. [10]. The sequential decays have a more elongated momentum distribution whereas the multifragmentation events are more spherical. Comparison between figures 14 and 15 indicates that our experimental results are qualitatively in agreement with a binary sequential decay mechanism.

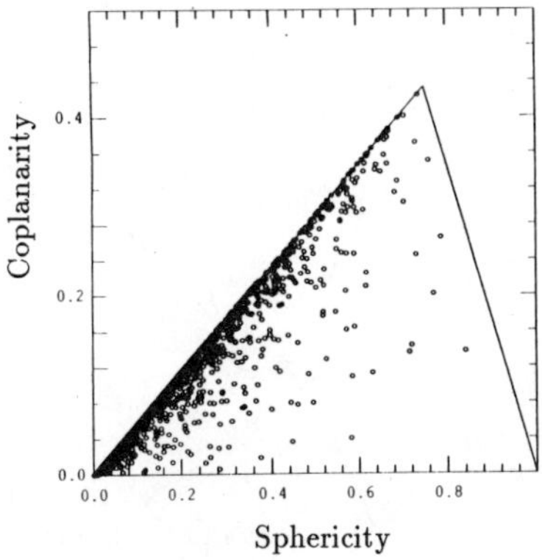


Figure **14** : *Experimental sphericity-coplanarity distribution.*

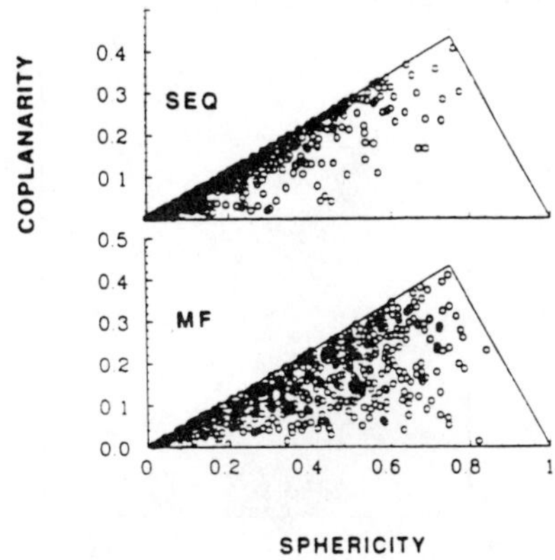


Figure **15** : *Sphericity-coplanarity distributions for sequential decay (top) and multifragmentation (bottom) predicted in reference [10].*

Another analysis of the shape of the momentum distribution has been performed using the sphericity tensor [11]. The eigenvalues of this tensor give the ellipsoid axis lengths from which are defined the symmetry axis and the eccentricity of the ellipsoid. The eccentricity value is 0 for a spherical shape, close to 1 for a cigar and -1/2 for a disk.

The eccentricity distribution is plotted in figure 16 as a function of the excitation energy. We observe both cigar and disk shaped events, the cigar configuration being the most probable at low excitation energy. When the excitation energy increases, the momentum distribution seems to become on average more spherical. However, this effect is probably correlated with the increase of the multiplicity with energy. Indeed, when not very large multiplicities are involved, the eccentricity parameter is rather sensitive to the multiplicity. It is interesting to notice that no significant difference is observed for events with three intermediate mass fragments. Thus, no conclusion can be made about a possible evolution towards a simultaneous multifragment breakup mechanism for high excitation energies.

An interesting piece of information concerning the effect of the angular momentum is provided by the angle $\theta_n$ between the longest axis of the ellipsoid and the normal to the reaction plane. This angle measures the average direction of matter flow with respect to the reaction plane. If the quantity $1 - cos(\theta_n)$ is equal to 0, the flow is perpendicular to the reaction plane ; if this quantity is equal to 1, the flow is in the reaction plane. The $1-cos(\theta_n)$ distribution extracted from the data shows that the fragments are preferentially emitted in the reaction plane (figure 17). This is a clear signature of a strong angular momentum effect on the breakup of the projectile. The concentration in the reaction plane is more pronounced for events with three intermediate mass fragments (figure 17), indicating that IMF emission is favoured for the highest angular momenta and associated with the first steps of the de-excitation.

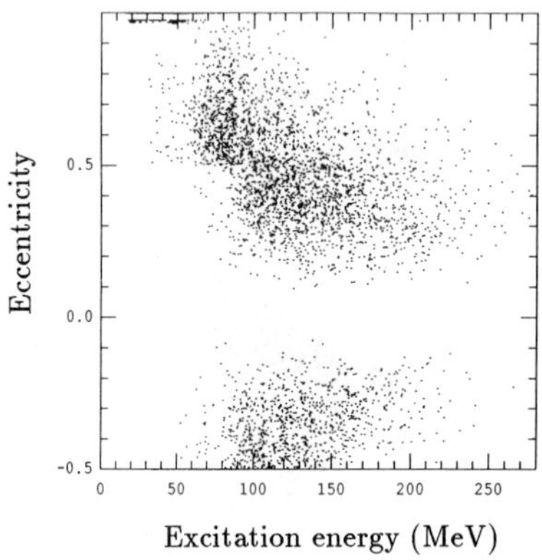


Figure **16** : *Eccentricity distribution deduced from experiment as a function of excitation energy.*

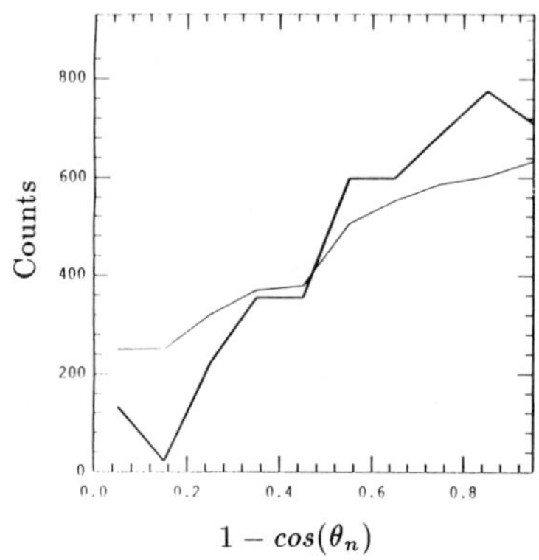


Figure **17** : *Angular distribution of the ellipsoid symmetry axis with respect to the normal to the reaction plane for all events (thin line) and for events with three IMF (thick line).*

## III. STUDY OF CENTRAL COLLISIONS

### 1) Selection of evaporation residues (ER):

We show here, as an example, the $^{32}S + {}^{58}Ni$ at 30 A·MeV [12] in which 4 trigger counters were mounted symmetrically around the beam axis at $\theta = 8°$ in order to detect evaporation residues. Figure 19 shows a mass versus velocity contour plot for residues detected in the four trigger detectors. The events enclosed in the triangle drawn in figure 19 originate from central collisions and are the so-called evaporation residues. However, since the system is rather symmetrical, these events may also include fission fragments (FF) (indeed a kinematic calculation of the fission of the composite system into two fragments of equal mass confirms that a fragment detected at 8° would have a velocity ranging between 2.8 and 3.8 cm/ns for a linear momentum transfer ranging from 50 % to 90 %) (figure 20). The $4\pi$ coverage of AMPHORA can be used to select events involving an ER and particles of charge Z=1 or Z=2 only and to reconstruct the total detected charge on an event-by-event basis. Figure 21 shows the distribution of the total charge $Z_t$ calculated as follows :

$$Z_t = \frac{A_{HI}}{2} + (2\,M_{He} + M_H)/\epsilon \tag{1}$$

where

$A_{HI}$ is the mass of the heavy fragment detected at 8°,
$M_{He}$ is the multiplicity of helium nuclei detected in AMPHORA,
$M_H$ is the multiplicity of hydrogen nuclei detected in AMPHORA,
$\epsilon$ is a factor that takes into account the efficiency of the device for detecting light particles, equal to 0.75.

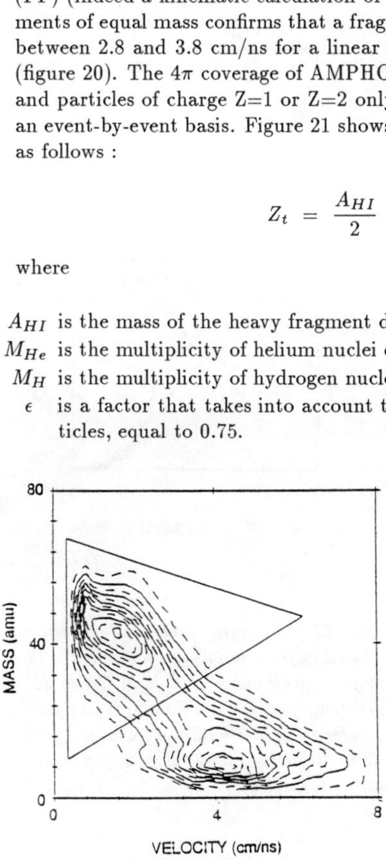


Figure **19** : *Mass versus velocity contour plots of fragments stopped in the Si detectors set at $\theta_{lab} = 8°$.*

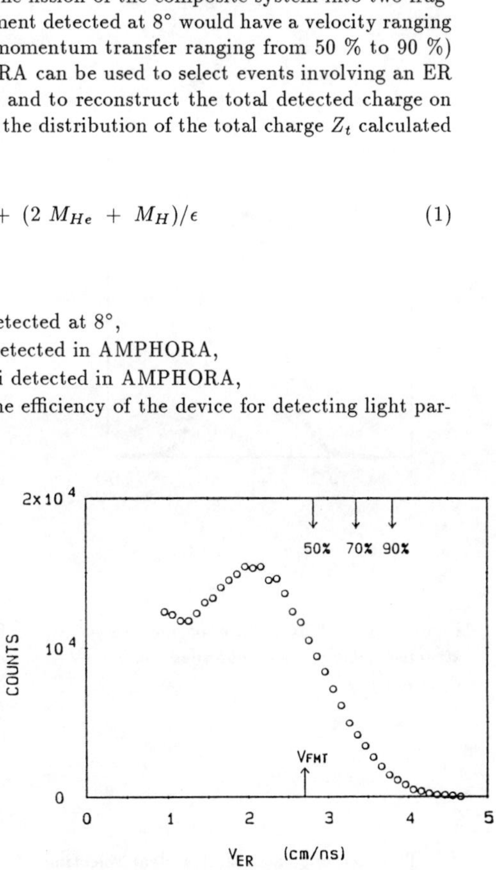


Figure **20** : *Velocity spectrum for events enclosed in the triangle drawn in figure 19.*

The region labelled "A" in figure 21 and defined by $Z_t > 40$ corresponds to genuine evaporation residues whereas the region labelled "B" refers to events in which the average undetected charge amounts to about the charge of the heavy fragment. When events are selected within class "A", we obtain the typical ER velocity spectrum shown in figure 22. When selecting within class of events "B" and additionally selecting either low or high total multiplicity of light particles $M_{LP}$, one selects either rather peripheral collisions in which the projectile-like fragment (PLF) is emitted at very small angle (within the beam pipe) and the detected one is the target-like fragment (TLF) or a fusion-fission mechanism. In the latter case, the velocity spectrum looks like the spectrum of the fragments detected at 8°.

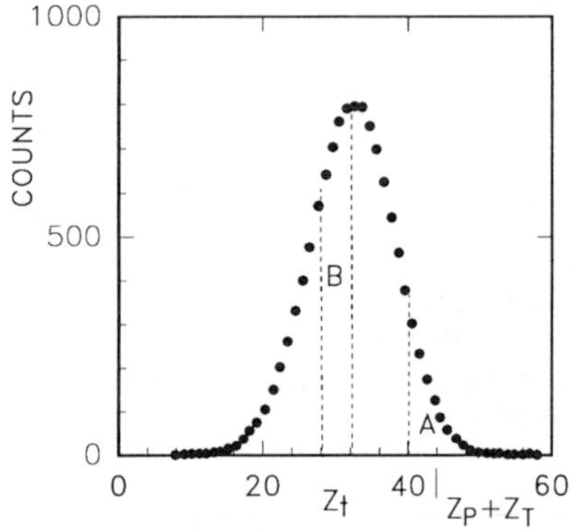


Figure **21** : *Distribution of the reconstructed charge $Z_t$ obtained using equation (1).*

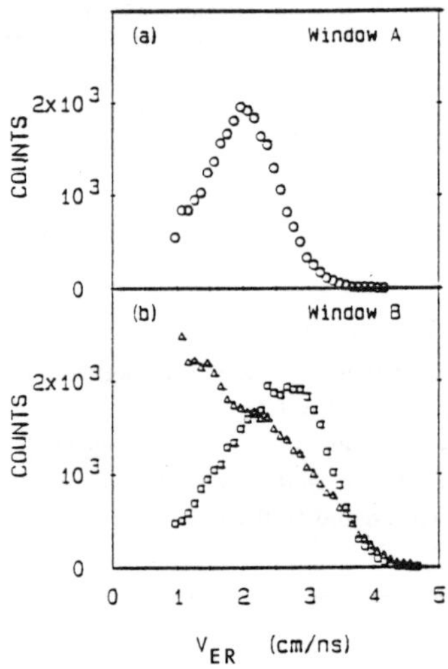


Figure **22** : *Velocity spectrum for events enclosed in the triangle drawn in figure 19 and for selections A and B in figure 21. Open triangles and open squares in figure (b) refer to events with light charged particle multiplicities lower than 4 and larger than 5, respectively.*

This demonstrates that a clear selection of evaporation residues in a symmetric system, like the one studied here, cannot be achieved simply by setting a gate on the heavy residue mass versus velocity contour plot measured at forward angle, but that an additional condition on $Z_t$ (class "A" events) is necessary.

**2) Study of IMF's emitted in central collisions :**

Class "A" events are produced in fusion-evaporation processes where probably no clusters heavier than $\alpha$-particles are emitted, neither as PLF in early stages of the reactions, nor as statistically emitted fragments from the composite system. Therefore, the excitation energy deposited in the primary compound nucleus can be rather accurately determined as :

$$E^*_{th} = E_b - \sum_{ij} < E_{ij} > (dM_{ij}/d\Omega_j)\Delta\Omega_j - E_{rot} - E_{kin} + Q \qquad (2)$$

where

$E^*_{th}$ is the thermal excitation energy deposited into the system,

$E_b$ is the incident energy,

i, j indicate the particle type and the angular position of the AMPHORA detectors,

$< E_{ij} >$ is the average energy of the pre-equilibrium component of the spectrum,

$\frac{dM_{ij}}{d\Omega_j}$ is the differential multiplicity of the pre-equilibrium component,

$E_{rot}$ is the rotational energy of the compound nucleus system estimated from the average value of the spin distribution for fusion,

$E_{kin}$ is the kinetic energy of the recoiling compound system and

Q is the reaction Q-value.

The values of the excitation energy per nucleon deposited into the system, $\epsilon^*_{th} = E^*_{th}/A_{CN}$, extracted with (2) are reported in table 1 a in MeV/a.m.u. Velocities $V_{ER}$ and $V_{CN}$ are listed in cm/ns, mass $A_{ER}$ in a.m.u., and temperatures $T^p_{EV}$ and $T^\alpha_{EV}$ in MeV.

| | a | b |
|---|---|---|
| $M^p_{PE}$ | $3.5 \pm 0.5$ | $2.6 \pm 0.5$ |
| $M^\alpha_{PE}$ | $2.7 \pm 0.8$ | $1.5 \pm 0.7$ |
| $V_{ER}$ | $2.0 \pm 0.09$ | $2.3 \pm 0.09$ |
| $V_{CN}$ | 1.98 | 2.31 |
| $A_{ER}$ | $49.0 \pm 2.4$ | $37.0 \pm 2.4$ |
| $\epsilon^*_{\mathbf{CN}}$ | $\mathbf{3.2 \pm 0.2}$ | $\mathbf{4.8 \pm 0.2}$ |
| $M^p_{EV}$ | $5.3 \pm 0.7$ | $3.7 \pm 0.7$ |
| $M^\alpha_{EV}$ | $3.2 \pm 1.3$ | $2.3 \pm 1.3$ |
| $T^p_{EV}$ | $5.0 \pm 0.6$ | $5.0 \pm 0.6$ |
| $T^\alpha_{EV}$ | $6.1 \pm 1.1$ | $6.0 \pm 1.4$ |

**Table 1**

The composite system is formed with an average excitation energy of about 3.2 A·MeV, whereas protons are emitted with a temperature of 5 MeV and $\alpha$'s with a temperature of 6.1 MeV. Such a value of the excitation energy is low compared to the total available center-of-mass energy of 6.9 A·MeV for compound nucleus formation in complete fusion. This originates from the particular selection of events for which one heavy fragment is detected at 8° in coincidence with particles of charge $Z = 1$ or $Z = 2$ in the AMPHORA array. This leads to a composite system of mass of the order of 75 with a great deal of preequilibrium light particle emission. A larger excitation energy could be obtained with a different selection on the AMPHORA data. The same analysis can be done by selecting one evaporation residue at 8° (with the additional constraint of no IMF at angles smaller than 30° in order to rule out events for which one IMF is emitted forward) and at least two IMF's originating in the evaporative stage ($Z > 3, \theta > 30°$). The class of events described now is compatible with the formation of a composite system of mass 80 (table 1 b) instead of 75 and an excitation energy of 4.8 A·MeV instead of 3.2 A·MeV in the previous selection. This demonstrates that the processes leading to emission of several intermediate mass fragments (two or more) are associated with higher excitation energies. It should be noted however that the temperatures of proton and $\alpha$ emission we observed are the same for both selections.

Therefore mainly two classes of events are found :

- up to excitation energies of the order of 3 A·MeV the compound nucleus decays by emission of light particles in a way that is well described by statistical calculations
- at higher excitation energies, one or more clusters are emitted on the average early in the de-excitation chain.

Although relative cross sections for these two processes cannot be deduced, the measured average excitation energies can be compared with the statistical multifragmentation model of Sa Ben-Hao [13] (figure 23).

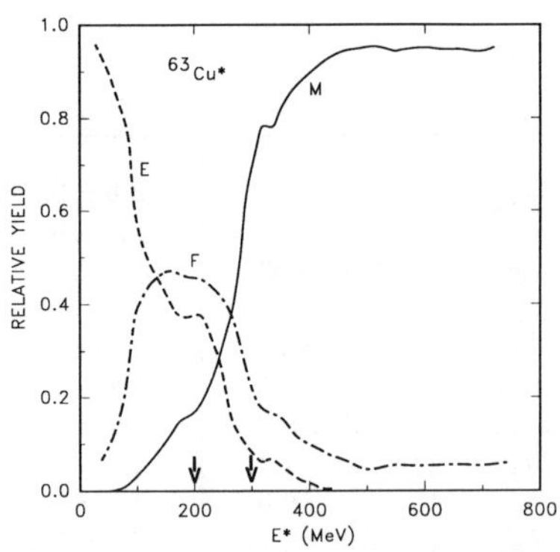


Figure **23** : *Relative yields for different partitions of a $^{63}Cu$ nucleus as a function of excitation energy as predicted in reference [13]. The two arrows indicate the average excitation energies deduced from table 1 a and b.*

In another experiment ($^{32}S$ + $Ag$ at 30 A·MeV) [6], the probability of having 1,2,3,4 IMF's has been investigated as a function of charged particle multiplicity (figure 24 a). In this way the IMF multiplicity has been deduced as a function of charged particle multiplicity (figure 24 b). The IMF multiplicity increases up to 1.4 for central collisions, associated with high AMPHORA multiplicities. From the data measured in AMPHORA and the results of a simulation, charge, linear momentum and energy balances can be reconstructed within three bins set on the heavy residue velocity (corresponding to $\approx 50, 75$ and 100% of the full linear momentum transfer). The missing mass and kinetic energy are estimated to be of the order of 13 to 20 mass units and 4 to 8 MeV depending of the velocity bin.

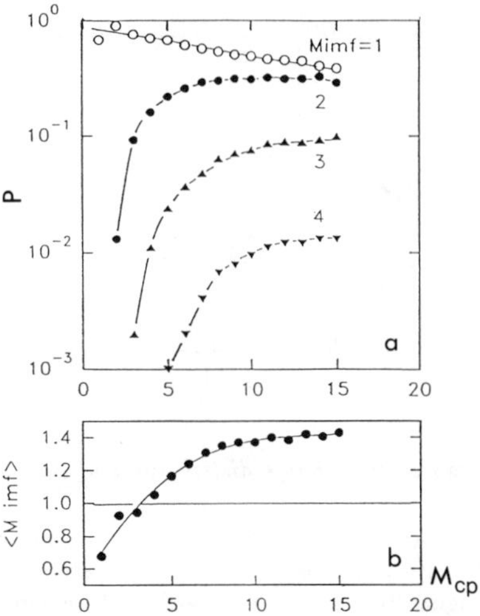


Figure **24** : *Probability of emission of one, two, three and four IMF's (a) and mean IMF multiplicity (b) as a function of charged particle multiplicity.*

The analysis of IMF's detected in ionisation chambers set at angles ranging from 8° to 120° confirms that point. It also seems to infer that, although IMF emission from DIC (deep inelastic collisions) cannot be ruled out, for large linear momentum transfers, IMF emission could be a statistical process. As seen in figures 25 and 26, the IMF angular distribution is more isotropic for high LMT's and at backward angles the charge distribution is of exponential type.

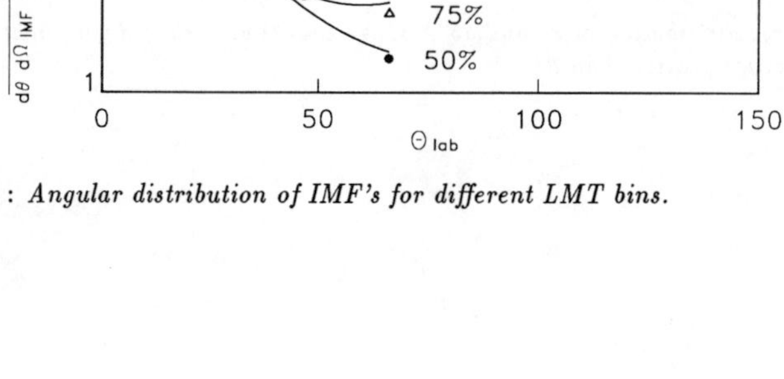


Figure **25** : *Angular distribution of IMF's for different LMT bins.*

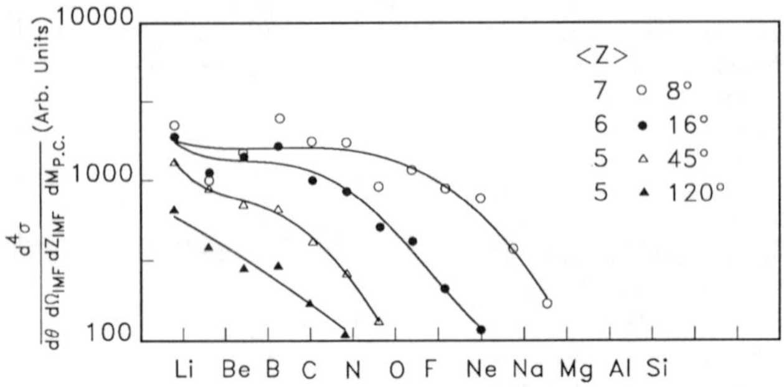


Figure **26** : *Charge distributions of IMF's at four laboratory angles.*

An attempt to find evidence for multifragmentation at SARA energies has been made by studying the $^{40}Ca$ + $^{27}Al$ system in central collisions at two incident energies [14]. Figure 27 shows that the yield of IMF's increases between 25 and 35 A·MeV as compared to the yield of fragments of higher masses. Figure 28 shows the total kinetic energy obtained from light particles $Z \leq 2$ for events of multiplicity greater than 8 (associated with central collisions) as compared to predictions of the evaporation code "EDMON" [15]. A good agreement is seen with experimental results with the assumption of an evaporative process although a nonevaporative component is present at high energies. Therefore no evidence is found for the onset of a multifragmentation process between 25 and 35 A·MeV.

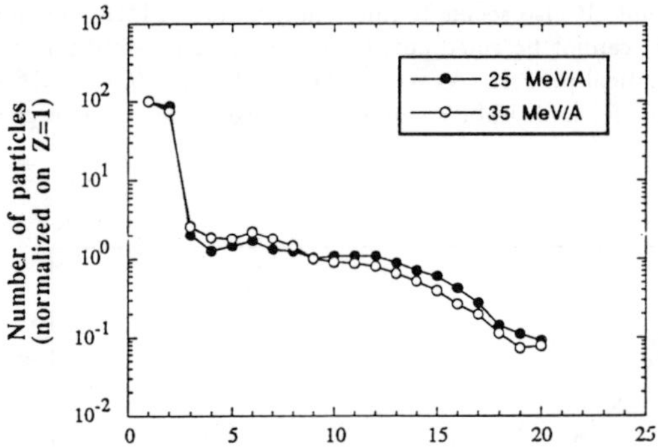


Figure **27** : *Charge distribution at 25 and 35 A·MeV. Events are selected with light charged particle multiplicities greater than 8.*

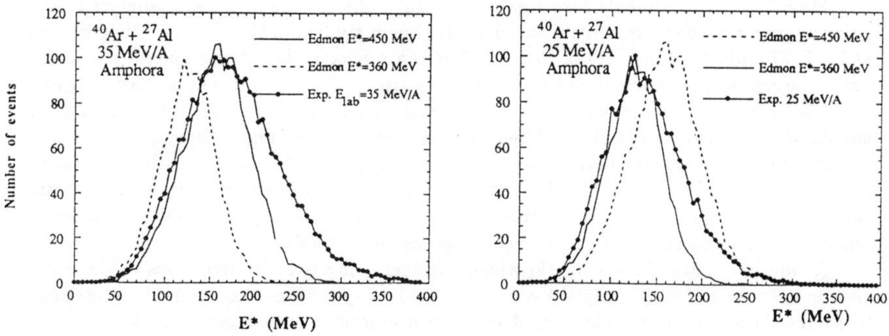


Figure **28** : *Total kinetic energies for light particles of charge lower than 3 as compared to the energies predicted by the evaporation code "EDMON" [15]. Events of light particle multiplicities larger than 8 have been selected.*

**3) Study of angular momentum effects :**

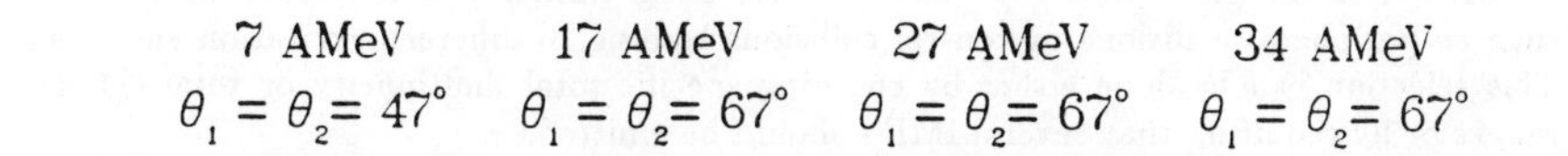


**Figure 29** : *Azimuthal angle correlation ($\theta$ and $\Delta\phi$ in laboratory) for $^4He - X$ pairs for the cases indicated. Solid lines are results of the calculation of references [17,18].*

Many other experiments investigating central collisions are now being analyzed. Angular momentum effects in central collisions are being studied in the $^{40}Ar$ + $^{nat}Ag$ system at 7, 17, 27 and 34 A·MeV [16]. Central collisions are selected by setting a condition on the light charged particle multiplicity and using as a trigger $^4He$ or Li nuclei at large angles.

The role of angular momentum in very excited nuclei can be studied by azimuthal angular distributions or correlations between emitted particles. Figure 29 shows a selection of such azimuthal angular correlations for $^4He - X$ pairs. No change in the qualitative pattern is observed from 7 to 34 A·MeV, while a change might be expected if the mechanism changes. This is just the pattern expected for evaporation from a thermalized composite nucleus i.e. favored emission perpendicular to the emitter spin.

A simulation calculation [17,18] which considers only two effective steps in the evaporation cascade and varys only the composite nucleus spin range, has be made to attempt to fit the data. It shows that indeed nuclei are produced with large momenta in these reactions, with initial excitations of up to 5 MeV per emitter nucleon and spins $\geq$ 100 $\hbar$.

## CONCLUSION

With the first results reported here, we have attempted to show that the $4\pi$ AMPHORA detector provides a good means of selecting various reaction types of interest, such as peripheral collisions or central collisions leading to different excitation energies. This selection can be done either by choosing specific total multiplicity or total charge ranges or by requiring that several IMF's should be emitted.

Most of the experiments involving central collisions deal with the questions : for highly excited nuclei is a thermalized composite nucleus still produced ? can a multifragmentation process appear at SARA energy ? The latter question is still open.

As far as the experimental set-up is concerned, an improvement of the IMF detection including charge identification is required. We are considering to replace the plastic scintillators by gaseous detectors which, besides good charge identification, would certainly provide lower detection thresholds. Scintillating gas chambers mounted in front of the CsI(Tl) crystals [19] would have the tremendous advantage of minimizing the changes to be made on the AMPHORA electronics currently used.

## REFERENCES


[1] D. FABRIS et al., Nucl. Phys. A471 (1987) 351.

[2] J.P. BONDORF et al., Nucl. Phys. A443 (1985) 321. and Nucl. Phys. A444 (1985) 460.

[3] D. DRAIN et al., Nucl. Instr. and Meth. in Phys. Res. A281 (1989) 528.

[4] P. DÉSESQUELLES, Thesis, University Joseph Fourier, Grenoble (1991).

[5] D.W. STRACENER et al., Nucl. Instr. and Meth. in Phys. Res. A294 (1990) 485.

[6] X.C. HU, Thesis, University Claude Bernard, Lyon (1991).

[7] J. PETER et al., Nucl. Phys. A519 (1990) 611.

[8] J.A. LOPEZ and J. RANDRUP, Nucl. Phys. A491 (1989) 477.

[9] M. GYULASSY et al., Phys. Lett. 110B (1982) 185.

[10] B.A. HARMON et al., preprint, LBL-27861 (1989).

[11] J. CUGNON et al., Phys. Lett. 109B (1982) 167.

[12] J.A. RUIZ et al., XXIX Winter Meeting on Nuclear Physics Bormio, Italy, January 1991.

[13] SA BEN-HAO et al., Nucl Phys. A509 (1990) 499.

[14] S. GROULT, Thesis, University of Caen, Caen (1991).

[15] F. AUGER et al., Phys. Rev. C35 (1987) 190.

[16] T. ETHVIGNOT et al., submitted to Phys. Rev.

[17] N.N. AJITANAND et al., Nucl Instr. and Meth. A243 (1986) 111.

[18] A. ELMAANI et al., preprint (1990)

[19] B. CHAMBON et al., submitted to Nucl. Instr. and Meth. in Phys. Res.

# THE RISE AND FALL OF MULTI-FRAGMENT EMISSION

C.A. Ogilvie[a], J.C. Adloff[f], M. Begemann-Blaich[a], P. Bouissou[e], J. Hubele[a], G. Imme[d], P. Kreutz[b], G.J. Kunde[a], S. Leray[e], V. Lindenstruth[a], Z. Liu[a], U. Lynen[a], R.J. Meijer[a], U. Milkau[a], W.F.J. Müller[a], C. Ngô[e], J. Pochodzalla[b], G. Raciti[d], G. Rudolf[f], H. Sann[a], A. Schüttauf[b], W. Seidel[c], L. Stuttge[f], W. Trautmann[a], and A. Tucholski[b]

[a] *Gesellschaft für Schwerionenforschung, D-6100 Darmstadt, Federal Republic of Germany*
[b] *Institut für Kernphysik, Universität Frankfurt, D-6000 Frankfurt, Federal Republic of Germany*
[c] *ZFK Rossendorf, Dresden, Federal Republic of Germany*
[d] *Dipartimento di Fisica dell' Universitá, I-95129 Catania, Italy*
[e] *Laboratoire National Saturne, CEN Saclay, F-91191 Gif-sur-Yvette, France*
[f] *Centre de Recherches Nucléaires, Strasbourg, France*

**Abstract**

We have studied multifragment decays of Au projectiles after collisions with C, Al, and Cu targets at a bombarding energy of 600 MeV/nucleon. We find that with increasing violence of the collision, measured via the associated multiplicity of light particles, the mean multiplicity of intermediate mass fragments first increases to a maximum $<M_{IMF}>\simeq 3$ and then decreases again. Calculations using the BUU model suggest that the fragmentation is governed by the energy $E_{dep}$ deposited into the Au nucleus and that $<M_{IMF}>$ reaches its maximum around $E_{dep}\simeq 8$ MeV/nucleon.

When a nucleus is excited to and beyond its total binding energy, it is predicted to decay into several intermediate mass fragments (IMFs). Competing models embody different decay mechanisms and experiments have yet to discriminate between several theoretical scenarios. These range from the near-simultaneous emission of fragments due to large fluctuations in regions of mechanical instability, [1],[2] to statistical multifragmentation models, [3],[4],[5] and sequential decay processes where equilibrium is re-established after each binary decay.[6],[7]

In order to constrain and confront these models, it is important to know under what conditions fragments are emitted. Is there an excitation energy at which multifragment emission begins to be a dominant process, and what happens if the excitation is further increased? Eventually the energy will be so large that the system will disassemble into nucleons and light fragments such as alpha particles. On this basis one expects a rise and fall of multi-fragment emission with excitation energy.

We use asymmetric nuclear collisions at relativistic energies to excite a large nucleus.[8] A schematic reaction mechanism is that the initial nucleon-nucleon collisions form a fireball and that some of these nucleons traverse through the larger nucleus. As these undergo secondary collisions they excite the large nucleus. In the second stage of the reaction, this excited nucleus decays into a variety of exit channels, some of which will be multi-fragment states. Previous work in this field is consistent with this reaction mechanism, and has shown that fragments are emitted nearly isotropically from an excited source that moves with a velocity near that of the larger nucleus.[9],[10],[11]

In this talk I report on exclusive data of fragments emitted in Au reactions on three light targets, C, Al, and Cu at a beam energy of 600 MeV/nucleon. Our aim is to answer the following questions. What type of fragments are emitted in the collision and how many? How does this change with the impact parameter and the size of the target nucleus? Can these changes be related to the energy deposited into the Au projectile, and is the rise and fall of multi-fragment emission observed experimentally?

The experiment was performed with the ALADIN forward spectrometer[12] at GSI, Darmstadt, with the beam accelerated by the SIS synchrotron. A schematic layout of the experimental setup is shown in figure 1. The inverse kinematics have the effect

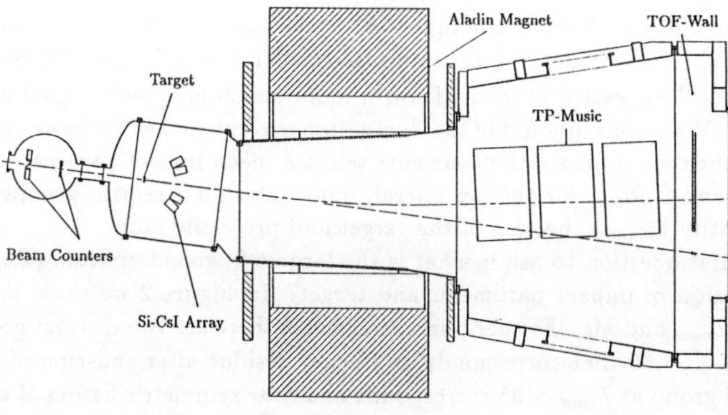


Figure 1: A schematic outline of the experimental apparatus used to measure the reaction Au at 600 MeV/nucleon on C, Al and Cu. The distance from target to TOF wall is approximately 6m.

that the fragments from the Au projectile are focused into the forward direction. A 64-element Si-CsI hodoscope measured the multiplicity of light charged particles from the mid-rapidity source and the ALADIN magnet accepted projectile fragments at forward angles. The TOF-wall was positioned approximately 6m from the target, and consisted of two layers, each of 40 vertical scintillator strips. The strips were read out from both the top and bottom enabling an amplitude reconstruction that is independent of the position in the scintillator. The two TOF layers were displaced from each other in the bending plane by one half of a strip's width. This leaves no in-active area between the strips. The discriminator threshold was set to be below alpha particles, and the TOF-wall determined the charge and multiplicity of nuclear fragments. The TP-MUSIC detector [12] was used in this analysis to calibrate the charge measured in the TOF-wall. Unit charge resolution was obtained for Z<9, increasing to a charge resolution of ±1.5 at Z=79. Further details of the experimental acceptance can be found in Ref. [13].

The TOF strips near $0^0$ provided a hardware interaction trigger and to eliminate background events a further software cut was applied of at least one charge particle in CsI hodoscope. Offline the events have been placed into three groups according to the multiplicity of light charged particles ($M_{lp}$) in the Si-CsI hodoscope. The multiplicity of charged particles is expected to be a measure of the size of the fireball formed in the initial stage of the reaction and hence a measure of the impact parameter (b). The borders of the three groups have been set such that the peripheral (lowest multiplicity), mid-central and central groups have 50%, 35%, and 15% of the events respectively. For the peripheral group, the minimum multiplicity is one charged particle. These cuts will select a fixed range of impact parameters if the following assumptions are true; that the cross-section is geometrically related to b, i.e. $d\sigma/db \sim b$, and that $M_{lp}$ depends on b monotonously with small fluctuations. Calculations with the Boltzmann-Uehling-Uhlenbeck (BUU) model, [14] Quantum Molecular Dynamics (QMD) model, [15] and FREESCO [5] suggest that the half-width-half-maximum of each impact distribution is $0.15b_{max}$. When this amount of the fluctuation is combined with the geometrical cross-section, then we estimate that our cuts select a mean impact parameter of $0.75b_{max}$, $0.5b_{max}$, and $0.3b_{max}$ for the peripheral, mid-central and central groups respectively. The quantity $b_{max}$ is the sum of the target and projectile radii.

The first question to ask is what is the largest fragment emitted ($Z_{max}$) per event as a function of impact parameter and target? In Figure 2 we show the correlation between $Z_{max}$ and $M_{lp}$. For peripheral collisions, there are two distinct groups of $Z_{max}$, one at $Z_{max} \sim 60$ that corresponds to a heavy residue after emission of some charge. A second group at $Z_{max} \sim 35$ corresponds to a near symmetric fission of the projectile. As $M_{lp}$ increases, i.e. as the violence of the collision increases, $Z_{max}$ becomes steadily smaller, until $Z_{max} \sim 25$ for the most violent collisions on the C target. For the Al and Cu targets there is an apparant sudden decrease in $Z_{max}$ at intermediate values of $M_{lp}$, which is followed by a steady reduction of $Z_{max}$ to small values. This may imply that $Z_{max}$ in these violent collisions can no longer be identified as the remnant from the projectile spectator. One should note that for the most central collisions on Al and Cu the maximum charge is $Z_{max} \sim 5 - 10$. This is a considerable reduction from an Au projectile!

In Figure 3 we present the charge distributions in three columns for the targets C, Al, and Cu. These distributions have been corrected for the loss of efficiency due to our finite coverage, with the correction being of the order 5% for Z=3 fragments and less than 2% for Z>6. The inefficiency is small because even the lighter fragments have a velocity distribution that is sharply peaked near the projectile velocity.

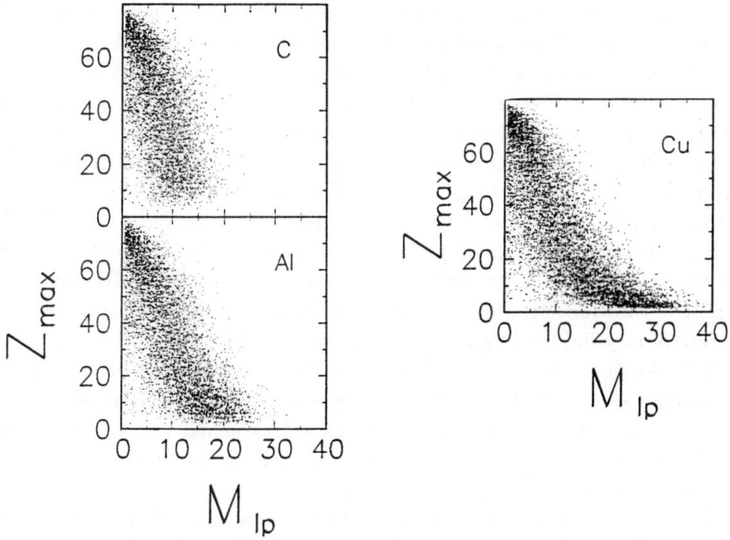


Figure 2: The correlation correlation between $Z_{max}$ and the multiplicity of charged particles in the CsI array $M_{lp}$ for the reactions Au at 600 MeV/nucleon on C, Al and Cu.

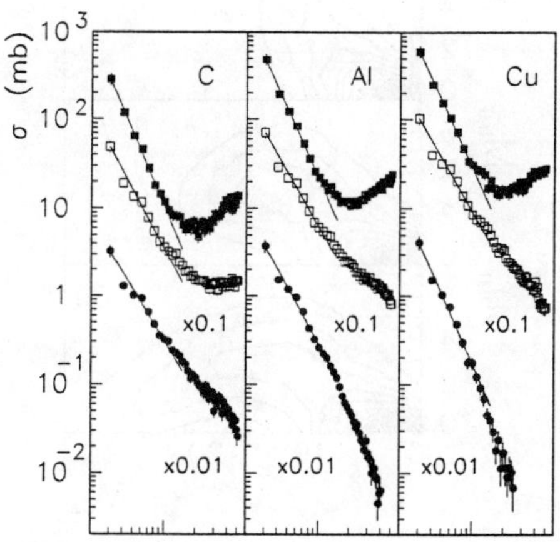


Figure 3: The measured charge distributions from Z=3 for Au 600 MeV/nucleon collisions on C, Al, and Cu targets are shown in columns from left to right. The solid square data points are for peripheral collisions, the open squares mid-central collisions and the solid circles central collisions. The lines are from the power law fit and are plotted from $3 \leq Z \leq 15$.

The absolute normalisation was determined using a scaled-down trigger of beam particles at the target; this normalisation has an uncertainty of 15%.

For all targets, there is a strong reduction of yield of fragments with Z>30 with increasing $M_{lp}$, i.e. as one goes from peripheral to central reactions. This is related to the reduction of $Z_{max}$ that was discussed above. For those fragments with Z<30, the distribution is steep in peripheral reactions, which is characteristic of light-fragment evaporation from a heavy system. On the C target, the Z<30 distribution is considerably broader for both mid-central and central collisions, i.e. the probability for emitting heavier fragments is greater. For reactions on Al, the mid-central distribution is similar to the distribution for central reactions on the C target. On the Cu target, the mid-central collisions produce a broad distribution, while the central reactions produce predominantly lighter fragments. These distributions were fitted with a power law ($\sigma \sim Z^{-\tau}$) for Z=3, $5 \leq Z \leq 15$ and the extracted $\tau$ parameters are listed in Table 1. The existence of a steadily decreasing charge distribution for central collisions on C, Al and Cu suggests that the emission of heavy fragments is governed by the same process as lighter fragments. This is consistent with $Z_{max}$ no longer being identifiable as the projectile remnant.

One of the more interesting questions is to ask how many fragments are emitted in each event. We define an IMF as a fragment with $3 \leq Z \leq 30$. In Figure 4 we plot the correlation between the multiplicity of IMFs and $M_{lp}$. For the C target, the IMF

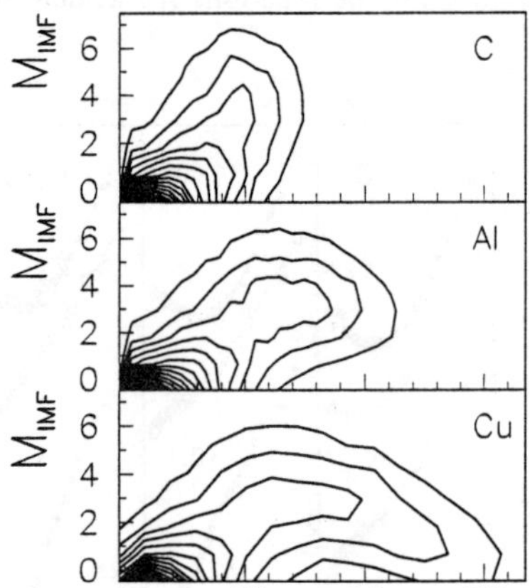


Figure 4: The measured correlation of the multiplicity of IMFs at forward angles with the multiplicity of light charged fragments in the CsI array. The results for 600 MeV/nucleon Au on C, Al and Cu are shown from top to bottom and the contours are drawn using a linear scale.

multiplicity increases with the violence of the collision and reaches a mean multiplicity of three to four for the most central collisions. This striking result reveals that emission of many fragments is the dominant exit channel for this class of collisions.

The reactions on Al behave similarly in that on average three to four IMFs are emitted, however this occurs over a broader range of $M_{lp}$. For the Cu target, the region of three to four IMFs is reached for lower relative $M_{lp}$. With more violent collisions, the mean multiplicity of IMFs reduces to one or two. This peak in the IMF multiplicity distribution on the Cu target is an observation of the rise and fall of multi-fragment emission. For each impact parameter group, we have evaluated the mean IMF multiplicity ($<M_{IMF}>$) and applied a 3 to 5% correction for the finite coverage of IMFs. The double-hit probability is of the order of 10% which leads to some mis-identification of fragments. This has been included in the systematic error of $<M_{IMF}>$.

We have summarized these results in Table 1, where we have also included the estimated range of impact parameters for each group. These results can be directly

| | $<b>$ (fm) | <Zmax> | $\tau$ | $<M_{IMF}>$ |
|---|---|---|---|---|
| | 7.5±1.5 | 56±2 | 2.8±0.2 | 0.93±0.03±0.09 |
| C | 5.5±2 | 38±1 | 2.2±0.1 | 2.31±0.06±0.21 |
| | 3.5±2 | 25±1 | 2.0±0.1 | 3.39±0.11±0.37 |
| | 8±1.5 | 53±2 | 2.6±0.1 | 1.13±0.03±0.10 |
| Al | 5±2 | 26±1 | 2.0±0.1 | 3.00±0.05±0.26 |
| | 3±1.5 | 12±0.5 | 2.1±0.1 | 3.30±0.08±0.17 |
| | 9±1.5 | 53±2 | 2.6±0.1 | 1.14±0.04±0.10 |
| Cu | 6±2 | 20±1 | 2.0±0.1 | 2.96±0.07±0.20 |
| | 3±1.5 | 6±0.5 | 2.6±0.1 | 1.97±0.07±0.09 |

Table 1: Summary of experimental results for Au 600 MeV/nucleon collisions on C, Al, and Cu targets. For each target, the average maximum charge in each event, the $\tau$ parameter from the charge distributions and the mean multiplicity of intermediate mass fragments are given for peripheral (top row), mid and central (bottom row) collisions. The uncertainties for $<M_{IMF}>$ correspond to the statistical and systematic errors respectively. The first column gives the estimated mean and the half-width at half-maximum of the impact parameter selection for each class of reaction.

compared to dynamical models that produce fragments, such as QMD. To facilitate comparisons with fragmentation models that do not include the reaction dynamics, and to help interpret these experimental observations, we have performed a series of calculations with the BUU model.[14] We have used the model to calculate the energy deposited into the Au projectile during the initial stage of the reaction. Similar approaches have been used in Refs [16],[17] and [18].

A projectile-like remnant is defined as all the nucleons within a sphere in coordinate space, where the position and size of the sphere is calculated from the distribution of projectile nucleons that have yet to undergo a nucleon-nucleon collision. The radius of this sphere is approximately that of a Au nucleus. The deposited energy is calculated from the momentum and potential energy of the nucleons within this sphere minus an estimate of the ground state energy. The deposited energy is relatively constant in time once the fireball-like nucleons have left the reaction zone, so an average deposited energy is calculated between 60 and 100 fm/c after the onset of the reaction. For central, mid-central and peripheral reactions a further average is taken over the impact parameter ranges that have been discussed earlier in this work. The number of nucleons within the sphere decreases from approximately 180 for peripheral collisions to 100 for central collisions on Cu.

In Figure 5 we plot the maximum charge in an event, the fitted $\tau$ parameter from the charge distribution, and the mean IMF multiplicity versus the calculated deposited energy/nucleon. For systems at low energy there is a heavy residue, a steep

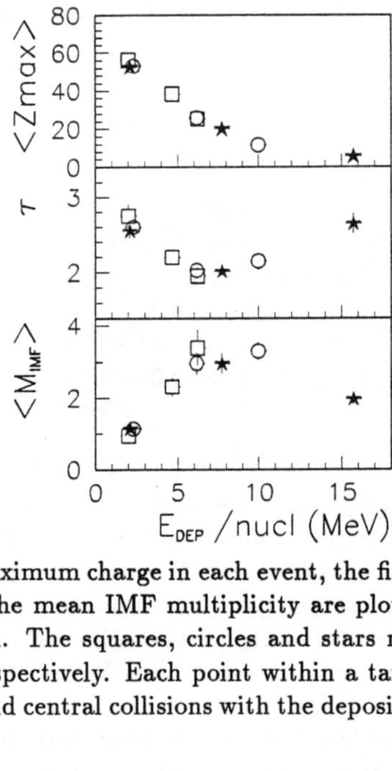


Figure 5: The average maximum charge in each event, the fitted $\tau$ parameter from the charge distribution and the mean IMF multiplicity are plotted versus the calculated deposited energy/nucleon. The squares, circles and stars represent collisions on the C, Al, and Cu targets respectively. Each point within a target group corresponds to peripheral, mid-central and central collisions with the deposited energy increasing with centrality.

charge distribution, i.e. large $\tau$ parameter, and small IMF multiplicity. As the energy increases, the number of emitted IMFs increases, the charge distribution broadens, and the maximum charge in an event decreases. At still higher energies, the multiplicity of IMFs decreases. In their place lighter fragments are produced as deduced from an increase in $\tau$, and the maximum charge in the event becomes quite small. There is a universal behaviour of the experimental results as a function of the calculated deposited energy. This suggests that this energy might be a significant parameter that drives the decay of the Au-like nucleus. The deposited energy is model dependent, but this dependence is expected to affect the scale of the deposited energy, and should not change the ordering of different targets and impact parameters. For reactions leading to excitations of up to ~3 MeV/nucleon, the role of the deposited energy in governing fragment emission has been experimentally established.[19]

In the low-energy region of Figure 5, the projectile spectator decays as expected for evaporation from a heavy nucleus. At high values of deposited energy, the small $Z_{max}$, high $\tau$ parameter and small $<M_{IMF}>$ are all consistent with the projectile nucleus disassembling nearly completely into its constituent nucleons. These two extremes respectively represent a nuclear liquid and a nuclear gas, so it is possible that somewhere imbetween is the liquid-gas nuclear phase transition.[1],[2]

One signature for such a phase transition would be the presence of large event-by-event fluctuations in the charge distribution of observed fragments. Recently it has been suggested that the scaled factorial moments of the charge distribution may reveal the presence of such large fluctuations.[20] These factorial moments are calculated for different sizes of charge bins($\Delta$Z) and it has been suggested that large fluctuations result in the scaled factorial moment increasing as the size of the charge bin $\Delta$Z decreases. The reader is referred to Ref [20] for definitions and more details.

In Figure 6 we plot the scaled factorial second moment for Au+Cu collisions at 600 MeV/nucleon as a function of the width of the charge bin $\Delta$Z. Each plot corresponds to an interval in $M_{lp}$ four multiplicity units wide, with the smallest $M_{lp}$ being in the top left-hand corner and $M_{lp}$ increasing across the page until the most violent collisions are shown in the bottom right-hand corner. The moments are observed to increase with decreasing $\Delta$Z for intermediate values of $M_{lp}$, i.e. middle impact parameters. For the peripheral and central collisions the scaled factorial moments are nearly independent of $\Delta$Z. This is consistent with statistical emission from a system in only one phase. The data in these plots are preliminary and more work has to be done before this can interpreted as evidence for a phase transition.

In summary, we have experimentally determined a set of optimum conditions for the multi-fragment decay of a large nucleus. For asymmetric reactions at 600 MeV/nucleon, multi-fragment emission is the dominant decay channel for the most violent collisions on a light target and for mid-central collisions on a heavy target. In peripheral collisions, a heavy residue and lighter fragments are observed. In central collisions on heavy targets, only light fragments are produced. It is now a challenge to confront the various models of multifragmentation with these systematics and to search for signals in the multi-fragment data that investigate the question of the nuclear liquid-gas phase transition.

The authors thank W. Bauer for the use of the BUU code and useful discussions. We acknowledge discussions with R.J. Lenk.

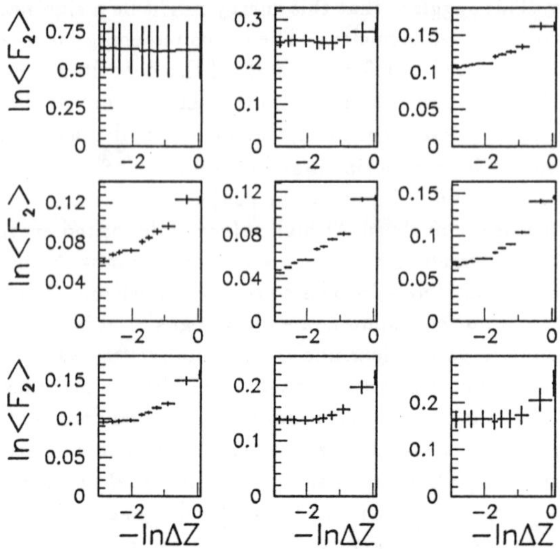


Figure 6: The scaled factorial second moment as a function of the size of the charge bin ΔZ for different $M_{lp}$ intervals in the reaction Au+Cu at 600 MeV/nucleon. The $M_{lp}$ intervals are in units of four light particles and the plots start with low multiplicity in the top left-hand side, with $M_{lp}$ increasing across the page until the most violent collisions are shown in the bottom right-hand corner.

# References


[1] G. Bertsch and P.J. Siemens, Phys. Lett. B **126**, 9 (1983).

[2] C.J. Pethick and D.G. Ravenhall, Nucl. Phys. A **471**, 19c (1987).

[3] D.H.E. Gross, Z. Xiao-ze, and, X. Shu-yan, Phys. Rev. Lett. **56**, 1544 (1986), D.H.E Gross and H. Massman, Nucl. Phys. A **471** 339c (1987).

[4] H.W. Barz et al., Phys. Lett. B **169**, 318 (1986).

[5] G. Fai and J. Randrup, Nucl. Phys A **381**, 557 (1982), Nucl. Phys. A. **404**, 551 (1983).

[6] W.A. Friedman and W.G. Lynch, Phys. Rev C **28**, 16 (1983).

[7] R.J. Charity et al., Nucl. Phys. A. **483**, 371 (1988).

[8] R. Stock et al., Phys. Rev. Lett. **44**, 1243 (1980).

[9] A.I. Warwick et al., Phys. Rev. C **27**, 1083 (1983).

[10] S.J. Yennello et al., Phys. Lett. B **246**, 26 (1990).

[11] U. Milkau et al., GSI preprint GSI-91-08.

[12] ALADIN Collaboration, GSI-Report 02-89 (1988).

[13] C.A. Ogilvie et al., GSI-Preprint 91-18 (1991).

[14] G.F. Bertsch and S. das Gupta, Phys. Rep. **160**, 189 (1988).

[15] G. Peilert et al., Phys. Rev. C **39**, 1402 (1989).

[16] E. Suraud et al., Nucl. Phys A **495**, 73c (1989).

[17] C. Cerruti et al., Nucl. Phys A **492**, 322 (1989).

[18] A.S. Botvina, A.S. Iljinov, and I.N. Mishustin, Nucl. Phys A **507**, 649 (1990).

[19] R. Trockel et al., Phys. Rev. C **39**, 729 (1989).

[20] M. Ploszajcak and A. Tucholski, Nucl. Phys A **523**, 651 (1991).

# II. Theoretical Aspects of Nuclear Collisions

# DENSE MATTER FROM CHIRAL AND SCALING LAGRANGIANS*

Mannque Rho

*Service de Physique Théorique, C.E. Saclay*
*91191 Gif-sur-Yvette, France*

## ABSTRACT

**I discuss dense hadronic matter from the point of view of an effective Lagrangian that incorporates approximate chiral and scale invariances of QCD. It is shown, within an effective Lagrangian that preserves the original (zero density) symmetries of QCD, that the change of vacuum structure due to the presence of matter is manifested in the scaling as a function of density of hadronic masses. The proposition is that this scaled Lagrangian be taken as a basis for a systematic chiral expansion for describing nuclear processes at and above the normal nuclear density. Some preliminary consequences of this approach in various nuclear processes are discussed.**

---

***Invited talk given at "First European Biennial Workshop on Nuclear Physics," Megève, France; March 25-29, 1991**

In this talk, I would like to describe a recent idea developed in collaboration with Gerry Brown on how to treat dense hadronic matter in a manner consistent with the basic premises of quantum chromodynamics. The ultimate goal of this approach is to address the issue of QCD phase transitions that are conjectured to take place at high density and/or temperature.

One of the most exciting new directions in nuclear physics is to study how nuclear phenomena change as the environment changes. Thus relativistic heavy-ion experiments are to address the state of nuclear matter at high temperature and/or density and high-energy high-duty-cycle electron machines are to probe the properties of individual hadrons in close encounter with other strongly interacting matter. These are the processes that reflect the change of the strong-interaction vacuum as density and temperature are "dialled." Given a fundamental theory of strong interactions, *i.e.*, QCD, one should, in principle, be able to calculate all the observables unambiguously as the environment is modified. It is possible that this feat will be eventually accomplished but it is highly unlikely that it will come soon enough to make contact with either on-going experiments or experiments to come in the near future. Here I would like to discuss, focusing on density effects, a recent proposal of Gerry Brown and myself for an economical and powerful approach to this problem, starting from low-energy effective theories based on spontaneously broken chiral symmetry that have been phenomenologically successful in describing low-energy and low-density hadronic interactions. Much of what I discuss here is to be published elsewhere [1].

I start with a known structure of effective Lagrangians at low energy and zero density (*i.e.*, free space), dictated by symmetries and other constraints of QCD (*e.g.*, chiral symmetry with its current algebra and anomaly, trace anomaly etc). We are interested in how this theory evolves as density (or temperature) is increased. Embedding a hadron in dense matter is equivalent to changing the vacuum, thereby modifying quark and gluon condensates in QCD variables. Our key assumption here is that as the condensates change, the symmetries of the Lagrangian remain more or less *intact*, while the relevant scale is changed in a prescribed way that we will explain below. This means that we will have, as density increases, the same Lagrangian but with the masses and coupling constants of the theory *modified according to the symmetry constraints of QCD*. This is really an assumption but I will give some evidence that it may be a valid one.

To illustrate our point, we start with the original Skyrme Lagrangian[2] consisting of the current algebra term characterized by a dimensional constant $f_\pi$ and the quartic stabilizing term characterized by a dimensionless constant $\epsilon$. (More precise definitions of these quantities will be given later.) It is convenient to work with physical quantities by taking $f_\pi$ to be the pion decay constant (experimentally $\approx 93$ MeV) and the axial vector coupling constant $g_A$ in place of $\epsilon^2$[3, 4]. If we accept that baryons arise as solitons from the Skyrme Lagrangian–which seems to be fairly well established by now, we can use simple scaling arguments[5, 4] to show that, modulo overall constants, the size and mass of the baryon go as

$$\langle r^2\rangle \sim g_A/f_\pi^2, \qquad M \sim \sqrt{g_A} f_\pi. \tag{1}$$

We now ask what happens to these quantities when the skyrmion is embedded in a dense medium. As in refs.[5, 4], we will argue that as long as there is no phase transition that changes symmetries of the "vacuum" (or more precisely the ground state), the leading modification in the theory is in the basic constants of the Lagrangian, *i.e.*,

$$\langle r^2(\rho)\rangle^\star \sim g_A^\star(\rho)/f_\pi^{\star 2}(\rho), \qquad M^\star(\rho) \sim \sqrt{g_A^\star(\rho)} f_\pi^\star(\rho). \tag{2}$$

Here we indicate the density-dependent quantities by stars. (From now on we will omit the density dependence in starred quantities, unless explicitly required.) We identify (2) as the *quasiparticle* size and mass, respectively, for single baryons in the medium. (Further (residual) interactions, say between quasiparticles (or *quasiskyrmions*), can be introduced in a way analogous to zero density case using a given effective Lagrangian.) Below, we will make this argument more explicit by introducing the trace anomaly of QCD into the effective Lagrangian.

So far, we have been concerned with baryons only. What about mesons? Chiral effective theories contain, in large $N_c$ limit, Goldstone bosons, ordinary mesons, and a $U_A(1)$ boson called $\eta'$, in addition to the baryons. The $\eta'$ decouples from the rest of the world in the large $N_c$ limit and does not concern us [6]. Now in considering the properties of mesons in dense medium, it is more convenient to introduce explicit degrees of freedom assocaited with other mesons than the Goldstone bosons. The most important of them all are the strong vector mesons (*e.g.*, $\rho$, $\omega$,...) and the glueballs. (In considering baryons, it is legitimate to "integrate out" these other meson degrees of freedom, as exemplified by the Skyrme model.) A convenient framework to incorporate the strong vectors is through the hidden gauge symmetry strategy advocated by Bando *et al*[7]. The glueballs that are needed for our purpose will be introduced through QCD anomalies.

Given a chiral Lagrangian which contains pseudoscalar (Goldstone) and vector (hidden gauge) bosons, what is the effect of changing vacuum structure by density? We will argue that similarly to the baryon case, the leading effect is in the effective masses of the mesons and their coupling constants.

The principal in-medium scaling law [8] conjectured in refs.[9] which we will justify is

$$m_\sigma^\star/m_\sigma \approx m_N^\star/m_N \approx m_\rho^\star/m_\rho \approx m_\omega^\star/m_\omega \tag{3}$$

where the unstarred masses stand for free-space values. To do this, we follow the approach of Campbell, Ellis and Olive[10] and introduce scale invariance into the effective Lagrangian. Their approach can be summarized as follows. The standard chiral Lagrangian lacks the trace anomaly which is an essential ingredient of QCD. (The axial anomaly, another ingredient of QCD, can be incorporated easily via the $\eta'$ field but we are not concerned with it here.) In order to understand what this is, we consider the scale transformation

$$x \rightarrow^{\lambda} x = \lambda^{-1} x \qquad \lambda \geq 0 \tag{4}$$

under which an arbitrary field $\phi$ transforms with the canonical mass dimension (called "scale dimension") $d_\phi$

$$\phi(x) \rightarrow \lambda^{d_\phi} \phi(\lambda x). \tag{5}$$

The conventional scale dimensions are 1 for scalar and vector fields and 3/2 for fermion fields. An action with a Lagrangian density with scale dimension 4 is clearly invariant under scale transformation. This means that the classical QCD action in chiral limit (*i.e.*, with zero quark mass) is scale-invariant. The quark mass term has scale dimension 3, thus breaks scale invariance. Quantum mechanics introduces a profound modification to this structure. A well-known property of QCD is that the pure Yang-Mills action is not scale-invariant at quantum level due to dimensional transmutaion. This can be stated in terms of the divergence of the dilatation current $D_\mu$ whose nonzero value signals the breaking of scale invariance or equivalently nonvanishing trace of energy-momentum tensor. Including the effect of quark masses, the trace anomaly is given by [11]

$$\partial^\mu D_\mu = \theta_\mu^\mu = \sum_q m_q \bar{q} q - \frac{\beta(g)}{g} Tr G_{\mu\nu} G^{\mu\nu} \tag{6}$$

where $\beta(g)$ is the usual beta function of QCD. The quantum nature of the trace anomaly is evident from the dependence on the strong fine-structure constant $\frac{g^2}{4\pi}$. In order to be consistent with QCD, effective Lagrangians must satisfy eq.(6) in terms of effective fields.

Now the chiral field $U = e^{i\pi/f_\pi}$ with $\pi = \vec{\tau} \cdot \vec{\pi}(x)$ has scale dimension zero [12, 13]. A derivative brings in a scale dimension 1 and hence the current algebra term $\frac{f_\pi^2}{4} Tr(\partial_\mu U \partial^\mu U^\dagger)$ has scale dimension 2, the mass term $cTr(MU + h.c.)$ with c a constant and $M$ the quark mass matrix has scale dimension zero and the Skyrme quartic term has scale dimension 4. Clearly this effective Lagrangian in its original form does not satisfy the trace condition (6). In order to restore consistency with the trace condition, it is convenient to define the scalar "glueball" field $\chi$

$$(\langle 0 | Tr G_{\mu\nu} G^{\mu\nu} | 0 \rangle)^{\frac{1}{4}} \propto \langle 0 | \chi | 0 \rangle \equiv \chi_0 \tag{7}$$

where $|0\rangle$ stands for the vacuum at zero density. (The "vacuum" at non-zero density or, more properly, ground state will be denoted by $|0^\star\rangle$.) As defined, the $\chi$ field has scale dimension 1. In terms of this field, the second term of eq.(6), *i.e.*, the trace anomaly, can be reproduced by a potential term of the form

$$V(\chi) \propto \chi^4 \ln \left( \chi / e^{\frac{1}{4}} \chi_0 \right). \tag{8}$$

(It is in practice convenient to add constant terms so that the minimum of the potential occurs at $\chi = \chi_0$ with $V(\chi_0) = 0$ when the density is zero.) With the help of the $\chi$ field, the

Skyrme Lagrangian can be rewritten so as to be consistent with the QCD scale property,

$$\begin{aligned}\mathcal{L} &= \frac{f_\pi^2}{4}\left(\frac{\chi}{\chi_0}\right)^2 Tr\partial_\mu U\partial^\mu U^\dagger + \frac{\epsilon^2}{4}Tr[U^\dagger\partial_\mu U, U^\dagger\partial_\nu U]^2 + \frac{1}{2}\partial_\mu\chi\partial^\mu\chi \\ &+c\left(\frac{\chi}{\chi_0}\right)^3 Tr(MU + h.c.) + V(\chi)\end{aligned} \tag{9}$$

where a scale-invariant kinetic energy term for the glueball field is introduced. As written, the scale-breaking resides entirely in the last line of eq.(9), all the rest being scale-invariant. It should be particularly noted that the Skyrme quartic term is scale-invariant by itself. This is a key point for later discussions [14].

One immediate consequence of the above discussion is that the quark condensate in dense medium scales as

$$\langle 0^\star|\bar{q}q|0^\star\rangle/\langle 0|\bar{q}q|0\rangle = (\chi_\star/\chi_0)^3, \qquad \chi_\star \equiv \langle 0^\star|\chi|0^\star\rangle. \tag{10}$$

This suggests defining an *effective* pion decay constant as

$$f_\pi^\star = f_\pi(\chi_\star/\chi_0) \tag{11}$$

so that

$$\langle 0^\star|\bar{q}q|0^\star\rangle/\langle 0|\bar{q}q|0\rangle = (f_\pi^\star/f_\pi)^3 \tag{12}$$

which is the relation deduced in [9]. Let us assume that as density increases, the potential develops a minimum at $\chi = \chi_\star$. This suggests expanding the $\chi$ field as

$$\chi = \chi_\star + \chi' \tag{13}$$

with $\chi'$ a fluctuating field. In the "vacuum" characterized by $\chi_\star$, the Lagrangian for the chiral field can be written as

$$\mathcal{L}_U = \frac{f_\pi^{\star 2}}{4}Tr\partial_\mu U\partial^\mu U^\dagger + \frac{\epsilon^2}{4}Tr[U^\dagger\partial_\mu U, U^\dagger\partial_\nu U]^2 + c\left(\frac{f_\pi^\star}{f_\pi}\right)^3 Tr(MU + h.c.) + \cdots \tag{14}$$

with

$$U(x) = exp(i\pi^\star/f_\pi^\star), \qquad \pi^\star \equiv \pi\chi_\star/\chi_0. \tag{15}$$

The ellipsis stands for other fields including $\chi'$ field with which we are not directly concerned here. We have thus effectively assigned a scale dimension 1 to the $f_\pi^\star$ and to the pion field $\pi^\star$. There is no explicit $\sigma$ field of linear $\sigma$ model (*i.e.*, the $\bar{q}q$ scalar in NJL). However a strong mixing is expected between the glueball field $\chi'$ and two pions in the scalar channel so as to give the effective $\sigma$ needed in nuclear physics. Thus in Nature we should identify the "observed" lowest-mass scalar to be a mixture of a quarkish scalar and the glueball ($f_0$ (1400)) scalar. This immediately leads to the scaling

$$m_\chi^\star/m_\chi \approx f_\pi^\star/f_\pi \rightarrow m_\sigma^\star/m_\sigma \approx f_\pi^\star/f_\pi \tag{16}$$

with the $\sigma$ field now understood to be the effective *nuclear physics* scalar.

The Lagrangian (14), together with (1) and (2), implies that the nucleon mass scales as

$$m_N^\star/m_N \approx \sqrt{g_A^\star/g_A} f_\pi^\star/f_\pi. \tag{17}$$

Since, modulo loop corrections, $g_A$ is scale-invariant (because the coefficient $\epsilon^2$ of the Skyrme quartic term to which $g_A$ is related is scale-invariant), we may set $g_A^\star/g_A \approx 1$. Alternatively we may invoke the phenomenological observation that the in-medium constant $g_A^\star$ saturates rapidly in light nuclei at 1 and stays constant as density (or mass number) increases[15]. In some problems [16] the density dependence of $g_A^\star$ is important but not at densities higher than nuclear matter $\rho_0$. The rapid variation in $g_A$ between $\rho = 0$ and $\rho = \rho_0$ is a loop effect as suggested in [5], indicating a possible new (lower) scale *induced* in nuclei. Accounting for such effects is an important subject for classical nuclear physics but not relevant at the mean-field order we are concerned with. We thus establish that

$$m_N^\star/m_N \approx f_\pi^\star/f_\pi. \tag{18}$$

This result was also obtained using *in-medium* QCD sum rules[17].

Before proceeding to complete the scaling relation (3), we pause to emphasize that eq.(18) is extremely practical. It tells us that knowledge of the effective mass of the nucleon $m_N^\star$ supplies the order parameter $f_\pi^\star$ of the broken symmetry regime at finite density. Essentially every nuclear physicist knows what $m_N^\star$ is at (at least) nuclear matter density although there is no consensus on its exact value. Quite conservatively (from the standpoint of the difference of $m_N^\star$ from $m_N$),

$$m_N^\star(\rho_0)/m_N \approx 0.8. \tag{19}$$

This, coupled with eqs.(18) and (12), has the remarkable consequence that

$$\langle 0^\star|\bar{q}q|0^\star\rangle/\langle 0|\bar{q}q|0\rangle \approx 1/2 \quad at \quad \rho = \rho_0; \tag{20}$$

that is, the quark condensate has dropped by $\sim 50$ % already in the middle of nuclei!

Let us now consider the vector-meson masses. In a generalized model with vector mesons (*i.e.*, the hidden gauge symmetric theory[7]), one has the KSRF relation

$$m_V^2 = 2g^2 f_\pi^2 \tag{21}$$

where $g$ is the hidden gauge coupling. We will ignore $O(1/N_c)$ correction and set $m_V = m_\rho = m_\omega$. In terms of the original Skyrme model, we can identify the gauge coupling $g^2$ with $(8\epsilon^2)^{-1}$ and hence find [18], thanks to the scale invariance of the Skyrme quartic term,

$$m_V^\star/m_V \approx f_\pi^\star/f_\pi. \tag{22}$$

This completes the relation (3). (A striking omission from eq.(3) is the scaling for Goldstone bosons, in particular for the pion which figures importantly in nuclear physics. The Lagrangian (14) implies that the pion mass $m_\pi^\star$ scales as $\sqrt{f_\pi^\star}$ which differs from eq.(3). The story turns out to be much more complex in the case of Goldstone bosons. As discussed in ref.[19], there is an intricate interplay between the explicit chiral symmetry breaking term and many-body effects which turns out to preserve *albeit approximately* $m_\pi^\star/m_\pi \approx 1$ in cold dense matter up to $\rho \approx \rho_0$. I will use this for applications.)

The key point of my discussion is that (14) with scaled masses be taken as the effective Lagrangian to calculate physical observables according to the chiral perturbation scheme as *e.g.*, discussed in Refs. [20, 21]. Let me conclude then by discussing a few examples of how the strategy works out.

Some consequences of these dropping masses are discussed in [16, 17, 22, 23, 24, 25]. More recently quite detailed analysis by Hosaka and Toki [26] who investigated effective matrix elements of the in-medium two-body interaction in the s,d-shell shows that a uniform decrease in various masses yields G-matrix elements which are consistent with empirical matrix elements.

A more striking case where the "dropping" of the masses seems to be required is found in "axial-charge" transitions in nuclei. It is known that in nuclear transitions, mediated by the time component of the weak axial current, of the type

$$(0^+, T = 0) \to (0^-, T = 1) \tag{23}$$

the pion-exchange currents play an important role [27]. In fact it has been recently shown [21] that the pion-exchange current is the only contribution to leading order in chiral expansion [20], with shorter-range exchange currents suppressed by two or more powers of the chiral expansion parameter. Incorporation of the approximate scale invariance as discussed above is simply to replace the free-space mass $m = m^\star(\rho = 0)$ by $m^\star(\rho)$. The net effect is that the single-particle matrix element is enhanced by

$$\mathcal{M}_1(m^\star) \approx (\frac{m}{m^\star})\mathcal{M}_1(m) \tag{24}$$

and the two-body matrix element is enhanced by

$$\mathcal{M}_2(m^\star) \approx (\frac{m}{m^\star})^2\mathcal{M}_2(m) \tag{25}$$

where $m$ is the generic mass involved, so the enhancement factor is $m/m^\star \approx m_N/m_N^\star$. The prediction for the ratio $R = (\mathcal{M}_1(m^\star) + \mathcal{M}_2(m^\star))/\mathcal{M}_1(m)$ comes out to be

$$\begin{aligned} R^{th} &= 1.5 \quad A = 16, \\ &= 2.2 \quad nuclear\ \ matter. \end{aligned} \tag{26}$$

This should be compared with the experimental values [28]

$$\begin{aligned} R^{exp} &= 1.4 \qquad\qquad A = 16, \\ &= (2.01 \pm 0.05) \quad A = 205 - -211. \end{aligned} \tag{27}$$

*I should note that this is a nice case where both chiral and scaling properties show up explicitly in many-body processes.*

Perhaps more significantly, the "dropping masses" are precursors to QCD phase transitions: From eqs.(1) and (2), combined with the preceding arguments, we see that the nucleon radius increases rapidly beyond $\rho = \rho_0$ [29], *i.e.*,

$$\langle r^2 \rangle^\star / \langle r^2 \rangle \approx f_\pi^2 / f_\pi^{\star 2}. \tag{28}$$

This rapid increase in radius with higher density may be interpreted as deconfinement as suggested in [10]. This basic notion was recently extended by Brown, Bethe and Pizzechero [30] to a physical description of the QCD phase transition from hadrons to a quark-gluon phase with a fascinating result on possible scenarios of the transition.

If we are living in a "swelled" world with all the masses dropping as density increases, how have the many calculations in conventional nuclear physics held up for so long in such a drastically different scenario? The answer to this puzzle is that to the extent that there is a common scale, it can be scaled out, leaving relatively small effects. To put more precisely, to the extent that $m_i^\star / m_i$ equals a common $\lambda(\rho)$ for any $i$, we have, to leading order (in mean field in the sense defined above),

$$H(m_i^\star, r) \approx \lambda(\rho) H(m_i, x) \tag{29}$$

with $m_i^\star r = m_i x$ defining $x$. Since $\lambda(\rho) \geq 0.8$ in nuclei, the energy is shifted little, say, less than 20%. But the distance swells by $\lambda(\rho)^{-1}$, *i.e.*, $r = (m_i / m_i^\star) x$. This "swelling" must of course show up in certain processes. Indeed one such case is the spin-orbit interaction in nuclei: It has been pointed out [31, 25] that the spin-orbit interaction in nuclei most sensitively displays the density-dependent masses, giving us a factor $(m/m^\star)^3$, as is obvious from its dimensionality in $r$.

## References

[1] G.E. Brown and M. Rho, Phys. Rev. Lett. , in press (1991)

[2] T.H.R. Skyrme, Nucl. Phys. **31**, 556(1962)

[3] G.E. Brown, A.D. Jackson, M. Rho and V. Vento, Phys. Lett. **B140**, 285(1984)

[4] I. Zahed and G.E. Brown, Phys. Repts. **142**, 1(1986)

[5] M. Rho, Phys. Rev. Lett. **54**, 767(1988)

[6] The $\eta'$ is connected with the flavor-singlet axial charge of the nucleon and figures importantly in what is (mistakenly) called "the proton spin crisis." However for our purpose, it is safe to ignore it.

[7] M. Bando, T. Kugo and K. Yamawaki, Phys. Repts. **164**, 217(1988)

[8] The Nambu-Jona-Lasinio model provides a microscopic description of the "dropping" masses as first shown by V. Bernard, U.-G. Meissner and I. Zahed, Phys. Rev. Lett. **59**, 966(1987). It predicts, in medium, $m_N^\star/m_\sigma^\star = 3/2$ *for all density* and hence $m_\sigma^\star/m_\sigma = m_N^\star/m_N$. Because of the low cut-off mass scale, however, it cannot describe reliably the in-medium behavior of vector mesons.

[9] G.E. Brown, Nucl. Phys. **A488**, 659c(1988); Z. Phys. **C38**, 291(1988)

[10] B.A. Campbell, J. Ellis and K.A. Olive, Nucl. Phys. **B345**, 325(1990)

[11] J. Collins, A. Duncan and S. Joglekar, Phys. Rev. **D16**, 438(1977); N.K. Nielsen, Nucl. Phys. **B120**, 212(1977)

[12] The pion field has scale-dimension zero in the nonlinear sigma model and one in the linear sigma model. See ref.[13].

[13] J. Ellis, Nucl. Phys. **B22**, 478(1976)

[14] The scale invariance of this term explains why the Skyrme Lagrangian gives, in high density limit, the same asymptotic property as noninteracting quarks.

[15] B. Buck and F. Perez, Phys. Rev. Lett. **50**, 1975(1983)

[16] G.E. Brown and M. Rho, Phys. Lett. **B222**, 324(1989)

[17] G.E. Brown, in *Proceedings of Akito Arima Symposium, May 1990*, Nucl. Phys. **A**, to be published

[18] This together with eq.(21) gives $\left(\frac{g^\star/m_V^\star}{g/m_V}\right)^2 \approx \left(\frac{f_\pi^\star}{f_\pi}\right)^{-2} \approx \left(\frac{m_N^\star}{m_N}\right)^{-2}$ as suggested in [22] to argue for the disappearance of tensor forces in nuclear matter.

[19] G.E. Brown, V. Koch and M. Rho, Stony Brook preprint (1990)

[20] S. Weinberg, Phys. Lett. **251**, 288(1990)

[21] M. Rho, Phys. Rev. Lett. **66**, 1275(1991)

[22] G.E. Brown and M. Rho, Phys. Lett. **B237**, 3(1990)

[23] G.E. Brown, C.B. Dover, P.B. Siegel and W. Weise, Phys. Rev. Lett. **60**, 2723(1988)

[24] T. Hatsuda, H. Hogaasen and M. Prakash, Phys. Rev. **C42**, 2212(1990)

[25] G.E. Brown, A. Sethi and N.M. Hintz, Phys. Rev. **C**, to be published

[26] A. Hosaka and H. Toki, University of Pennsylvania preprint-0449T, November 1990

[27] K. Kubodera, J. Delorme and M. Rho, Phys. Rev. Lett. **40**, 755(1978)

[28] E.K. Warburton, private communication from D.H. Wilkinson.

[29] Up to $\rho = \rho_0$, the nucleon radius is roughly constant because of the interplay between the factor $g_A^\star/g_A$ which decreases and $f_\pi/f_\pi^\star$ which increases. This effect is important for $y$ scaling in nuclei as shown in ref.[16].

[30] G.E. Brown, H.A. Bethe and P. Pizzechero, "The Hadron to Quark/Gluon Transition in Relativistic Heavy Ion Collisions," Caltech preprint (1991)

[31] G.E. Brown, H. Müther and M. Prakash, Nucl. Phys. **A506**, 565(1990)

# Chaos and Order in Excitation Functions of Heavy-Ion Collisions

S. Yu. KUN

*Max-Planck Institut für Kernphysik, 6900 Heidelberg 1, FRG*

**Abstract.** Chaotic and ordered quasiperiodic structures in fluctuating excitation functions of dissipative heavy-ion collisions are found from a Regge-pole representation for the $S$-matrix elements which allows for residual angular momentum coherence of the reaction amplitudes. The correspondence is established between the existence of ordered structures and degree of the time-orientation localization of a decaying dinuclear system: stronger localization produces more ordered structures. The results are applied for the analysis of energy autocorrelation functions in $^{19}F + ^{89}Y$ dissipative reaction.

**1.** Ever since the discovery of fluctuations in the excitation functions of dissipative heavy-ion collisions (DHIC) [1-3] one has asked whether it is possible to analyze these fluctuations in the framework of Ericson's fluctuation theory for compound nuclear decay [4]. The considerations [5,6] resulted in a Lorentzian shape of energy auto correlation function of the cross section (EAF) in agreement with compound nucleus theory. The deviation from Ericson's theory was obtained in [7], where EAFs display oscillating behaviour due to taking into account correlation between $S$-matrix elements with different angular momenta. Such a correlation is absent in compound nucleus theory. In this contribution we follow to consideration [7]. Especial attention is paid to the ordered long-range structures in DHIC excitation functions which appear in addition to compound nucleus like short-range chaotic fluctuations.

**2.** For the sake of simplicity, we consider all spins to be zero and we omit the indices of the entrance and exit channels. Then the reaction cross section and the statistical amplitude are

$$\sigma(E,\theta) = 2\pi \sin\theta |F(E,\theta)|^2, F(E,\theta) = \sum_l (2l+1) S_l(E) \exp[i\varphi(l)] P_l(\cos\theta), \qquad (1)$$

where $E$ is the incident beam energy, $S_l(E)$ are $S$-matrix elements with the orbital momentum $l$, $\varphi(l)$ are the phase shifts due to elastic (coherent) scattering, and $\theta$ is the CM scattering angle. If the reaction fragments have spins the formulas (1) and (2) are also valid under the condition of planar reaction mechanism [8].

We model the DHIC process by taking $S$-matrix in the Regge-pole form [6,7]

$$S_l(E) = i[W(l)]^{1/2} \sum_\mu b_\mu^l / [E^*(l) - E_\mu(l) + \tfrac{1}{2} i \Gamma_\mu(l)]. \qquad (2)$$

where $E^*(l) = E - E_{rot}(l)$ is the energy of intrinsic excitation of intermediate complex, $E_{rot}(l) = \hbar^2 l^2 / 2\mathcal{J}$ is its rotation energy, $\mathcal{J}$ is a moment of inertia, and $E_\mu(l)$, $\Gamma_\mu(l)$ are the positions and total widths of the overlapping resonance levels of intermediate dinuclear system (IDS). Assuming the DHIC is peripheral, we take the average reaction probability $W(l)$ to have the $l$-window form $W(l) = \exp[-(l-L)^2/d^2]$, where $L \simeq (L_{cr} + Lgr)/2$ ($L_{cr}$ and $L_{gr}$ are the critical and grazing angular momenta), and $d$ is the $l$-window width. To allow for the residual angular momentum correlations between $S_l(E)$ and $S_{l'}(E)$ we assume $l$-coherence for the residues $b_\mu^l$ and consider these to be $l$-independent inside of the $l$-window – $b_\mu^l = b_\mu$. This can be expected for peripheral collisions ($d \ll L$) under the condition of relative stability of the intrinsic wave function of a rotating IDS at the small variations of its angular velocity $\Delta\omega \leq \omega d/L \ll \omega$ around its average value $\omega = \hbar L/\mathcal{J}$. For the same reason we take $\mu$-average level spacing $D$ and total width $\Gamma$ in $l$-independent form, while $E_\mu(l)$ and $\Gamma_\mu(l)$ are allowed to depend on $l$.

Using Eq.(2) and the condition $L >> d$ we have

$$< S_l(E+\varepsilon) S_{l'}^*(E) >= iB[W(l)W(l')]^{1/2} / [\epsilon - (l - l') + i\gamma], \qquad (3)$$

where $B = 2\pi < |b_\mu|^2 >_\mu / \hbar\omega D$, $\gamma = \Gamma/\hbar\omega$, and $\epsilon = \varepsilon/\hbar\omega$. For $\epsilon = 0$ we get angular correlation function introduced in Ref. 9 where $\gamma$ was considered as a free parameter.

Due to the peripheral character of DHIC we may expand $\varphi(l)$ into the series

$$\varphi(l) = \varphi(L) + \Phi(L)(l-L) + \tfrac{1}{2}\dot{\Phi}(L)(l-L)^2, \tag{4}$$

where $\Phi(L) = \dot{\varphi}(L)$ is a deflection function and $\dot{\Phi}(L)$ is its derivative both taken at $l = L$. We shall adopt the notation $\Phi(L) = \Phi,\ \dot{\Phi}(L) = \dot{\Phi}$.

**3.** The cross section EAF normalized to 1 at $\varepsilon = 0$ has the form

$$C(\varepsilon,\theta) = |\rho(\varepsilon,\theta)|^2/[\rho(\varepsilon=0,\theta)]^2,\ \rho(\varepsilon,\theta) = 2\pi\sin\theta < F(E+\varepsilon,\theta)F^*(E,\theta) >, \tag{5}$$

where $\rho(\varepsilon,\theta)$ is the amplitude EAF.

Using Eqs. (3) and (4), the asymptotic form of the Legendre polynomials and replacing the sums over the angular momenta by integrals with the help of the Poisson formula we have

$$\rho(\varepsilon,\theta) = 2(2\pi)^{3/2}BdL\{\sum_{n=-\infty}^{\infty} R(a,b_n^{(-)},\gamma,\epsilon) + \sum_{m=-\infty}^{\infty} R(a,b_m^{(+)},\gamma,\epsilon) - iq\exp[i(2L+1)\theta - d^2\theta^2]\sum_{\nu=-\infty}^{\infty} R(a,c_\nu^{(+)},\gamma,\epsilon) + iq\exp[-i(2L+1)\theta - d^2\theta^2]\sum_{\lambda=-\infty}^{\infty} R(a,c_\lambda^{(-)},\gamma,\epsilon)\}, \tag{6}$$

where

$$b_n^{(\pm)} = \Phi \pm \theta + 2\pi n,\ c_\nu^{(\pm)} = \Phi + 2\pi\nu \pm i\dot{\Phi}\theta d^2,$$

$$a = (d^{-2} + \dot{\Phi}^2 d^2)/4,\ q = (\hbar\omega/\Delta E\theta)\sin(\Delta E\theta/\hbar\omega),$$

$$R(a,b,\gamma,\epsilon) = \exp[a(\gamma - i\epsilon)^2 - b(\gamma - i\epsilon)]\mathrm{erfc}[a^{1/2}(\gamma - i\epsilon) - \tfrac{1}{2}ba^{-1/2}],$$

and $\mathrm{erfc}(z)$ is the Error Function. The factor $q$ appear in (6) from the additional energy averaging from $\bar{E} - \frac{1}{2}\Delta E$ to $\bar{E} + \frac{1}{2}\Delta E$ due to energy dependence of $L(E) \simeq L + (E-\bar{E})/\hbar\omega$, where $L = L(\bar{E})$.

Having $\rho(\varepsilon,\theta)$ we can calculate the time power spectrum (TPS)

$$
\begin{aligned}
P(t,\theta) \sim \int_{-\infty}^{\infty} d\varepsilon \exp(-i\varepsilon t/\hbar)\rho(\varepsilon,\theta) \sim (4\pi a)^{-\frac{1}{2}} H(t) \exp(-\Gamma t/\hbar) \\
\{\sum_{n=-\infty}^{\infty} \exp[-(\Phi - \theta + 2\pi n - \omega t)^2/4a] + \sum_{m=-\infty}^{\infty} \exp[-(\Phi + \theta + 2\pi m - \omega t)^2/4a] \\
+2q \exp(-\theta^2/4a) \sum_{\lambda=-\infty}^{\infty} \exp[-(\Phi + 2\pi\lambda - \omega t)^2/4a] \sin[(2L+1)\theta - \\
(\Phi + 2\pi\lambda - \omega t)\theta\dot{\Phi} d^2/2a]\},
\end{aligned}
\tag{7}
$$

where $H(t)$ is the Heaviside function.

For $n, m \geq 0$ the terms of the first sums in Eqs. (6) and (7) represent the near side contribution to the decay after $n$ revolutions of IDS while the terms of the second sums correspond to the far side contribution after $m$ revolutions. The third and fourth sums in (6) as well as the $\lambda$-summation in (7) come from the interference between the near and far side collision amplitudes. These sums have a wave origin and display high frequency $t$ and $\theta$ oscillations.

Eq. (7) shows that the rotation of IDS has a macroscopic character. Indeed, the terms in the $n$-sum and $m$-sum in $P(t,\theta)$ represent *wave packets* rotating with angular velocity $\omega$.

To obtain an approximate form for $\rho(\varepsilon,\theta)$ we consider the backward angular range $\theta \geq |\Phi| + 2(a)^{1/2}$. Under the additional condition $a^{1/2}\gamma << 1$, which is always fulfilled for the long-living dinuclear configurations on which we concentrate, one can find [7]

$$
\begin{aligned}
\rho(\varepsilon,\theta) \sim \exp[a(\gamma - i\epsilon)^2]\{\exp[(\pi - \theta)(\gamma - i\epsilon)] + \\
\exp[-(\pi - \theta)(\gamma - i\epsilon)]\}\{1 - \exp[-2\pi(\gamma - i\epsilon)]\}^{-1},
\end{aligned}
\tag{8}
$$

while $< \sigma(E,\theta) >= \rho(\varepsilon = 0, \theta) \sim \cosh[(\pi - \theta)\gamma]$, which is the formula for the angular distribution of a classical model of a rotating IDS having a lifetime $\hbar/\Gamma$ and an angular velocity $\omega$ [10].

For the cross section EAF we have

$$
C(\epsilon,\theta) = \underbrace{\exp(-2a\epsilon^2)}_{\mathrm{I}} \underbrace{\frac{\cos[2(\pi - \theta)\epsilon] + \cosh[2(\pi - \theta)\gamma]}{1 + \cosh[2(\pi - \theta)\gamma]}}_{\mathrm{II}} \underbrace{\frac{[1 - \exp(-2\pi\gamma)]^2}{|1 - \exp[-2\pi(\gamma - i\epsilon)]|^2}}_{\mathrm{III}} . \tag{9}
$$

The analysis of Eq. (9) shows that $C(\varepsilon,\theta)$ contains four types of energy dependence:

1) The monotonically decreasing exponential factor I has the coherent length $\Delta\varepsilon_1 = (2a)^{-1/2}\hbar\omega = \hbar/\Delta t_1$, where $(2a)^{1/2}$ is the angular dispersion of the orientation of IDS and $\Delta t_1$ is the corresponding time uncertainty.

2) A periodic long-range structure with period $\Delta\varepsilon_2 = \pi\hbar\omega/(\pi-\theta) = 2\pi\hbar/\Delta t_2$ comes from factor II, where $\Delta t_2 = t_{f.s.} - t_{n.s.} = 2(\pi-\theta)/\omega$ and $t_{f.s.}(t_{n.s.})$ are the times IDS needs to reach the orientations which give decays into scattering angle $\theta$ after dinucleus was formed on far(near) side. This energy structure is the result of interference between the near and far side wave packets, and, hence, it disappears if the contribution from only one side is present. The corresponding criterion for the existence of the near-far side interference is $2(\pi-\theta)\gamma \leq 1$.

3) The factor III at $\varepsilon \leq 2-3\Gamma$ gives Ericson type energy structure with a width of order $\Delta\varepsilon_3 \simeq \Gamma = \hbar/\Delta t_3$ , where $\Delta t_3$ is the average lifetime of IDS.

4) The factor III also generates a periodic energy structure with the period $\Delta\varepsilon_4 = \hbar\omega = 2\pi\hbar/T(2\pi)$ , where $T(2\pi)$ is the time for one complete revolution of IDS. This periodic structure is the result of an interplay between the contributions from a different number of revolutions of IDS. This structure disappears under the condition $2\pi\gamma \geq 1$, which means that the average number of IDS revolutions is less than 1.

Consider long-living dinuclear configurations such that $2\pi\gamma \leq 0.5$. The average number of IDS revolutions is then more than 2, and the angular distributions of the reaction products are practically isotropic – $\cosh[\gamma(\pi-\theta)] = 1$. In this case, Eq. (9) takes the simpler form

$$C(\varepsilon,\theta) = \sum_{n=0}^{\infty} \exp(-2an^2)\cos^2(\theta n)[1+(\varepsilon-n\hbar\omega)^2/\Gamma^2]^{-1}. \tag{10}$$

where $\cos^2(\theta n)$ appears due to the near-far side interference. Eqs. (6-10) have been obtained assuming a good angular resolution of $\Delta\theta < \pi/L$. For poor angular resolution, $\Delta\theta > \pi/L$, the near-far side interference is completely destroyed and $C(\varepsilon,\theta)$ becomes independent of $\theta$ and corresponds to Eq. (10) with $\theta = 0$.

Examples of $C(\varepsilon,\theta)$ calculated from Eq.(10) are depicted in Fig. 1 and show the transition from the destructive near-far side interference at $\theta = 90°$ to the constructive one at $\theta = 175°$. The lowest curve in Fig. 1 corresponds to a poor angular resolution and is quite similar to $C(\varepsilon,\theta)$ at $\theta = 175°$ for a precise angular resolution. From values $\gamma = 0.055$ and $a = 0.1$, the angular dispersion of the IDS orientation is $(2a)^{1/2} = 25°$, and the average number of rotations is $n = 1/2\pi\gamma \simeq 3$.

It is seen that $n$=1,2,... terms in Eq.(10) generate additional to Ericson type long-range quasiperiodic structure. However, destruction of the time-orientation localization of IDS by making the angular dispersions of wave packets large enough $(2a)^{1/2} \geq (2)^{1/2}$ leaves only a Ericson type contribution $n = 0$, while $P(t,\theta)$ becomes to be close to the angular independent exponential form. This shows an intimate link between the existence of long-range quasiperiodic energy structures in excitation functions and degree of the time-orientation localization of IDS : stronger localization produces more ordered structures.

Disappearance of the ordered structures due to time-orientation delocalization can be demonstrated by analyzing TPS of IDS. We pass from the representation of $P(t,\theta)$ (7) through the number of rotations to the summation over the differences of angular momenta $\mu = l - l'$. Using the Poisson formula we have

$$P(t,\theta) \sim H(t)(4\pi a)^{-\frac{1}{2}} \exp(-\Gamma t/\hbar) \sum_{-\infty}^{\infty} \exp(-a\mu^2)[\exp(i\mu(\Phi+\theta-\omega t))+\exp(i\mu(\Phi-\theta-\omega t))]. \tag{11}$$

One can see that at $a \geq 1$ the contributions $|\mu| \geq 1$ become to be small and TPS takes angular independent exponential form $P(t,\theta) \sim \exp(-\Gamma t/\hbar)$, which immediately gives Ericson's result for EAF. Hence the criterion for disappearance of ordered structure in excitation functions can be obtained without any use of expression (10) for EAF itself but using the condition of time-orientation delocalization of IDS.

Another aspect we would like to point out is that the width of wave packets $(2a)^{1/2} =$

$[(d^{-2}+\dot{\Phi}^2d^2)/2]^{1/2}$ is smaller when the deflection function $\Phi(l)$ varies smoothly with impact parameter $l$, i.e. $\dot{\Phi} \leq (2)^{1/2}/d$. This is the condition for the existence of the ordered long-range correlated structures in DHIC fluctuating excitation functions, in addition to the Ericson type random fluctuations. On the contrary, if $\Phi(l)$ has a strong dependence on impact parameter $l$ (orbiting regime), any ordering in the excitation functions disappears and we obtain the Ericson short-range correlated picture. If we take $\Phi(l)$ to be completely independent of impact parameter, packet width goes to zero as $d$ becomes large. Then excitation functions show maximal ordering – periodicity, while $P(t,\theta)$ is given by $\delta(\Phi\pm\theta+2\pi m-\omega t)$, corresponding to classical rotation motion of pointed objects along deterministic trajectories.

**4.** In order to apply our results for the analysis of data we use measurements of excitation functions in the reaction $^{89}$Y($^{19}$F,x)y between 135 and 143.25 MeV incident energy in 250 keV steps [3]. We consider the reaction channels Z=7,8,10 and 11 at $\theta_{lab} = 160°$. To fit the behaviour of experimental EAFs for all Z at $\varepsilon \leq 500$ keV we take $\Gamma$=100 keV. The comparison of the experimental EAFs (Fig. 2) with the calculations using Eq. (10) gives $\hbar\omega$=1.7MeV which corresponds to $\gamma = 0.07$, $T(2\pi) = 2.2\times10^{-21}s$. The average number of IDS revolutions before its decay is 2.5. The parameter $\gamma$ extracted by analyzing EAFs is in agreement with the angular distribution having an isotropic character from $\theta = 120°$ to $\theta = 170°$ [11]. Note that estimates [3,11] for $L$=67 give $\hbar\omega$=2 MeV which is close to the value extracted by analyzing the shape of $C(\varepsilon,\theta)$.

To model Z-dependence of $a$(Z) we suppose a linear decreasing of $\dot{\Phi}$ as the atomic number or charge of the reaction fragments increases: $\dot{\Phi}$(Z)=$\dot{\Phi}$(Z=11)+(11-Z)$k$, where $\dot{\Phi}$(Z=11) and $k$ are unknown parameters. For $a$(Z) we have

$$a(\mathrm{Z})=\tfrac{1}{4}[d^{-2} + d^2(\dot{\Phi}(\mathrm{Z}=11)+(11-\mathrm{Z})k)^2].$$

Putting $d$=8 and fitting amplitudes of the first bumps at $\varepsilon \simeq 1.7$MeV in EAFs for Z=11 and 10 by using Eq.(10) we find $\dot{\Phi}$(Z=11)=0.13, $k$=0.014 which allows us to plot also EAFs for Z=7 and 8 (Fig. 2).

Fig. 1.

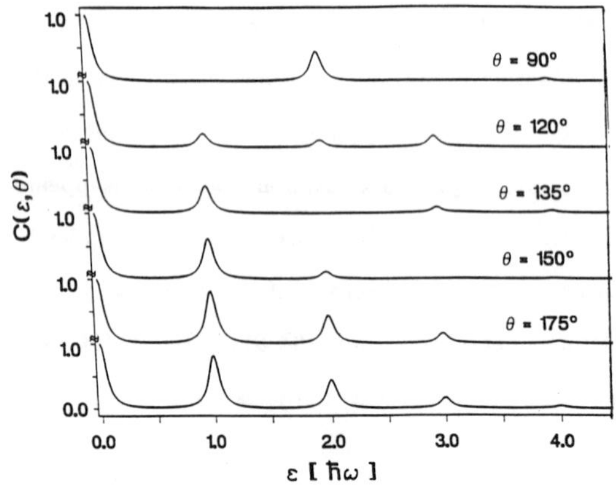


Fig. 2.

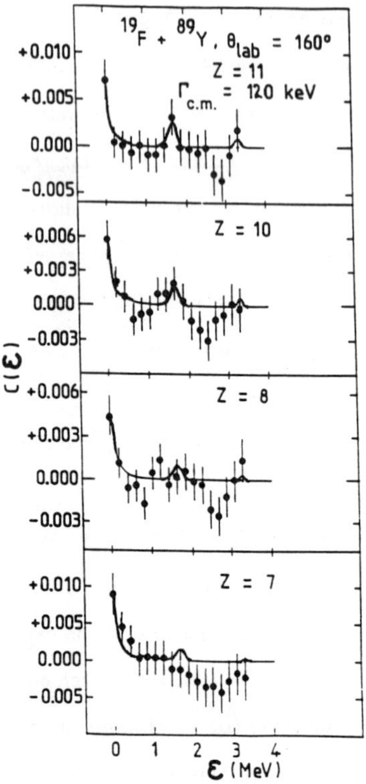


Fig. 3.

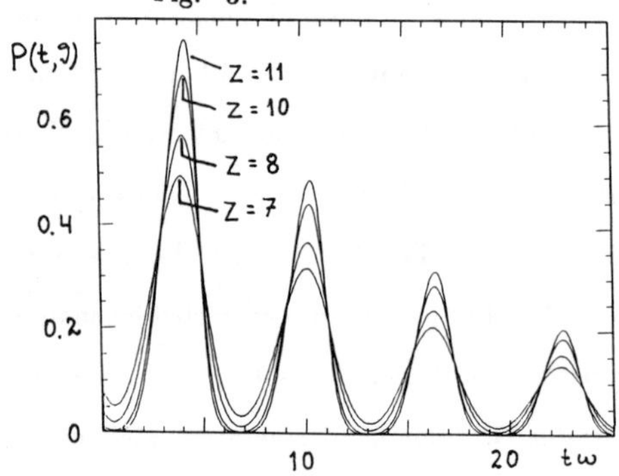


**Fig.** 1. $C(\varepsilon, \theta)$ at the different $\theta$ for pure angular resolution. Lowest part: angular independent EAF for poor angular resolution. Parameters are $\gamma = 0.055, a = 0.1$.

Fig. 2. EAFs for Z=7,8,10 and 11 reaction fragments. Theoretical EAFs (solid lines) is normalized to experimental one at $\varepsilon = 0$ and calculated in accordance with Eq. (10). Experiment is from ref.[3].

Fig. 3. $P(t, \theta)$ for a Z=7,8,10 and 11 reaction fragments.

In Fig. 3 we depict TPSs (Eq.7) for Z=7,8,10 and 11 putting $\Phi = 0.5$[11]. Contributions from near and far sides are not given separately due to their practically complete overlapping. We see the correlation between EAF and TPS behaviours. Namely more pronounced time-orientation localization for Z=11 produces more strong bumps at $\varepsilon = \hbar\omega, 2\hbar\omega$. For Z=7 contributions from consistent rotations in TPS become to overlap and bumps are close to complete disappearance.

**5.** In conclusion we have found the coexistence of both chaotic and ordered quasiperiodic structures in excitation functions of DHIC under the condition of long-living configurations of IDS. The excitation functions display short-range chaotic structures providing energy interval of investigation $\Delta E < \hbar\omega$, while at $\Delta E > \hbar\omega$ additional quasiperiodic structures appear. The good agreement of the data with our calculations supports the coexistence of both structures and the interpretation of quasiperiodic one in terms of time-orientation distribution of IDS.

The possible origin of regular quasiperiodic structures is regularity (periodicity) of underlying classical model of rotating dinucleus. However the classical picture appears from the quantum mechanical formulation of the problem, where the crucial element is residual correlation between amplitudes of reaction with different angular momenta.

This work was initiated during the author's stay at MPI für Kernphysik, Heidelberg by Humboldt Fellowship.

**References**

[1] A. De Rosa, G. Inglima, V. Russo, M. Sandoli, G. Fortuna, G. Montagnoli, C. Signorini, A.M. Stefanini, G. Cardella, G. Pappalardo and F. Rizzo, Phys. Lett. **160B** (1985) 239.

[2] A. Glaesner, W. Dünnweber, W. Hering, D. Konnerth, R. Ritzka, R. Singh, and W. Trombik, Phys. Lett. **169B** (1986) 153.

[3] T. Suomijärvi, B. Berthier, R. Lucas, M.C. Mermaz, J.P. Coffin, G. Guillaume, B.

Heusch, F. Jundt, and F. Rami, Phys. Rev. **36C** (1987) 181.

[4] T. Ericson and T. Mayer-Kuckuck, Ann. Rev. Nucl. Sci. **16** (1966) 183.

[5] R. Bonetti and M.S. Hussein, Phys. Rev. Lett. **57** (1986) 194.

[6] D.M. Brink and K. Dietrich, Z. Phys. **A326** (1987) 7.

[7] S.Yu. Kun, Phys. Lett. **B**, in press.

[8] V.M. Strutinsky and S.M. Vydrug-Vlasenko, Z. Phys. **A294** (1980) 281.

[9] V.M. Strutinsky, Z. Phys. **A286** (1978) 77.

[10] J. Barrette, P. Braun-Munzinger, C.K. Gelbke, H.E. Wegner, B. Zeidman, A. Gamp, H.L. Harney, and Th. Walcher, Nucl. Phys. **A279** (1977) 125.

[11] M.C. Mermaz, T. Suomijärvi, R. Lucas, B. Berthier, J. Matuszek, J.P. Coffin, G. Guillaume, B. Heusch, F. Jundt, and F. Rami, Nucl. Phys. **A456** (1986) 186.

# Symmetry adapted non-spurious wavefunctions for multi-cluster systems

Jacob Katriel

Department of Chemistry

Technion - Israel Institute of Technology, Haifa 32000, Israel

and

Akiva Novoselsky

The Racah Institute of Physics

The Hebrew University of Jerusalem, Jerusalem 91104, Israel

**Abstract**

**An algorithm for the construction of multi-cluster non-spurious harmonic oscillator wavefunctions with arbitrary permutational symmetry is developed. Each cluster is presented in terms of a harmonic oscillator wavefunction, expressed in an appropriate set of Jacobi coordinates. The single-cluster wavefunctions are constructed recursively, using a new type of harmonic oscillator coefficients of fractional parentage. These coefficients are the eigenvectors of the two-cycle class operator of the intra-cluster symmetric group, in the appropriate basis. The matrix elements of the class operator are evaluated by using a specific version of the harmonic oscillator brackets.**

**The cluster wavefunctions are coupled into non-spurious multi-cluster states with an overall well-defined angular momentum and permutational symmetry. This is achieved by diagonalizing an appropriate set of single-cycle class operators of the symmetric group involving all the constituent particles. The technique for the evaluation of the necessary matrix elements is developed. It provides an interesting generalization of the corresponding technique for a system consisting of a single cluster of identical particles.**

**The results are immediately applicable to the study of nuclear cluster models, nuclei as quark clusters, nuclear fusion, fission and multifragmentation, as well as in atomic and molecular physics.**

## I. Introduction

The treatment of a finite system of identical particles within non-relativistic quantum mechanics starts with the construction of a many-particle basis set with appropriate symmetry properties. We shall deal with spherically symmetrical systems, for which the wavefunctions are characterized by their total angular momentum as well as by their behaviour with respect to the symmetric group. There may be additional symmetries, such as the symplectic symmetry which gives rise to the seniority scheme, which we shall not discuss explicitly; they can always be incorporated in order to take care of whatever degeneracies remain after the classification in terms of the symmetric group and the total angular momentum has been carried out.

An important distinction is the one between the system of electrons in an atom on the one hand and the system of nucleons in a nucleus, on the other hand. In the former case a natural origin exists, namely the atomic nucleus which is much heavier than the system of electrons anchored to it. In the latter case the internal motions should be referred to the center-of-mass (c.m.) of the particles themselves. An entirely analogous situation occurs in the van-der-Walls clusters of identical rare-gas atoms or in alkali-metal clusters.

One of the technical difficulties that have to be faced in the study of systems of the second kind involves the construction of a basis set in which the spurious c.m. motion is excluded. The common procedures involve the construction of a basis set contaminated by spurious states, which have than to be eliminated. This is usually done by adding an appropriate interaction to the hamiltonian, which pushes up the spurious states in such a way that they do not affect the eventual calculations[1]. However, these states are still present, yielding large, inefficient basis sets.

A more serious difficulty is encountered in the calculation of spectroscopic factors in nuclear reactions[2]-[3]. These involve non-spurious two-cluster wavefunctions. Several methods have been proposed to eliminate two-cluster spurious states, but each one of them is only applicable within a small class of special cases[1].

The present contribution consists of a review of the group theoretical foundations and of the recursive construction of single-cluster and multi-cluster symmetry adapted wavefunctions in an arbitrary single-particle basis. This is followed by a discussion of the construction of fully non-spurious many-particle wavefunctions by using the special transformation properties of the harmonic oscillator (h.o.) basis set.

## II. Group theoretical preliminaries

Since the theory of angular momentum coupling is so fully developed and well known, we shall in all the cases considered below start from a set of $n$-particle functions corresponding to a well-defined total angular momentum. This set of functions has to span an invariant subspace with respect to the $n!$ permutations of the $n$ particle indices. To split this subspace into irreducible subspaces, each one of which carries a well-defined irreducible representation (irrep) of the symmetric group $S_n$, we have to diagonalize simultaneously the single-cycle class-operators $[(2)]_n$, $[(3)]_n$, $\ldots$, $[(n)]_n$, each one of which is the sum of all the permutations consisting of a single cycle of the appropriate length. In fact, it turns out that for any given $n$ a small subset of this set of single-cycle class-operators is sufficient to fully characterize the irreps of the symmetric group. A further reduction in the number of class-operators that need to be diagonalized is achieved if the space constructed corresponds to a well-defined irrep of a symmetric subgroup. In the important special case in which all basis functions correspond to a particular irrep of $S_{n-1}$, the class of transpositions of $S_n$ provides a complete label. The irrep labels are determined by the corresponding eigenvalues. In those instances in which one can choose to diagonalize simultaneously the class-operators corresponding to a chain of subgroups of the form $S_n \supset S_{n-1} \supset \ldots \supset S_2$, the states are eventually labeled in terms of the sequence of irreps $\Gamma_n, \Gamma_{n-1}, \ldots, \Gamma_2$, or, equivalently, in terms of a Yamanouchi symbol $Y_n$[4].

## III. A single cluster in an arbitrary basis

A basis set of one-particle wavefunctions centered at the nucleus, which span an invariant subspace with respect to the three dimensional rotation group, is appropriate for the atomic case. If the invariant subspace is irreducible the situation encountered is known as the single-shell case. The many-particle symmetry-adapted functions are linear combinations, with appropriate coefficients, of products of these one-particle functions. A comprehensive scheme can be formulated in terms of a set of coefficients of fractional parentage (cfps) which relate the $n$-particle state corresponding to some total angular momentum $J_n$ and a Yamanouchi symbol $Y_n$ to the set of states obtained by coupling the angular momenta of all $(n-1)$-particle states with the appropriate Yamanouchi symbol $Y_{n-1}$ to an $n$'th particle state. For a single shell $j^n$ we denote the states by

$$|j^n Y_n \alpha_n \vec{J}_n > \tag{1}$$

where $j$ is the single-particle angular momentum, $\vec{J}_n$ is the total angular momentum and $\alpha_n$ labels any remaining degeneracy. We will use the abbreviated notation $\Phi_n \equiv \alpha_n \vec{J}_n$. The scheme just described is specified by the relation

$$|j^n Y_n \Phi_n >= \sum_{\Phi_{n-1}} [(j^{n-1} Y_{n-1} \Phi_{n-1}) j \vec{J}_n | \} j^n Y_n \Phi_n] |(j^{n-1} Y_{n-1} \Phi_{n-1}) j \vec{J}_n > \tag{2}$$

The cfps appearing in eq. (2) are the eigenvectors obtained upon diagonalizing $[(2)]_n$ in the basis specified on the r.h.s. of that equation. Full details, including the incorporation of the seniority scheme and the application to LS-coupling, are presented in ref. [4].

## IV. Multi-cluster states in an arbitrary basis

In many physical situations involving clustering, fission, fusion or fragmentation it is desirable, indeed necessary, to express the $n$-particle wavefunction in terms of subsystems with, say, $n_1$ and $n_2$ particles, respectively. It will be convenient to denote the set of indices belonging to the first cluster by $[n_1]'$ or, explicitly, by $\{ 1, 2, \ldots, n_1 \}'$, those belonging to the second cluster by $[n_2]''$ or $\{ 1, 2, \ldots, n_2 \}''$, and $[n] = [n_1]' \cup [n_2]''$. The group $S_{[n]}$ commutes with $S_{[n_1]'} \supset S_{[n_1-1]'} \supset \ldots \supset S_{[2]'}$ as well as with $S_{[n_2]''} \supset S_{[n_2-1]''} \supset \ldots \supset S_{[2]''}$ so that one can use a mutually consistent set of labels consisting of the Yamanouchi symbols $Y_{[n_1]'}$ and $Y_{[n_2]''}$ of the two clusters, the irrep label of $S_{[n]}$ (i.e., $\Gamma_{[n]}$) and the total angular momentum $\vec{J}_{[n]}$. Note that the individual cluster angular-momenta do not commute with the operators of $S_{[n]}$ and are consequently not conserved. Therefore, the two-cluster wavefunction is fully specified as

$$|j^n(Y_{[n_1]'}Y_{[n_2]''})\Gamma_{[n]}\alpha_{[n]}\vec{J}_{[n]}> \tag{3}$$

where $\alpha_{[n]}$ labels any remaining degeneracies.

We construct this two-cluster wavefunction in terms of the set of wavefunctions obtained by coupling the angular momenta of all the pairs of single-cluster states with the desired Yamanouchi symbols $Y_{[n_1]'}$ and $Y_{[n_2]''}$, into the desired total angular momentum $\vec{J}_{[n]}$. This yields

$$\begin{aligned} &|j^n(Y_{[n_1]'}Y_{[n_2]''})\Gamma_{[n]}\alpha_{[n]}\vec{J}_{[n]}> \\ &= \sum_{\Phi_{[n_1]'}\Phi_{[n_2]''}} \left[\left(j^{n_1}Y_{[n_1]'}\Phi_{[n_1]'}\right)\left(j^{n_2}Y_{[n_2]''}\Phi_{[n_2]''}\right)\vec{J}_{[n]}|\}j^n\left(Y_{[n_1]'}Y_{[n_2]''}\right)\Gamma_{[n]}\alpha_{[n]}\vec{J}_{[n]}\right] \\ &\qquad |\left(j^{n_1}Y_{[n_1]'}\Phi_{[n_1]'}\right)\left(j^{n_2}Y_{[n_2]''}\Phi_{[n_2]''}\right)\vec{J}_{[n]}> \end{aligned} \tag{4}$$

The coefficients in this expression can be referred to as the two-cluster coefficients of fractional parentage. They are obtained by diagonalizing an appropriate set of class-operators of $S_n$ in the basis appearing on the r.h.s. of Eq. (4). Full details, including the treatment of systems consisting of an arbitrary number of clusters, are presented in ref. [5]. In anticipation of Section VI we point out that the resulting states contain the spurious c.m. motion in an uncontrolled and inseparable way.

## V. Single-cluster non-spurious states

In order to construct non-spurious states for $n$ identical isotropic three-dimensional h.o.s we have to use a set of coordinates where the c.m. is separated from the $n-1$ internal coordinates. Among the various sets of coordinates satisfying this requirement the normalized Jacobi coordinates

$$\vec{\rho}_{[n]} = \frac{1}{\sqrt{n}}\sum_{j=1}^{n}\vec{r}_j \quad ; \quad \vec{\rho}_{\{i\},[i-1]} = \sqrt{\frac{i-1}{i}}\left(\vec{r}_i - \frac{1}{i-1}\sum_{j=1}^{i-1}\vec{r}_j\right) \quad i = 2, 3, \ldots, n \tag{5}$$

were found to be preferable because each internal coordinate $\vec{\rho}_{\{i\},[i-1]}$ $i = 2, \ldots, n$ depends on the first $i$ single particle coordinates only. This property enables the formulation of an efficient recursive procedure for constructing the h.o. non-spurious states [6].

The h.o. wavefunctions expressed in Jacobi coordinates are naturally separated into an internal and a c.m. wavefunction. Since the c.m. wavefunction is totally symmetric, the permutational symmetry of the total wavefunction is completely determined by that of the internal wavefunction. Our aim is to construct internal wavefunctions, consisting of $n-1$ h.o.s, which belong to an irrep of the symmetric group, $S_n$.

With the present choice of coordinates and basis functions it is found that in order to span an invariant subspace with respect to the symmetric group $S_n$ one has to include all the $n$-particle internal wavefunctions corresponding to given total energy and angular momentum.

The internal wavefunction of an $n$-particle single cluster with a well defined permutational symmetry is[7]

$$|Y_n\Phi_n;\vec{\rho}_{\{2\},[1]}\vec{\rho}_{\{3\},[2]}\cdots\vec{\rho}_{\{n\},[n-1]}> \tag{6}$$

where $[i]\equiv\{1,2,\ldots,i\}$, $i=1,2,\ldots,n$. The symbol $\Phi_n$ stands for the good quantum numbers

$$\Phi_n=\vec{\Lambda}_n\epsilon_n\alpha_n \tag{7}$$

where $\vec{\Lambda}_n$ is the overall internal angular momentum, $\epsilon_n$ is the total internal energy and $\alpha_n$ is an additional label that takes care of any remaining degeneracy.

We shall denote by $\eta_{\{i\}}$ the individual h.o. radial and angular quantum numbers $N_i$ and $L_i$ corresponding to the $i$'th Jacobi coordinate.

For two particles the internal wavefunction can be written as

$$|\Gamma_2\,\vec{\Lambda}_2\,\epsilon_2\,;\vec{\rho}_{\{2\},\{1\}}>=|\Gamma_2\,\Phi_2\,;\vec{\rho}_{\{2\},\{1\}}>=\Psi_{N_2\vec{L}_2}(\vec{\rho}_{\{2\},\{1\}})=|\eta_{\{2\}};\vec{\rho}_{\{2\},\{1\}}> \tag{8}$$

where $\epsilon_2=2N_2+L_2$, $\vec{\Lambda}_2$ is the internal angular momentum and $\vec{L}_2=\vec{\Lambda}_2$. $\Gamma_2$ is determined by $L_2$: $\Gamma_2=[2]$ for even $L_2$ and $\Gamma_2=[1^2]$ for odd $L_2$.

To obtain three-particle internal wavefunctions labeled by the sequence of Young frames $\Gamma_2\Gamma_3$ and internal angular momentum $\vec{\Lambda}_3$, we first form angular momentum coupled direct product states of the form

$$|\Gamma_2\Phi_2\eta_{\{3\}}\vec{\Lambda}_3;\vec{\rho}_{\{2\},\{1\}}\vec{\rho}_{\{3\},[2]}> \tag{9}$$

The permutation operators transform these states within invariant subspaces, each of which is characterized by a given internal energy $\epsilon_3=\epsilon_2+2N_3+L_3$ and internal angular momentum $\vec{\Lambda}_3$. The linear combinations which belong to well defined irreps of the symmetric group are written in the form

$$\begin{aligned}&|\Gamma_2\Gamma_3\Phi_3;\vec{\rho}_{\{2\},\{1\}}\vec{\rho}_{\{3\},[2]}>=\\&\sum_{\substack{\Phi_2\eta_{\{3\}}\\(\epsilon_3=\epsilon_2+2N_3+L_3)}}\left[\Gamma_2\Phi_2\eta_{\{3\}}\vec{\Lambda}_3|\}\Gamma_2\Gamma_3\Phi_3\right]|\Gamma_2\Phi_2\eta_{\{3\}}\vec{\Lambda}_3;\vec{\rho}_{\{2\},\{1\}}\vec{\rho}_{\{3\},[2]}>\end{aligned} \tag{10}$$

Let us assume that the $(n-1)$-particle wavefunctions, symmetry adapted to $S_{n-1}$, have already been constructed. The general expression for the $n$-particle internal wavefunction, symmetry adapted to $S_n$, can than be written in the form

$$|Y_n\Phi_n;\vec{\rho}_{\{2\},\{1\}}\vec{\rho}_{\{3\},[2]}\cdots\vec{\rho}_{\{n\},[n-1]}>=\sum_{\substack{\Phi_{n-1}\eta_{\{n\}}\\(\epsilon_n=\epsilon_{n-1}+2N_n+L_n)}}$$

$$\left[Y_{n-1}\Phi_{n-1}\eta_{\{n\}}\vec{\Lambda}_n|\}Y_n\Phi_n\right]|Y_{n-1}\Phi_{n-1}\eta_{\{n\}}\vec{\Lambda}_n;\vec{\rho}_{\{2\},\{1\}}\vec{\rho}_{\{3\},[2]}\cdots\vec{\rho}_{\{n\},[n-1]}> \qquad (11)$$

where $Y_n = Y_{n-1}\Gamma_n$.

The coefficients appearing on the r.h.s. of Eq. (11) will be referred to as harmonic oscillator coefficients of fractional parentage (hocfps). They satisfy orthogonality and completeness relations[6] similar to those satisfied by the single shell cfps for arbitrary permutational symmetry, defined in ref. [4]. On the other hand, the $n$'th particle state $|\eta_{\{n\}};\vec{\rho}_{\{n\},[n-1]}>$ is not unique in the present case. We have to sum over all the different single particle states consistent with the angular momentum coupling $\vec{\Lambda}_{n-1}+\vec{L}_n=\vec{\Lambda}_n$ and the energy relation $\epsilon_n=\epsilon_{n-1}+2N_n+L_n$, since the elements of $S_n$ couple all those states. This is the price paid for using the Jacobi coordinates, which explicitly eliminate the c.m., rather than using single particle coordinates.

The hocfps are evaluated by diagonalizing $[(2)]_n$ within the set of functions specified by the r.h.s. of eq. (11). Details of the computations that this entails are presented in ref. [6] where it is shown that the problem reduces to the evaluation of the matrix elements of the transposition $(n-1,n)$ with respect to the wavefunctions corresponding to the last two Jacobi coordinates, coupled to a total angular momentum $\vec{\lambda}$. This is the only step which we wish to recall, for a reason which will become apparent at the end of the present section.

The Jacobi coordinates $\vec{\rho}_{\{n-1\},[n-2]}$ and $\vec{\rho}_{\{n\},[n-1]}$ defined in Eq. (5) are neither symmetric nor antisymmetric in the coordinates $\vec{r}_{n-1}$ and $\vec{r}_n$. Therefore, the application of the transposition $(n-1,n)$ to the wavefunction on the r.h.s. of Eq. (11), expressed in these coordinates, is very cumbersome. A passage to a new set of coordinates

$$\vec{\rho}_{\{n\},\{n-1\}}=\frac{1}{\sqrt{2}}(\vec{r}_n-\vec{r}_{n-1})$$

and

$$\vec{\rho}_{\{n-1,n\},[n-2]}=\sqrt{\frac{2(n-2)}{n}}\left[\frac{\vec{r}_{n-1}+\vec{r}_n}{2}-\frac{\vec{r}_1+\vec{r}_2+\ldots+\vec{r}_{n-2}}{n-2}\right]$$

which are antisymmetric and symmetric, respectively, in the coordinates $\vec{r}_{n-1}$ and $\vec{r}_n$, is required. This passage is achieved by means of an appropriate rotation.

To take advantage of the symmetry properties of the new coordinates we have to express the wavefunction on the r.h.s. of Eq. (11) in terms of wavefunctions in the new coordinates using the h.o. brackets for two particles with different masses. These brackets were first introduced by Talmi[8] and were then calculated explicitly for equal masses[9]-[10] and for different masses[11].

The internal h.o. states with arbitrary symmetry, derived above, are expressed in terms of the Jacobi coordinates (Eq. (5)). However, in many calculations in atomic and nuclear physics it is desirable to have expressions for the wavefunctions in terms of the single particle coordinates. This is particularly important when the h.o. states are used as a basis set in a calculation involving non-harmonic potentials, which are not easily expressible in Jacobi coordinates[12]. In fact, for potentials which depend on the two-body distance, the expression of the wavefunction in terms of the coordinates $\vec{\rho}_{\{n\},\{n-1\}}$ and $\vec{\rho}_{\{n-1,n\},[n-2]}$ enables a convenient evaluation of the matrix elements. The expression to be evaluated reduces into a one body integral over the relative coordinate $\vec{\rho}_{\{n\},\,\{n-1\}}$.

## VI. Multi-cluster non-spurious states

We now introduce a general procedure to construct non-spurious multi-cluster states with arbitrary permutational symmetry. For the sake of simplicity we shall deal with two clusters only, since the extension required to deal with multi-cluster systems is rather straightforward. The two-cluster states are expressed in terms of two single-cluster internal wavefunctions coupled to a h.o. relative wavefunction. The c.m. is assumed to be at rest ($0s$ state). Consequently, the overall angular momentum is the same as the internal angular momentum. As for the single cluster case, the c.m. wavefunction is totally symmetric with respect to permutations of the particle coordinates and therefore does not affect the overall permutational symmetry. In view of the above, we will consider only the internal wavefunction.

The good quantum numbers of a two-cluster wavefunction are the Yamanouchi symbols of the two clusters $Y_{[n_1]'}$ and $Y_{[n_2]''}$, the total permutational irrep symbol $\Gamma_{[n]}$, angular momentum $\vec{\mathcal{L}}_{[n]}$ and internal energy $\epsilon_{[n]}$ (cf., [5]). The two-cluster wavefunction is constructed in terms of a basis set consisting of states of the form

$$| \left( Y_{[n_1]'}\Phi_{[n_1]'}Y_{[n_2]''}\Phi_{[n_2]''} \right) \vec{\Lambda}_{[n]}\Phi_{[n_2]'',[n_1]'}\vec{\mathcal{L}}_{[n]}\epsilon_{[n]} > \tag{12}$$

The intercluster h.o. wavefunction

$$|\Phi_{[n_2]'',[n_1]'} >= |N_{[n_2]'',[n_1]'}\vec{L}_{[n_2]'',[n_1]'};\vec{\rho}_{[n_2]'',[n_1]'} > \tag{13}$$

is expressed in terms of the intercluster relative coordinate

$$\vec{\rho}_{[n_2]'',[n_1]'} = \left( \frac{\vec{r_1}'' + \vec{r_2}'' + \ldots + \vec{r}_{n_2}''}{n_2} - \frac{\vec{r_1}' + \vec{r_2}' + \ldots + \vec{r}_{n_1}'}{n_1} \right) \sqrt{\frac{n_1 n_2}{n_1 + n_2}} \tag{14}$$

$N_{[n_2]'',[n_1]'}$ and $L_{[n_2]'',[n_1]'}$ are the corresponding radial and angular quantum numbers.

The total internal energy is

$$\epsilon_{[n]} = \epsilon_{[n_1]'} + \epsilon_{[n_2]''} + \epsilon_{[n_2]'',[n_1]'} \tag{15}$$

where

$$\epsilon_{[n_2]'',[n_1]'} = 2N_{[n_2]'',[n_1]'} + L_{[n_2]'',[n_1]'} \tag{16}$$

is the internal energy associated with the intercluster relative motion. The angular momenta are coupled in the two-cluster states (12) in the following order: first we couple the angular momenta of the two clusters $\vec{\Lambda}_{[n_1]'} + \vec{\Lambda}_{[n_2]''} = \vec{\Lambda}_{[n]}$ and then we couple the resultant angular momentum with the angular momentum of the intercluster relative motion. This yields the overall internal angular momentum, i.e., $\vec{\Lambda}_{[n]} + \vec{L}_{[n_2]'',[n_1]'} = \vec{\mathcal{L}}_{[n]}$.

The states (12) do not have a well defined $n$-particle permutational symmetry. We now construct a set of states which, in addition to the good quantum numbers mentioned above, (i.e., the Yamanouchi symbol of each cluster, the total energy and angular momentum) can be labeled in terms of their overall permutational symmetry.

As explained in Section IV, an appropriate set of class-operators of the symmetric group should be diagonalized within the subspace of multi-cluster states with a given choice of single-cluster Yamanouchi symbols and a given total angular momentum. The common eigenstates of these class-operators belong to irreps of the symmetric group that are fully specified by the corresponding eigenvalues.

Since the functions given by Eq. (12) are non-spurious, so will be any linear combination of such functions. Choosing any desirable values for $Y_{[n_1]'}$, $Y_{[n_2]''}$, $\vec{\mathcal{L}}_{[n]}$ and $\epsilon_{[n]}$, and noting that the corresponding operators commute with the class-operators of the symmetric group $S_{[n]}$, we diagonalize the latter and obtain linear combinations belonging to irreps denoted by $\Gamma_{[n]}$

$$\begin{aligned} &| \left(Y_{[n_1]'}Y_{[n_2]''}\right) \Gamma_{[n]}\vec{\mathcal{L}}_{[n]}\epsilon_{[n]}\alpha_{[n]} >= \sum_{\Phi_{[n_1]'}\Phi_{[n_2]''}\Phi_{[n_2]'',[n_1]'}\vec{\Lambda}_{[n]}} \\ &\left[\left(Y_{[n_1]'}\Phi_{[n_1]'}Y_{[n_2]''}\Phi_{[n_2]''}\right) \vec{\Lambda}_{[n]}\Phi_{[n_2]'',[n_1]'}\vec{\mathcal{L}}_{[n]}\epsilon_{[n]}|\} \left(Y_{[n_1]'}Y_{[n_2]''}\right) \Gamma_{[n]}\vec{\mathcal{L}}_{[n]}\epsilon_{[n]}\alpha_{[n]}\right] \\ &\qquad | \left(Y_{[n_1]'}\Phi_{[n_1]'}Y_{[n_2]''}\Phi_{[n_2]''}\right) \vec{\Lambda}_{[n]}\Phi_{[n_2]'',[n_1]'}\vec{\mathcal{L}}_{[n]}\epsilon_{[n]} > \end{aligned} \tag{17}$$

where $\alpha_{[n]}$ labels any remaining degeneracies.

The coefficients appearing on the r.h.s. of Eq. (17) are the overlap integrals between the states defined in Eq. (12) and the permutational symmetry adapted states specified on the l.h.s. of Eq. (17). They are in fact the common eigenvectors of the matrices representing the class-operators of $S_{[n]}$ in the basis set specified in Eq. (12).

It is shown in ref. [5] that the matrix elements that need actually to be evaluated involve a small number of inter-cluster permutations. In order to evaluate the elements of these matrices we propose to transform the two-cluster wavefunctions (15) into three-cluster wavefunctions in which the particles appearing in the permutation of interest are placed in the third cluster. In the case of the transposition $(\{n_1\}', \{n_2\}'')$ the third cluster consists of the two particles $\{n_1\}'$ and $\{n_2\}''$.

The third cluster is formed recursively, one particle at a time. Assume that at a certain stage the first cluster consists of the particles $[n_1]' = \{1, 2, \ldots, n_1\}'$, the second consists of $[n_2]'' =$

$\{1, 2, \ldots, n_2\}''$ and the third consists of $[k]''' = \{1, 2, \ldots, k\}'''$. Say that we are now interested in moving the last particle of the first cluster into the third cluster. The coordinates involving this particle originally are its Jacobi coordinate within the first cluster, $\vec{\rho}_{\{n_1\}',[n_1-1]'}$ (Eq. (5)), as well as the coordinate of the c.m. of the second cluster relative to that of the first, $\vec{\rho}_{[n_2]'',[n_1]'}$, and the coordinate of the c.m. of the third cluster relative to that of the first two, $\vec{\rho}_{[k]''',[n_1]'\cup[n_2]''}$

After the transformations the particle originally referred to as $\{n_1\}'$ is renamed $\{k+1\}'''$ and it appears in an appropriate Jacobi coordinate within the third cluster as well as in the coordinate of that (augmented) cluster relative to the c.m. of the (new) first and second clusters. In addition, we have to introduce a new relative coordinate between the (new) first cluster and the (unaffected) second one.

The transformation is achieved by means of two consecutive Talmi-Moshinsky type rotations[8]-[11], along with appropriate angular-momentum recoupling transformations. For full details we refer to [13].

The transfer of a particle from the second cluster into the third one can be effected with obvious minor modifications.

Once all the relevant particles have been moved into the third cluster the evaluation of the matrix element of the intercluster class-operator becomes trivial, since the third cluster is expressed as a linear combination of terms, each one of which belongs to an irrep of the symmetric group $S_{[k]'''}$. Each such term is an eigenstate of the relevant intercluster class-operators.

The generalization of the algorithm to more than two clusters involves some further angular-momentum recoupling transformations, but is otherwise rather straightforward.

## VII. Conclusions

A computationally efficient procedure for constructing non-spurious h.o. states with arbitrary symmetry is presented. The explicit elimination of the spurious states results in a very significant reduction in the size of the basis employed. This is a very crucial point for calculations involving a very large number of states. Moreover, the states constructed in our method have a definite permutational symmetry. This property is essential for calculations involving multiple angular momentum quantum numbers, such as non-relativistic quark calculations.

The procedure is extended to the construction of permutational symmetry adapted non-spurious multi-cluster wavefunctions, which can be used in several very different physical contexts. They provide a very straightforward means for the evaluation of nuclear spectroscopic factors and enable the study of interacting clusters of identical particles ranging from quarks through nucleons to rare-gas atoms. Among the many applications we find the study of fission and fusion of clusters of nuclei or atoms particularly exciting. The latter are crucial steps in the process of nucleation and droplet formation in the gas phase. The coordinate transformations discussed in the present article are equally useful in the study systems of identical particles within classical mechanics, such as in molecular dynamics simulations of the behaviour of gas-phase atomic clusters.

## References


[1] R. D. Lawson, *Theory of the Nuclear Shell Model* (Clarendon, Oxford, 1980).

[2] Yu. F. Smirnov and Yu. M. Tchuvil'sky, Phys. Rev. C **15**, 84 (1977).

[3] V. G. Neudatchin, Yu. F. Smirnov and N. F. Golovanova, *Clustering Phenomena and High-Energy Reactions*, in *Advances in Nuclear Physics* Vol. 11, edited by J. W. Negele and E. Vogt, (Plenum, New York, 1979).

[4] A. Novoselsky, J. Katriel and R. Gilmore, *J. Math. Phys.* **29** (1988), 1368.

[5] J. Katriel and A. Novoselsky, *Multi-cluster Wavefunctions With Arbitrary Permutational Symmetry* , Ann. Phys. (N. Y.) (in press).

[6] A. Novoselsky and J. Katriel, Ann. Phys. (N. Y.) **196**, 135 (1989).

[7] A. Novoselsky and J. Katriel *Non-Spurious Harmonic Oscillator States for Many Body Systems*, in *Proceedings of the VI International Conference on Recent Progress in Many body Theories*, edited by Y. Avishai, (Plenum, New York, 1991).

[8] I. Talmi, Helv. Phys. Acta **25**, 185 (1952).

[9] M. Moshinsky, Nucl. Phys. **13**, 104 (1959).

[10] T. A. Brody and M. Moshinsky, *Tables of Transformation Brackets* (Gordon and Breach, 1967).

[11] A. Gal, Ann. Phys. (N. Y.) **49**, 341 (1968).

[12] A. Novoselsky and J. Katriel, J. Math. Phys. **31**, 1164 (1990).

[13] A. Novoselsky and J. Katriel, *Non-Spurious Two-Cluster Harmonic Oscillator Wavefunctions*, Phys. Rev. Lett. (submitted).

# THEORETICAL INVESTIGATION OF TRANSVERSE COLLECTIVE MOTION IN HEAVY-ION REACTIONS BELOW 100 MeV PER NUCLEON

F. SEBILLE*, V. DE LA MOTA**, M. FARINE*, B. REMAUD*, P. SCHUCK***.

* Laboratoire de Physique Nucléaire, IN2P3-CNRS/ Université de Nantes, 2 rue de la Houssinière - F- 44072 Nantes Cédex 03, FRANCE

** GANIL, BP 5027, F - 14021, Caen Cédex, France

*** ISN, F - 38026, Grenoble Cédex, France

## Abstract

The behavior of transverse nucleon flow is studied in heavy-ion collisions below 100 MeV per nucleon. In connection with recent experiments, a theoretical analysis of the flow inversion is made within the Landau -Vlasov semi-classical approach. Comparisons between theoretical results and experimental data are performed taking into account the experimental constraints. The sensitivity of the flow observables on the nuclear matter properties are discussed, underlying the role of non locality effects.

## I. Introduction

An active field of investigations is the search for informations on the nuclear Equation Of State (EOS) in heavy-ion collisions. To achieve this goal, many theoretical and experimental works are currently performed in a wide range of energies. A special effort has been devoted to the study of transverse collective motion of nuclear matter created in heavy-ion collisions. This transverse nucleon flow results from a balance between the attractive and repulsive components of the nuclear force. At incident energies above 200 MeV per nucleon, where this collective motion has been firstly investigated, experimental observations [1] and theoretical calculations [2] clearly indicate that the nuclear dynamics is dominated by the repulsive part of the nucleon-nucleon interaction. In this range of energies, unambiguous informations on nuclear matter properties, have not been yet extracted, and possible signatures of the nucleonic EOS are expected to vanish with increasing production of internal kinetic pressure. At incident energies below 200 MeV per nucleon, recent experiments [3-4] show that a net transverse flow still remains. In addition they put in evidence an inversion in the transverse direction below incident energies where the flow vanishes, these energies will be referred to as Energies of Vanishing Flow (E .V .F.). The flow inversion underlines that the nuclear dynamics is dominated by the attractive mean field at lower energies, where a greater sensitivity to the E.O.S. could be expected.

We present in this paper an analysis of theoretical and experimental results on collective momentum transfers around E.V.F., we shall discuss the experimental constraints which have to be taken into account in order to lead to meaningful comparisons between experimental data and theoretical predictions. A study of some collective observables is performed, and their sensitivity to E.O.S. parameters and two-body correlations is analyzed.

## II. Outlines of the theoretical approach.

Our theoretical approach is based on the Landau-Vlasov equation governing the time evolution of the semiclassical one body distribution $f(\vec{r}, \vec{p}, t)$ :

$$\frac{\partial f}{\partial t} + \{f, H\} = I_{coll}(f) \qquad (2\text{-}1)$$

The symbol { } stands for the Poisson bracket and $H(\vec{r}, \vec{p})$ is the Hartree-Fock Hamiltonian. The residual interactions, created by binary collisions between nucleons, are taken into account with the Uehling-Uhlenbeck collision term $I_{coll}(f)$.

In this approach, the one body distribution function is described using a basis of coherent states. The method treats the nuclear statics and dynamics within the same point of view. It gives a well defined self-consistent description of the ground state with realistic smooth phase space distributions, and proper initial conditions for the dynamics. The analytical treatment of the nuclear mean field ensures that the conservation laws of the Vlasov equation are exactly satisfied. Without entering the details of the model [5], we will just underline some features relevant to our study.

In practical calculations, different effective interactions have been considered to define the one body Hamiltonian appearing in equation (2.1). Some of them are extensively used in simulations at intermediate energies, due to their simple form ; they are the Zamick forces whose Hartree-Fock potential in the phase space representation reads :

$$V_{HF}(\vec{r}, \vec{p}) = a\frac{\rho}{\rho_0} + b\left(\frac{\rho}{\rho_0}\right)^{1+v} \qquad (2.2)$$

Here $\rho_0$ is the saturation density of nuclear matter and a, b and v are parameters extracted from fundamental properties of nuclear matter, namely the saturation density, the binding energy, and the incompressibility modulus at saturation density. To investigate the influence of the incompressibility modulus on the collision dynamics, the following two sets of parameters have been used :

Zamick parameters

| Zamick interaction | a | b | ν | $k_\infty$ | m*/m |
|---|---|---|---|---|---|
| soft | -356 | 303 | 1/6 | 200 | 1 |
| stiff | -123 | 70 | 1 | 380 | 1 |

Although they are oversimplified forces of Skyrme type, comparisons [6] with experimental data demonstrate that they are able to give some valuable insights on reaction mechanisms. Nevertheless, they present the disadvantage of neglecting the well established non-local property of the nuclear force. To avoid this drawback, we have used in this work, a realistic finite range force, called the Gogny force [7]. This force reproduces many nuclear properties, satisfies the stability conditions and sum rules of the Landau parameters, and has a functional form particularly well suited to an analytical treatment of the mean-field [5] in the Landau-Vlasov model. Its phase space representation, without spin-orbit and isospin dependent terms, reads :

$$V_{HF}(\vec{r},\vec{p}) = \frac{7}{8}\, t_3\, \rho^{4/3}(r) + \sum_{i=1}^{2} \alpha_i \int d^3r'\, e^{-(\vec{r}-\vec{r}\,')^2/\mu_i^2}\, \rho(\vec{r}\,') - \sum_{i=1}^{2} \beta_i \int d\vec{p}\,'\, e^{-(\mu_i^2/4\hbar^2)(\vec{p}-\vec{p}\,')^2}\, f(\vec{r},\vec{p}\,')$$

$$\alpha_i = W_i + \frac{B_i}{2} - \frac{H_i}{2} - \frac{M_i}{4}$$

$$\beta_i = \left(\frac{W_i}{4} + \frac{B_i}{2} - \frac{H_i}{2} - M_i\right) x\, (\sqrt{\pi}\ \mu_i)^3$$

Gogny interaction parameters

| Gogny interaction | i | $\mu_i$ | $W_i$ | $B_i$ | $H_i$ | $M_i$ | $t_3$ |
|---|---|---|---|---|---|---|---|
| D1-G1 | 1 | 0.7 | - 402.40 | -100.00 | - 496.20 | - 22.36 | 1350 |
| | 2 | 1.2 | - 21.30 | -11.77 | - 32.27 | - 68.81 | 1350 |

This set of parameters commonly used is referred to as the Gogny D1-G1 force, its incompressibility modulus is $K_\infty$ = 228 MeV. The depth of the real optical potential fits well the experimental estimated one for energies below 200 MeV, nevertheless there is no genuine energy dependence, and the resulting effective mass fails to reproduce the enhancement near the Fermi energy. To close this section, let us emphasize that in our model the nucleon-nucleon cross section differs from the free one, it has isospin and energy dependence fitted to Brueckner nuclear matter calculations [ 8 ].

## III. Comparisons between experimental data and theoretical predictions.

We now turn to detailed comparisons with the experimental data ; however let us recall that our theoretical approach describes the overall dynamics in terms of one body distribution functions. We will first single out the transverse collective motions analyzing the mean transverse momentum in the reaction plane as a function of the longitudinal rapidity, according to the so called transverse momentum analysis [9]. Our investigation is connected with recent experiments [4] performed at GANIL with the Ar + Aℓ system for incident energies ranging from 45 to 85 MeV/u. These inverse kinematic reactions of an asymmetric system appear to provide an efficient way [4] to select the impact parameter ; we specialize our to study semi-central collisions with an impact parameter of 3fm.

Meaningful comparisons between theoretical and experimental results require to introduce different filters. A first step will be to simulate the detector acceptance. According to ref. [4], a fit of this acceptance is to remove the nucleonic distributions having velocities $\vec{v}$ in the laboratory defined by $|\vec{v}|/c \leq 0.1$. The main consequence is an increase of the apparent transverse momentum at low rapidity, since nucleons with low transverse and longitudinal momenta are not taken into account.

For reliable comparisons with experimental data, a second step is needed. In experiments, a finite number of charged particles is detected, inducing an indetermination on the estimation of the reaction plane over which the mean transverse momentum is calculated. To take into account this fundamental effect in the Landau-Vlasov model, according to ref. [10], each element of the initial coherent state basis has been considered as a nucleonic degree of freedom, and the finite number effect is analyzed from a statistical point of view, associating at random a coherent state to each nucleon of the final distribution. The latter is obtained from the total system distribution by removing heavier fragments and using the filter previously mentioned. Each random selection in the statistical ensemble representing an event, the reaction plane is defined in agreement with the experimental procedure [4].

The corrected results of the Landau - Vlasov model, are plotted in Fig. 1 in comparison with experimental results. The theoretical results (circles) exhibit similar behavior as the experimental ones : a plateau near the initial projectile rapidity and a quasi linear behavior at lower rapidities, due to the filter introduced to simulate the detectors acceptance. In the lower rapidity region, the detection threshold having less influence with increasing energy, our theoretical results tend to reach intermediate values between those displayed for Z = 1 and Z = 2. This is coherent with the fact that theoretically we are dealing with one-body distributions, unable to discriminate individual nucleons and nucleons in clusters. The Z = 1 and Z = 2 experimental results present net different values in this domain, it is likely due to a higher detection threshold for heavier fragments, underestimated with our filter. In the higher rapidity region, a weaker fragment mass dependence still remains. Our theoretical results overestimate slightly the experimental data. A tentative explanation can be advanced, in the corrected theoretical results, only the heavy fragments have been removed, and the one body distribution should include Z > 2 fragment-mass, which contribute with slightly greater transverse momenta.

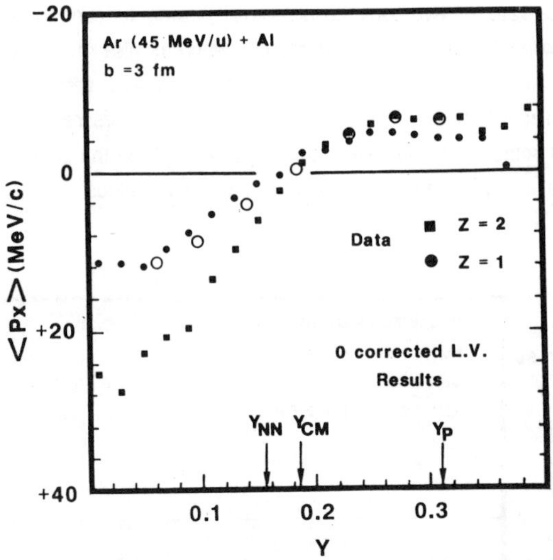


*Fig. 1 : Mean transverse momentum as a function of the longitudinal rapidity. The effects of various experimental filters are theoretically simulated (see text for details). The arrows indicate on the abscissas, the rapidities of respectively, the projectile $Y_P$, of the center of mass $Y_{CM}$ and of the nucleon-nucleon frame $Y_{nn}$.*

The momentum transfer from longitudinal to transverse components in the mid-rapidity region is characterized by the flow observable. Our definition of the flow F is the following

$$F = \left.\frac{d < p_x >}{d\tilde{y}}\right]_{\tilde{y}=0} \qquad (3-1)$$

where $\tilde{y} = y_z / y_{proj}$ is the rapidity normalized to the rapidity of the projectile. The mean transverse momentum $< p_x >$ is given by :

$$< p_x > = \int d^3r \int dp_x\, dp_y\, p_x f(\vec{r},\vec{p}) \int d^3r \int dp_x dp_y\, f(\vec{r},\vec{p}) \qquad (3-2).$$

In Fig. 2, theoretical flow values versus energy are plotted for the Nb + Nb, Ca + Ca and C + C systems, keeping nearly constant the ratio of the impact parameter to the linear dimension of the system. This constraint aims at preventing the relative weight of mean-field and collision term to change due to geometrical scalings. The E.V.F. values are indicated by arrows. A clear increase of this quantity is observed for decreasing masses, following approximatively a

$A^{-1/3}$ law since the rate of two-body collisions is growing function of A. For lighter nuclei, nucleon-nucleon collisions decrease very rapidly whereas the mean-field stays rather well defined. Following this simple picture, one can understand why zero collective flow occurs at higher beam energies when the mass of the colliding nuclei decreases. For comparison, we have also indicated some experimentally estimated E.V.F. [3] on the abscissa of Fig. 2. Although experimental errors are quite large, we observe that mean values are in good agreement with theoretical results.

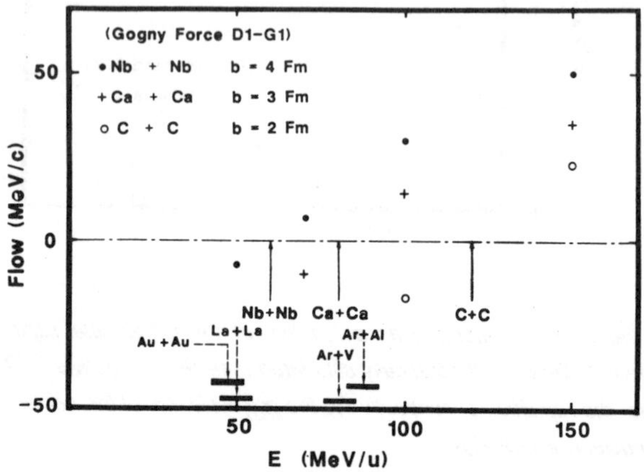


*Fig. 2. : Collective sidewards flow, calculated with the Gogny D1-G1 force, as a function of incident energy per nucleon, for three different symmetric systems. Experimental values of the energy balance are also indicated : La + La, Ar + V, Ar + Al and Au + Au. Horizontal bars indicate experimental uncertainties.*

## IV. Analysis of the flow observables sensitivity to the nuclear matter properties.

Let us first analyze to what extent the E.V.F. depend on the E.O.S. We therefore present in Fig. 3 the flow F as a function of beam energy per particle for three different forces : local stiff and soft Zamick forces with $K_\infty$ = 380 MeV and $K_\infty$ = 200 MeV respectively, and the finite range Gogny force D1-G1 with $K_\infty$ = 228 MeV. We investigated the system Nb + Nb at an impact parameter b = 4fm. For beam energies well above the E.V.F we see that the stiff Zamick force efficiently simulates the finite range Gogny force, whereas the soft Zamick force yields appreciably lower values for the flow. Let us remark that the E.V.F. almost occurs at the same energy for all three forces. A qualitative understanting of this fact is the following. The stiff force induces a higher pressure than the soft one, giving raise to higher transverse momenta in the entrance phase. In counterpart the reaction time becomes longer, then the attractive mean-field can act longer and pulls the flow to negative values.

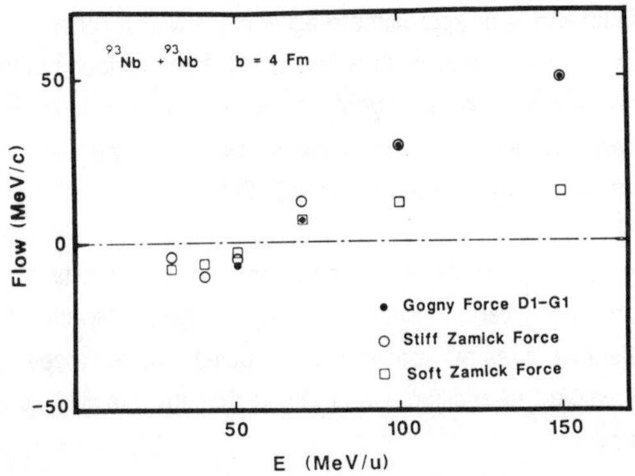


*Fig. 3. : Collective sidewards flow versus some typical values of incident energies, calculated with three different effective forces.*

In Fig. 3, we have seen that the E.V.F. depends very little on the E.O.S., in Fig. 2, on the contrary the two-body collisions seem to have a very strong influence. We now want to make a fine tuning of this later effect in slightly scaling up and down the effective nucleon-nucleon cross-section of the collision integral. For numerical convenience we choose the relativety light system Ca + Ca but we think that qualitatively the conclusions will stay the same for heavier systems. In Fig. 4. we present the variation of the E.V.F. with a slight up/down scaling of $\sigma_{NN}$ ; we see that 20% variation of $\sigma_{NN}$ induces a shift of roughly 10 MeV/u of the E.V.F..

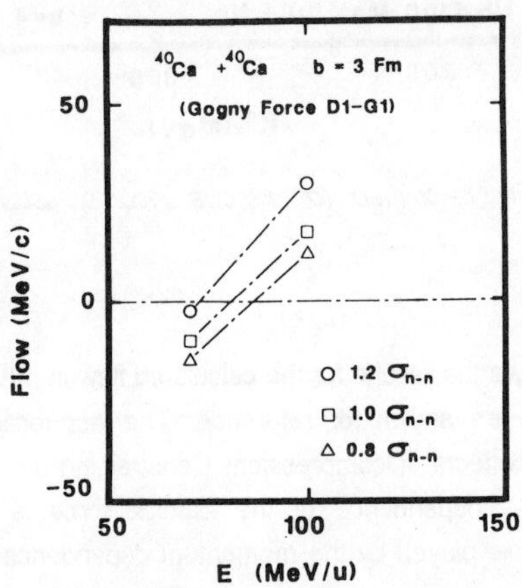


*Fig. 4. Collective sidewards flow as a function of incident energy per nucleon. For the Ca + Ca system at an impact parameter of 3 Fm. Calculations were done with the Gogny D1-G1 force, for different scalings of the nucleon-nucleon cross-section.*

This finding is in approximate agreement with a recent investigation by Olgivie et al [3] if one takes into account that in their work, the free nucleon-nucleon cross-section was used ; our in-medium corrected one is roughly a factor of two smaller. From Fig. 4 one also should note that scaling slightly $\sigma_{NN}$ leaves the slope of the flow near the E.V.F. practically unaltered ; this slope may be more sensitive to the E.O.S.

In order to understand better the role of non-locality in the flow creation, we have made calculations using various sets of Gogny D1 type interactions with different values for $K_\infty$ and for the effective mass m*. Basic infinite nuclear matter properties and ranges of the force were kept at D1 values. In addition, we checked that the stability conditions of the Landau parameter were verified.

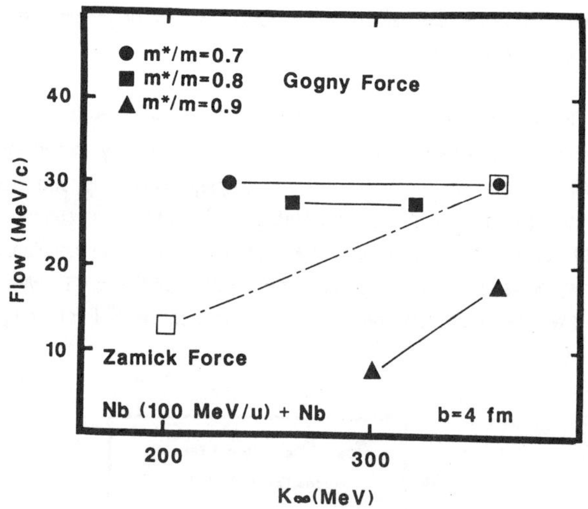


*Fig. 5. : Values of the flow calculated for various incompressibility modulus and average effective masses.*

Fig. 5 shows the results for the calculated flow in a typical reaction ; the results for the local Zamick force are shown for reference. The non-locality of the effective force completely washes out the effects of compression. Constraining the Gogny force to be almost local (m*/m = 0.9), the $K_\infty$ dependence of the Zamick force is recovered. These results underline the fundamental role played by the momentum dependence of the optical potential in the flow creation.

**V. Conclusions.**

In this work, we studied the behavior of transverse nucleon flow in an energy region where a transition from positive to negative deflections occurs. It is related with a current field of active investigations looking for informations on nuclear matter properties in heavy-ion collisions. This search goes beyond the mere determination of the nuclear incompressibility modulus since one must take into account other fundamental characteristics of nuclear matter such as the momentum and energy dependence of the optical potential.

We have shown that the implementation of experimental constraints in our model allows quantitative confrontations between theoretical results and experimental data. These comparisons show that the EVF values are sensitive to small variations of the nucleon-nucleon cross section and pratically insensitive to effective force parametrizations. Information on the EOS parameters could be looked for in the slope of the flow around the EVF, but statistical and quantal fluctuations absent in the present study may spoil this expectancy.

Our microscopic investigations established that the flow is created very soon in the reaction, in an out of equilibrium situation. The non local character of the nuclear interaction plays a preponderant role, washing out the effects of compression. The flow observables may bring informations on nuclear matter, but the connection with the cold one at equilibrium is certainely not direct.

## REFERENCES.


1. H. A. Gustafsson et al, Phys. Rev. Lett. 52 (1984) 1590
   H.G. Ritter et al, Nucl. Phys. A447 (1985) 3C
   K.G.R. Doss et al, Phys. Rev. Lett. 57 (1986) 302
2. H. Stocker and W. Greiner, Phys. Rep. 137 (1986) 277
   G.E. Bertsch and S. Das Gupta, Phys. Rep. 160 (1988) 190
   C. Gale, G. Bertsch and S. Das Gupta, Phys. Rev. C35 (1987) 1666
   J. Aichelin, A. Rosenhauer, G. Pielert, H. Stocker and W. Greiner, Phys. Rev. Lett. 58 (1987) 1926.
   G.F. Bertsch, W.G. Lynch, M.B. Tsang, Phys. Lett. B 189 (1987) 387
3. D. Krofcheck et al, Phys. Rev. Lett. 63 (1989)
   C.A. Olgivie et al, Phys. Rev. C40 (1989) 2592
   C.A. Olgivie et al, Phys. Lett. B231 (1989) 35
   C.A. Olgivie et al, Phys. Rev. C42 (1990) 10
   W.H. Zhang et al, Phys. Rev. C42 (1990) 491

4. J. Peter et al, Phys. Lett. B. 237 (1990) 187
K. Hagel et al, Phys. Lett. B. 229 (1989) 20
J.P. Sullivan et al, Phys. Lett. B. 249 (1990) 10
J. Peter et al. XXIX Winter Meeting on Nuclear Physics. Bormio (1991)
5. C. Grégoire, B. Remaud, F. Sébille, L. Vinet and Y. Raffray, Nucl. Phys. A 485 (1987) 317
B. Remaud, C. Grégoire, F. Sébille and P. Schuck. Nucl. Phys. A 488 (1988) 423
F. Sébille, G. Royer, C. Grégoire, B. Remaud and P. Schuck, Nucl. Phys. A 501 (1989) 137
P. Schuck et al, Prog. Part. Nucl. Phys. 22 (1989) 181
6. M.F. Rivet, B. Borderie, C. Grégoire, D. Jouan and B. Remaud, Phys. Lett. B215 (1988) 55
D. Jacquet, Thèse d'Etat, Orsay (1987)
D. Ardouin et al, Z. Phys. A. 329 (1988) 505.
7. J. Decharge and D. Gogny, Phys. Rev. C21 (1980) 1568
D. Gogny and R. Padjen, Nucl. Phys. A 2931 (1977) 365
8. J. Cugnon, A. Lejeune and P. Grange, Phys. Rev. C35 (1987) 861
9. P. Danielewicz and G. Odyniec, Phys. Lett. B 157 (1985) 146
10. H. Zielinska-Pfabé and C. Grégoire, Phys. Rev. C37 (1988) 2594

# FERMIONIC MOLECULAR DYNAMICS FOR GROUND STATES AND COLLISIONS OF NUCLEI

H. FELDMEIER and K. BIELER
Gesellschaft für Schwerionenforschung mbH, Darmstadt, Germany

**Abstract**

Fermionic Molecular Dynamics has been proposed as an approximation scheme to solve the many-body problem of interacting identical fermions with spin 1/2. In this model the interacting system is represented by an antisymmetrized many-body wave function consisting of single-particle states which are localized in phase space. The equations of motion for the parameters characterizing the many-body state (e.g. positions, momenta and spins of the particles) are derived from a quantum variational principle. This contribution gives first results for ground states of nuclei and collisions between light nuclei.

## 1 INTRODUCTION

The large fluctuations observed in final states of heavy ion collisions demand a theoretical description which allows very different types of reactions for identical initial conditions. For a given impact parameter and beam energy the nuclei may fuse or undergo a deeply inelastic collision or at higher energies they may fragment. These kind of fluctuations cannot be described by a one-body mean-field theory which for one initial set-up evolves to only one final state with very little fluctuations around the mean result.

Molecular dynamics models [1-8,15-17] are in principle capable of producing large fluctuations as they evolve in time individual members of an ensemble which represents the initial configuration (i.e. two nuclei in their ground state with a relative velocity). If the dynamics of the system is such, that small (quantal) fluctuations which are initially present want to grow in time, molecular dynamics lets these fluctuations develop because it is a microscopic time evolution of a many-body state. On the other hand if the system is stable against fluctuations molecular dynamics will give the same result as a mean field theory since it contains the mean field aspect.

In heavy ion collisions, however, the nonquantal molecular dynamics known from atomic collisions is not applicable. Unlike atoms or atomic molecules the nucleons are indistinguishable fermions which inside a nucleus are closely packed in phase space. The de Broglie wave length is of the order of the inverse Fermi momentum (about 1 fm) which does not allow to localize the particles well enough within the scale of typical variations of the two-nucleon potential (about 0.1 fm). Due to this lack of localization one has to take the large width of the single-particle wave packets into account and it is necessary to antisymmetrize the many-body state to obtain proper Fermi-Dirac statistics which is essential for the properties of bound nuclei and fragments.

The Fermionic Molecular Dynamics (**FMD** ) model proposed in Ref. [9] is an attempt to incorporate the fermionic nature of the molecules (here the nucleons) into a molecular dynamics picture in which, firstly, the complex interaction between the molecules is approximated by a two-body potential, and secondly, the molecules are described by parameters which characterize wave packets. These packets which are localized in phase space are used to construct an antisymmetric many-body state.

## 2 THE FERMIONIC MOLECULAR DYNAMICS MODEL (FMD)

In the **FMD** model [9, 10] the equations of motion for the parameters are obtained from the time-dependent quantum variational principle [23–28]

$$\delta \int_{t_1}^{t_2} dt \, \langle\, Q(t) \,|\, i\frac{d}{dt} - \underset{\sim}{H} \,|\, Q(t) \,\rangle \;=\; 0 \;, \tag{2.1}$$

where the antisymmetrized many-body state $|\, Q(t) \,\rangle$ is parametrized by a set of time-dependent parameters $Q(t) = \{\, q_\nu(t),\ \nu = 1, 2, \cdots \}$ which are regarded as the generalized coordinates of the system. The variation has to be performed with respect to all $q_\nu(t)$. $\underset{\sim}{H}$ denotes the total hamiltonian of the system and $t$ the time. If we define a Lagrange function by

$$\mathcal{L}\left(Q(t), \dot{Q}(t)\right) := \langle\, Q(t) \,|\, i\frac{d}{dt} - \underset{\sim}{H} \,|\, Q(t) \,\rangle \tag{2.2}$$

in which $\dot{Q}(t) = \{\, \dot{q}_\nu \equiv dq_\nu/dt,\ \nu = 1, 2, \cdots \}$ is the set of generalized velocities, the variation (2.1) leads to the Euler-Lagrange equations

$$\frac{d}{dt}\frac{\partial \mathcal{L}}{\partial \dot{q}_\nu} = \frac{\partial \mathcal{L}}{\partial q_\nu} \;, \tag{2.3}$$

which are the **FMD** equations of motion for the parameters. They have the following general structure

$$\sum_\mu \mathcal{A}_{\nu\mu} \frac{dq_\mu}{dt} = -\frac{\partial}{\partial q_\nu} \langle\, Q(t) \,|\, \underset{\sim}{H} \,|\, Q(t) \,\rangle \;, \tag{2.4}$$

with the skew symmetric matrix $\mathcal{A}_{\nu\mu}$ defined by

$$\mathcal{A}_{\nu\mu} = \frac{1}{i}\left( \langle \frac{\partial}{\partial q_\nu} Q(t) \,|\, \frac{\partial}{\partial q_\mu} Q(t) \rangle - \langle \frac{\partial}{\partial q_\mu} Q(t) \,|\, \frac{\partial}{\partial q_\nu} Q(t) \rangle \right) \;. \tag{2.5}$$

The task is to find a parametrization for the many-body state $| Q(t) \rangle$ which contains the essential degrees-of-freedom of the system. Classical molecular dynamics describes the motion of point-like particles under the influence of mutual interactions. Fermionic Molecular Dynamics extends this picture in a natural way towards quantum mechanics by introducing for each molecule (here each nucleon) a wave packet $|q(t)\rangle$ which is localized in phase space. For convenience we take a gaussian form

$$\langle \vec{x}|q(t)\rangle = \exp\left\{ -\frac{(\vec{x} - \vec{b}(t))^2}{2a(t)} \right\} \begin{pmatrix} \cos \frac{\chi(t)}{2} \\ \sin \frac{\chi(t)}{2} e^{i\phi(t)} \end{pmatrix} . \tag{2.6}$$

The time dependent parameters $q(t) = \left\{ \vec{b}(t), a(t), \chi(t), \phi(t) \right\}$ define the wave packet uniquely. The parameters

$$a(t) = a_R(t) + i\, a_I(t) \quad \text{and} \quad \vec{b}(t) = (\vec{r}(t) - a_I(t)\vec{p}(t)) + i\,(a_R(t)\vec{p}(t)) \tag{2.7}$$

are complex and contain the information about the spreading, the mean position $\vec{r}(t)$ and the mean momentum $\vec{p}(t)$. The spin of the fermion is represented by a two-component spinor where the real polar angles $\chi(t)$ and $\phi(t)$ denote the direction of the spin.

For identical fermions the most simple ansatz for the parameterized many body wave function $| Q(t) \rangle$ is a single Slater determinant built from N different single-particle wave packets $| q_k(t) \rangle$. A single determinant together with a suitable G-matrix interaction, which takes into account the short range hard-core correlations is appropriate for stationary ground state properties. In heavy-ion collisions, however, the phase space distributions of the two nuclei are time dependent and due to their relative shift in momentum space they allow for hard scatterings between nucleons belonging to different nuclei because the final states are no longer Pauli-blocked. It seems questionable if these time dependent hard-core collisions can still be described by a stationary G-matrix picture in which the time dependence is only in the momentum distribution which enters the Pauli-blocking and not in the intermediate two-body states. Therefore it has been proposed in ref. [10] to employ a time-dependent Jastrow ansatz [21] for the trial wave function instead of a single determinant. In coordinate representation the correlated many-body state is given by

$$\{ \langle \vec{x}_1 | \otimes \cdots \otimes \langle \vec{x}_N | \} | \hat{Q} \rangle =$$

$$\frac{1}{N!} \sum_{all\ P} sgn(P) \langle \vec{x}_1 | q_{P(1)} \rangle \otimes \cdots \otimes \langle \vec{x}_N | q_{P(N)} \rangle \prod_{k<l}^{N} f(c_{k,l}; |\vec{x}_k - \vec{x}_l|), \tag{2.8}$$

which then is normalized

$$| Q \rangle = \frac{1}{\langle \hat{Q} | \hat{Q} \rangle^{\frac{1}{2}}} | \hat{Q} \rangle . \tag{2.9}$$

The two-body correlation function $f(c_{k,l}; |\vec{x}_k - \vec{x}_l|)$ is quenching the wave function where the hard core of the interaction between particle $k$ and $l$ is located. It is parametrized by the correlation parameters $c_{k,l}(t)$ which in general may be regarded as additional variables belonging to the set $Q(t)$. At present we are working out approximation schemes to solve eqs. (2.3) for the trial wave function (2.8).

In the following we restrict ourselves to the case of a single Slater determinant (i.e. $f(c_{k,l}; |\vec{x}_k - \vec{x}_l|) \equiv 1$ for all $k, l$) so that the trial many-body state contains correlations induced by the Pauli principle and long range correlations due to the two-body interaction.

### 2.1 GROUND STATES IN FMD

Since the Fermi motion and the Pauli principle are included in a quantum mechanical fashion, namely in the momentum distribution of the wave function, the **FMD** model contains true quantum ground states. This is at variance with other molecular dynamics models where the Fermi motion is replaced by random motion of the classical particles. In a conventional model the clusters created during the heavy ion collision will cool by particle emission to a classical crystal where all the remaining particles are at rest. If one wants, for example, to look for possible liquid-gas phase transitions one cannot learn much from models which obey Boltzmann instead of Fermi-Dirac statistics which is the only appropriate one at energies where one expects these transitions.

In **FMD** the total energy of a system is given by the expectation value of the hamiltonian $\underset{\sim}{H}$ calculated with the normalized trial many-body state $|\, Q(t) \,\rangle = |\, q_1(t), \cdots, q_N(t) \,\rangle$ which in the present investigation is chosen to be a single Slater determinant built from gaussian-shaped single-particle states. The ground state in **FMD** is defined as the state which has the lowest energy among the trial wave functions. Therefore at the ground state all derivatives of the energy with respect to the single-particle parameters $q_\nu$ vanish.

$$\frac{\partial}{\partial q_\nu} \langle\, Q(t) |\; \underset{\sim}{H} \;| Q(t) \,\rangle \;=\; 0 \tag{2.10}$$

The minimum of the energy, $E_{GS}$, is found by the method of steepest descent for the mean positions and the width of the gaussians and the spin directions. From eqs. (2.10) and (2.4) it is obvious that the ground state is absolutely stationary, all positions, momenta, widths and spins are time independent. This seemingly trivial fact causes a lot of problems in conventional molecular dynamics, including QMD [15–17], because there the zero point motion is mocked up by classical motion of the center of mass coordinates of the nucleons and since the system has Boltzmann statistics the nucleus will evaporate nucleons even in the state which is considered to be the ground state. This is especially worrisome in the final stage of a collision where the fragments are formed.

Mass distributions of fragments and small clusters are an important part of the information obtained in nucleus-nucleus collisions. For that one first has to make sure that within the **FMD** model light nuclei are described in a realistic way. In a first step we are investigating [12] four central interactions proposed in the literature [13, 14]. These two-body interactions consist of a Wigner and a Majorana term with a radial part which has a soft repulsive core. Since there are no spin and isospin exchange terms these interactions favour the $\alpha$-cluster picture as can be seen in Fig. 1 from the density contour plot for carbon. For $^5$He we obtain three single-particle wave packets at the center, two protons with opposite spin and a neutron. The remaining two neutrons with equal spin, which is opposite to the spin of the middle neutron, are shifted out of the center because the Pauli principle does not allow them to be at the same location. In the $^6$Li case the crosses indicate the mean positions of three proton-neutron pairs. Again the outer protons and neutrons have equal spins, respectively.

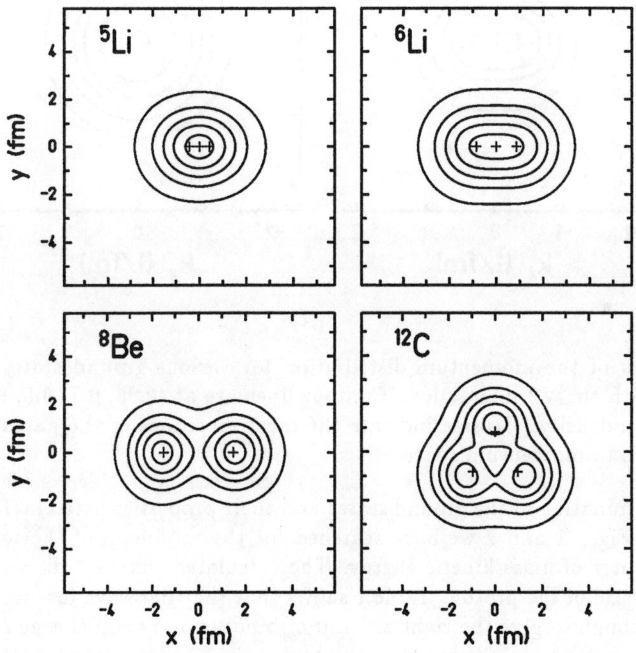


Figure 1
Density contour plots for various ground states in **FMD** calculated with the V2 interaction[13]. Contour lines are at (0.9, 0.7, 0.5, 0.3, 0.1) of the maximum density. Crosses indicate mean positions of the gaussian single-particle states in coordinate space.

The momentum distribution for $^6$Li and $^{12}$C displayed in Fig. 2 demonstrates clearly the Fermi motion of the particles although for a ground state all mean momenta vanish, $\vec{p}_k(t) = 0$. In **FMD** the Fermi motion is a quantal zero-point motion and not a random motion of classical particles. The Pauli principle causes the momentum distribution of $^6$Li to become deformed which reflects the broken symmetry in coordinate space. In the distribution for $^{12}$C the effect of the antisymmetrization is a flatening in the central region compared to the gaussian distribution which each single-particle state contributes.

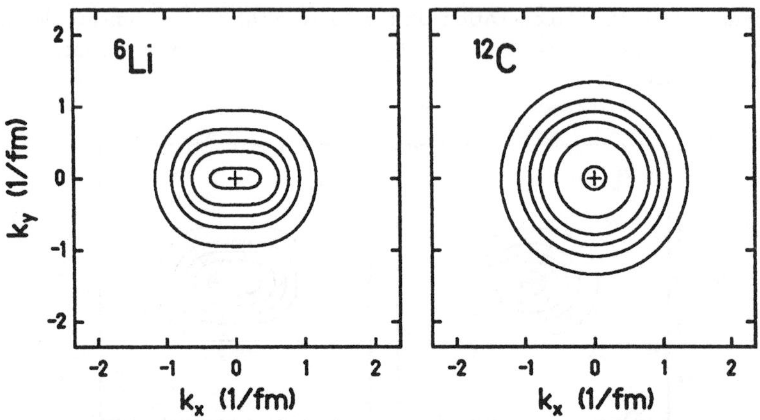


Figure 2
Contour plots of the momentum distribution for various ground states in **FMD** calculated with the V2 interaction. Contour lines are at (0.95, 0.7, 0.5, 0.3, 0.1) of the maximum density. Crosses indicate the mean positions of the gaussian single-particle states in momentum space.

For the determination of the ground states and their properties listed in Table 1 and displayed in Figs. 1 and 2 we have searched for the minimum of the total energy minus the center of mass kinetic energy. The calculated charge radii are corrected for the finite size of the proton. Table 1 shows that the trial wave functions seem to be flexible enough to give the right amount of binding and good charge radii. Even for the deuteron (not included in the table) where one needs an exponential instead of a gaussian tail in the wave function we get a local minimum, which is, however, at a small positive energy of $E_{GS} = 0.58$MeV for the V2 interaction. Within **FMD** the deuteron cannot decay out of this minimum, so it is stable. Its rms radius is 2.25fm not much different from the experimental value of $2.116 \pm .006$fm.

| | $^4$He | | $^5$Li | |
|---|---|---|---|---|
| | $E_{GS}$/MeV | rms-rad/fm | $E_{GS}$/MeV | rms-rad/fm |
| B1 | -28.19 | 1.65 | -18.72 | 3.40 |
| C1 | -28.20 | 1.57 | unbound | |
| V1 | -27.92 | 1.61 | -19.40 | 3.61 |
| V2 | -28.79 | 1.61 | -20.54 | 3.65 |
| Exp | -28.3 | 1.671±.014 | -26.3 | |
| | $^8$Be | | $^{12}$C | |
| | $E_{GS}$/MeV | rms-rad/fm | $E_{GS}$/MeV | rms-rad/fm |
| B1 | -44.9 | 2.60 | -67.4 | 2.77 |
| C1 | -43.7 | 2.50 | -65.3 | 2.68 |
| V1 | -46.0 | 2.39 | -76.8 | 2.53 |
| V2 | -49.0 | 2.39 | -83.5 | 2.52 |
| Exp | -56.5 | 2.46± .09 | -92.2 | 2.464±.012 |

Table 1
Ground state energies and charge radii for different interactions. B1, C1 from Ref. [14], V1, V2 from Ref. [13], experimental binding energies from Ref. [31] and radii from Ref. [32].

## 2.2 TIME EVOLUTION FOR TWO COLLIDING NUCLEI

In Fig. 3 we display the time evolution of two $\alpha$ particles which are colliding head-on with a relative velocity of 0.02c. In order to study the effect of the Pauli principle on the generalized trajectories of the particles we compare the correct calculation which uses an antisymmetrized many-body state (r.h.s.) with a calculation where a simple product state is assumed (l.h.s.). In both cases the ground state of the $\alpha$ particle is the same because the two neutrons and the two protons have opposite spins. This way all four particles are distinguishable and the antisymmetrization is not needed. However, when the two nuclei begin to overlap the Pauli principle becomes active because then identical particles with the same spin are interacting. Besides the contour plots for the single-particle density, at the times indicated, we show the mean positions $r_1$ and $r_2$ for a nucleon from the left and the right $\alpha$ particle, respectively. Furthermore the mean momentum $p_2$ which is equal to $-p_1$ and the real and imaginary part of the width parameters, $a_R, a_I$, which due to symmetries are the same for all eight nucleons. The graph exhibits clearly the repulsive effect of the Pauli principle on the trajectories. In the case of distinguishable particles the motion is completely different. The oscillation of the two $\alpha$ particles is much faster and the amplitudes for the momentum and the width parameters are larger. It remains to be seen in how far the effects of the Pauli principle can be simulated with a momentum dependent addition to the hamiltonian [30] together with a simple product ansatz for the many-body state. A similar result has been obtained in Ref. [9] where the relative motion of two identical fermions has been considered.

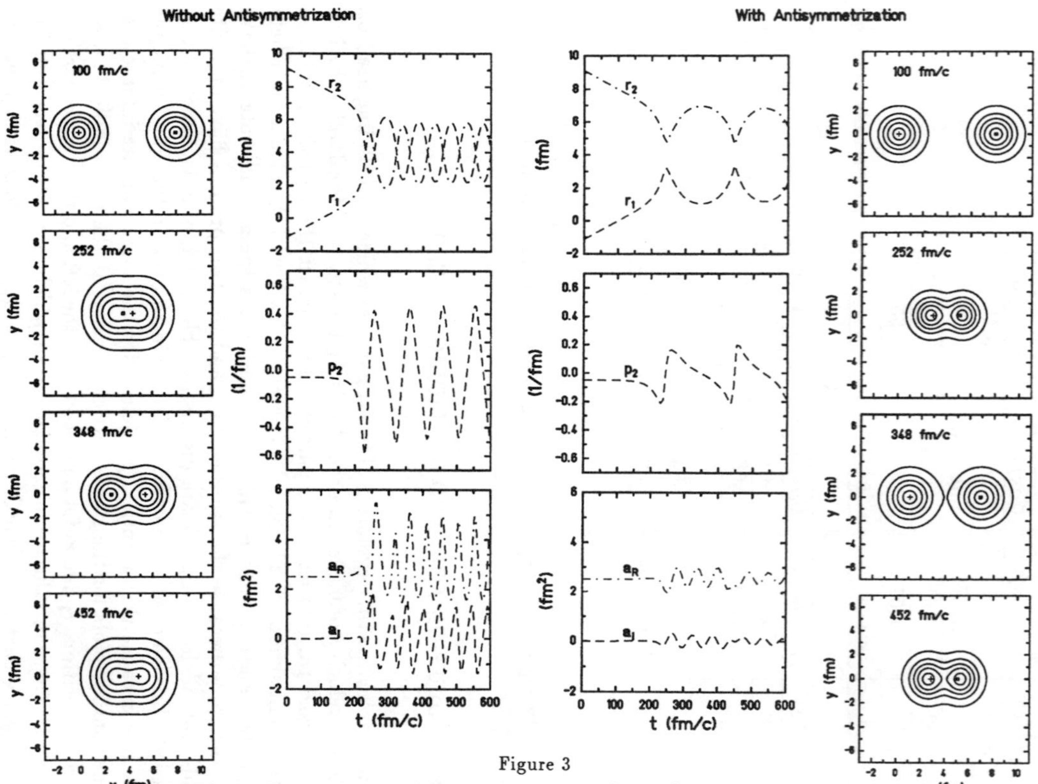


Figure 3

Densities and parameters as function of time for a ${}^4$He + ${}^4$He head-on collision. Left hand side: without Pauli principle, right hand side: proper treatment with antisymmetrization. Contour lines of single particle densities are at (0.9,0.7,0.5,0.3,0.1) of the maximum density. For details see text.

## 3 OUTLOOK

In case the model can be tackled numerically also for heavy nuclei, **FMD** will provide a dynamical event generator which should be capable of covering a wide range of phenomena like fusion at the barrier, deep inelastic collisions, incomplete fusion, or fragmentation reactions with subsequent deexcitation by particle or cluster emission. There is no need for another model after freeze-out as **FMD** will describe the ground state properties on the level of Hartree Fock (plus short range correlations when Jastrow correlations are included). Even without solving the **FMD** equations explicitly the **F**ermionic **M**olecular **D**ynamics model may serve as a foundation of existing molecular dynamics models and to improve those.

The first numerical results are very promising but the quest for an approximation scheme is obvious as the present equations for N particles couple all particles like a N-body interaction. The reason is that the Pauli principle is a N-body correlation. In Ref. [10] an expansion in terms of one- and two-body clusters has been proposed which we shall investigate next. As far as the Pauli principle is concerned this expansion is equivalent to a low phase-space density approximation. If it is successful for the ground states it will be good for any other state too, including high excitations and large spatial densities because the phase-space density for fermions has an upper limit of one, which is reached only in the ground state.

First attempts are in progress to study the relation between Quantum Molecular Dynamics [15] and a trial wave function which is just a product state without antisymmetrization [30].

It is also a very interesting question in how far the **FMD** equations (2.4) can be approximated by Hamilton equations with a "Pauli potential",

$$\frac{d\vec{r}_k}{dt} = \frac{\partial \mathcal{H}_{Pauli}}{\partial \vec{p}_k} \quad \text{and} \quad \frac{d\vec{p}_k}{dt} = -\frac{\partial \mathcal{H}_{Pauli}}{\partial \vec{r}_k}, \tag{3.1}$$

as proposed by Wilets[1, 2] and persued by others[18, 19, 20]. This problem is also related to the question of the existence of canonical variables investigated by Coriano et al. [29].

# 4 REFERENCES

[1] L. Wilets, A.D. MacKellar and G.A. Rinker, Jr., Proc. IV Int. Workshop on Gross Properties of Nuclei and Nuclear Excitations (Hirschegg, Austria, 1976), ISSN 0720-8715, p. 111

[2] L. Wilets, E.M. Henley, M. Kraft and A.D. MacKellar, Nucl. Phys. **A282** (1977) 341, and L. Wilets, Y. Yariv and R. Chestnut, Nucl. Phys. **A301** (1978) 359

[3] A.R. Bodmer and C.N. Panos, Phys. Rev. **C15** (1977) 1342, and A.R. Bodmer, C.N. Panos and A.D. MacKellar, Phys. Rev. **C22** (1980) 1025

[4] J.J. Molitoris, J.B. Hoffer, H. Kruse and H. Stöcker, Phys. Rev. Lett. **53** (1984) 899

[5] T.J. Schlagel and V.R. Phandharipande, Phys. Rev. **C36** (1987) 162

[6] E. Betak, preprint DUBNA E-86-701

[7] G.S. Anagnostatos, Phys. Rev. **C39** (1989) 877

[8] R. Schmidt, B. Kämpfer, H. Feldmeier and O. Knospe, Phys. Lett. **B229** (1989) 197

[9] H. Feldmeier, Nucl. Phys. **A515** (1990) 147

[10] H. Feldmeier, in Proc. CORINNE 90, Int. Workshop on Particle Correlations and Interferometry in Nuclear Collisions, Nantes 1990, ed. D. Ardouin, World Scientific, p.248

[11] H. Feldmeier, Proceedings of the NASI conference on 'The Nuclear Equation of State', Peniscola, Spain (1989), eds. W. Greiner and H. Stöcker, (NATO ASI Series B, Physics; Vol. 216A), Plenum Press, New York (1989) p. 375

[12] K. Bieler, Diplomathesis 1991

[13] A.B. Volkov, Nucl. Phys. **74** (1965) 33

[14] D.M. Brink, E. Boeker, Nucl. Phys. **A91** (1967) 1

[15] J. Aichelin, A. Rosenhauer, G. Peilert, H. Stöcker and W. Greiner, Phys. Rev. Lett. **58** (1987) 1926, and Phys. Rev. **C37** (1988) 2451

[16] G. Peilert, H. Stöcker, W. Greiner, A. Rosenhauer, A. Bohnet and J. Aichelin, Phys. Rev. **C39** (1989) 1402 and Nucl. Phys. **A488** (1988) 437c

[17] T. Maruyama, A. Ohnishi, H. Horiuchi, Phys. Rev. **C42** (1990) 386

[18] D.H. Boal and J.N. Glosli, Phys. Rev. **C38** (1988) 1870 and 2621

[19] C. Dorso, S. Duarte and J. Randrup, Phys. Lett. **B188** (1987) 287, and C. Dorso and J. Randrup, Phys. Lett. **B215** (1988) 611, and Phys. Lett. **B232** (1989) 29

[20] G. Peilert, J. Randrup, H. Stöcker and W. Greiner, UFTP preprint 247/1990, to be published in Phys. Lett. B

[21] R. Jastrow, Phys. Rev. **98** (1955) 1479

[22] J.W. Clark, Progress in Particle and Nuclear Physics, Vol. **2**, ed. D. Wilkinson, Pergamon Press (1979), p. 89

[23] F. Arickx, J. Broeckhove, E. Kesteloot, L. Lathouwers and P. van Leuven, Chem. Phys. Lett. **128** (1986) 310

[24] E. Krotschek and K. Takahashi, Phys. Lett. **61B** (1976) 393

[25] A.K. Kerman and S.E. Koonin, Annals of Phys. **100** (1976) 332

[26] E. Caurier, B.Grammaticos and T. Sami, Phys. Lett. **B109** (1982) 150, and S. Drozdz, J. Okolowicz and M. Ploszajczak, Phys. Lett. **B109** (1982) 145, and S. Drozdz, M. Ploszajczak and E. Caurier, Ann. Phys. **171** (1986) 108

[27] M. Saraceno, P. Kramer and F. Fernandez, Nucl. Phys. **A405** (1983) 88

[28] H. Horiuchi, T. Maruyama, A. Ohnishi and S. Yamaguchi, Kyoto University preprint (1990)

[29] C. Coriano, R. Parwani and H. Yamagishi, Nucl. Phys. **A522** (1991) 591

[30] J. Konopka, H. Stöcker, W. Greiner and L. Neise Proc. XXIX Int. Winter Meeting on Nuclear Physics, Bormio (Italy), January 1991, ed. E. Iori

[31] C.M. Lederer and V.S. Shirley, Table of isotopes, 7th edition (1978), John Wilet and Sons, INC, New York

[32] H. de Vries, C.W. de Jager and C. de Vries, Atomic Data and Nuclear Data Tables **36** (1987) 495

# III. Nuclear Structure and Exotic Nuclei

# EXOTIC BEAMS

**R. BIMBOT**
Institut de Physique Nucléaire, F-91406 Orsay Cedex

**Abstract**

*Beams of radioactive nuclei are produced by two different techniques based on i) the acceleration of radioactive nuclei produced in thick targets using proton induced spallation, or similar reactions, and ii) the fragmentation of intermediate or high energy heavy ion beams. The first method implies the use of an isotope separator for selecting the nuclei of interest, and in the second method a magnetic selection is performed with a recoil spectrometer, and associated to a beam purification step using a degrader placed in an intermediate focal plane of the spectrometer.*

*Although the intensities of such secondary beams are still low, several types of experiments have been performed, and are described in this paper.*

*The first series of experiments concern the cross-section measurements of the astrophysical reaction* $^{13}N(p,\gamma)^{14}O$ *at very low energy, which have been made, either directly, from an accelerated* $^{13}N$ *beam, or through the Coulomb break-up of an intermediate energy* $^{14}O$ *beam, which gives access to the inverse reaction* $^{14}O(\gamma,p)^{13}N$.

*The second series of experiments concern the determination of the strong interaction radius for exotic nuclei from total reaction cross-section measurements, and the study of neutron halos in very neutron rich light isotopes, and particularly in* $^{11}Li$. *This last study involved inclusive experiments in which* $^{9}Li$ *and other nuclei from* $^{11}Li$ *fragmentation were detected, and also an exclusive experiment in which the angular distribution of neutrons was measured in coincidence with* $^{9}Li$ *fragments from* $^{11}Li$ *break-up.*

## 1 INTRODUCTION

In the early sixties, a major evolution of Nuclear Physics was induced with the development of Heavy Ion Beams, which progressively led to the possibility of accelerating ions of all elements up to Uranium (Berkeley SuperHilac 1972, Darmstadt Unilac 1976). At that time, all the stable isotopes of the periodic chart had become potential projectiles, and Heavy Ions Physics contributed extensively to the progress of Nuclear Physics in many domains, from reaction mechanisms (micro and macrophysics) to nuclear structure (high spins, superdeformation), including exotic nucleus production. The next step in such an evolution was to use unstable nuclei (i.e. radioactive isotopes) as projectiles, and the interest in this new way of research has been increasing during the last 10 years. After 3 workshops devoted to this topics in the USA [1,2] (1981,1984) and in Canada [3] (1985), the First International Conference on Radioactive Nuclear Beams [4] was organised at Berkeley in 1989, and the second one will take place at Louvain-la-Neuve (Belgium) in August 1991. A review article on radioactive beams was written in 1987 by Tanihata [5].

As soon as radioactive nuclei, produced through a nuclear reaction, are ejected from the target and driven towards a detector, for example, one may speak of a secondary beam. Such beams have been produced from many years, at very low energy, in isotope separators such as ISOCELE (Orsay) or ISOLDE (CERN). On the other hand, since the first experiments (Berkeley, 1979) [6], high and intermediate energy fragment beams have been produced and transported through beam lines on long distances, for purposes of identification and study of new isotopes. Without any doubt, these pioneering experiments have been the first steps in the direction of intense exotic beam production. However, in the present talk, the subject will be limited to the production of exotic beams[1], and their use for inducing nuclear reactions in a secondary target.

The two basically different methods which can be used for secondary beam production (i.e. the production of radioactive nuclei at rest and their post-acceleration on the one hand, and the fragmentation of high energy projectiles on the other hand) will be described in the first part of the talk. In the second part, typical examples of the exotic beam physics will be given.

# 2 EXOTIC BEAM PRODUCTION

## 2.1 Accelerated radioactive beams

The basic principle of this method consists in producing radioactive nuclei at rest, through a nuclear reaction induced with a primary beam in a primary target, in selecting the proper nuclei using an isotope separator, and accelerating these nuclei to the desired energy. This technique is particularly suitable for producing low energy beams, for example for astrophysical purposes, when subbarrier energies are needed.

Among the most efficient methods to produce large amounts of radioactive nuclei, one can mention the spallation reactions induced with high energy protons. This is the reason why a number of projects [7] are centered on high intensity hadron machines such as the CERN-ISOLDE facilities in Switzerland, the TRIUMF machine in Vancouver (Canada), the Japanese Hadron project (KEK, Tsukuba), the Moscow Meson Factory, and the ISIS facility at the Rutherford Appleton Laboratory (UK).

However, the first radioactive beam ever produced through this method was a $^{13}$N beam obtained at Louvain-la-Neuve (Belgium) by coupling a low energy (15-30 MeV) proton cyclotron delivering a high intensity beam (up to 500 $\mu$A), with a heavy ion cyclotron (CYCLONE) [8]. The general layout of this facility is shown in Fig. 1.

[1] the term exotic being here synonymous of radioactive, and not necessarily far from stability

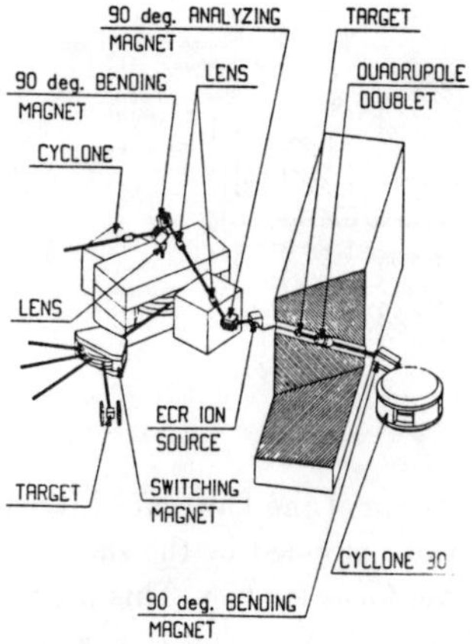


Figure 1: General lay-out of the radioactive ion beam facility at Louvain-la-Neuve [8].

The $^{13}$N nuclei are produced through the (p,n) reaction in a $^{13}$C target, and extracted from this target using a small nitrogen support gas flow. The $^{13}$N-$^{14}$N molecules are thus driven to a single stage ECR ion source especially designed to optimize the production of $N^{1+}$ ions. The $^{13}N^{1+}$ ions extracted from this source using a 7 KV voltage are selected with a 90° magnet and axially injected into the CYCLONE accelerator. In order to obtain the very low energy (0.65 MeV/u) required for the designed astrophysics experiment (see sect. 3.1), the cyclotron is operated in the $6^{th}$ harmonic mode.

Using a 100 $\mu$A primary beam of protons, the first experiments led to the production of $1.5.10^8$ $^{13}$N/s at 0.65 MeV/u, with the following intermediate yields [8] :

$^{13}$N/proton produced in the primary target : $1.6.10^{-3}$

Extraction yield (from primary target) : 9 %

Ionization yield (from primary target) : 4 %

Acceleration yield rom primary target) : 4 %

But this rather satisfactory production was associated with a very severe C contamination, the contaminant being 1000 times more abundant that the $^{13}$N beam. This huge contamination was recently eliminated, and the intensity of $^{13}$C in the $^{13}$N beam is now less than 5 %. Simultaneously, the $^{13}$N intensity was increased and amounts presently to $4.5.10^8$ pps (70 ppA) [9].

The first experiment carried out with this $^{13}$N beam was the measurement of the cross section for the nuclear reaction $^1$H($^{13}$N,$\gamma$)$^{14}$O of astrophysical interest. The motivation of this experiment and the preliminary results will be given in section 3.1, together with other approaches of the same problem.

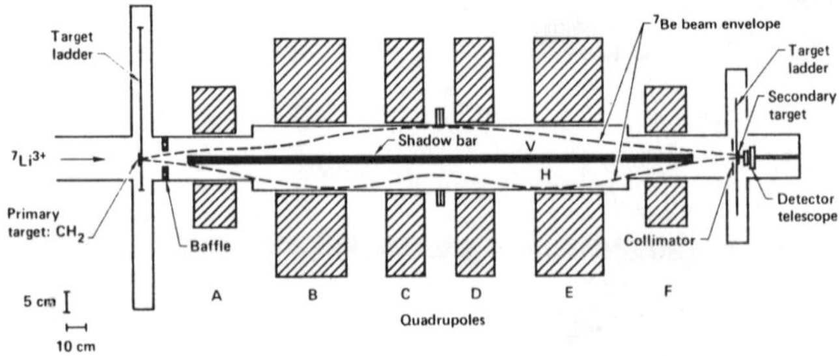


Figure 2: The Quadrupole Sextuplet beam transport system at the Lawrence Livermore National Laboratory [10]. Most of the primary $^7$Li beam is stopped by the shadow bar. The quadrupoles A B C are tuned to get an intermediate focus for $^7$Be. This procedure (with a central focus) reduces significantly the $^7$Li background. Note that this separation is obtained without any dipole.

## 2.2 Secondary-fragment beams

This method is particularly suitable when the primary beam is an intermediate energy or a high energy heavy ion beam. At such energies, the projectile fragmentation reactions lead to a wide distribution of nuclei emitted forward with velocities not far from that of the primary beam. Taking advantage of this effect, the secondary beams may be selected and driven to a secondary target using magnetic transport elements.

### 2.2.1 Low energy beams

Before describing in detail the various existing devices, it seems worth mentioning that similar techniques may be used at low energy for very specific purposes. For example, at the Lawrence Livermore National Laboratory, secondary beams of $^7$Be, $^{13}$N and $^{15}$O (1-3 MeV/u) were obtained in 1983 from the interaction of primary beams of $^7$Li, $^{12}$C and $^{14}$N respectively, with proton and deuteron targets [10]. Here, the reactions involved were not projectile fragmentation, but (p,n) or (d,n) reactions in inverse kinematics. The production efficiency was of $3.10^{-6}$ for the $^7$Be/$^7$Li intensity ratio, and of $10^{-7}$ for the $^{13}$N/$^{12}$C and $^{15}$O/$^{14}$N intensity ratios. The system used for beam selection and transport, based on a quadrupole sextuplet is shown in Fig. 2, and the bidimensional matrix for $^{13}$N identification is shown in Fig. 3. Recently, the device was improved by adding an electrostatic deflection [11]. These low energy beams of light particles have been used for nuclear astrophysics research.

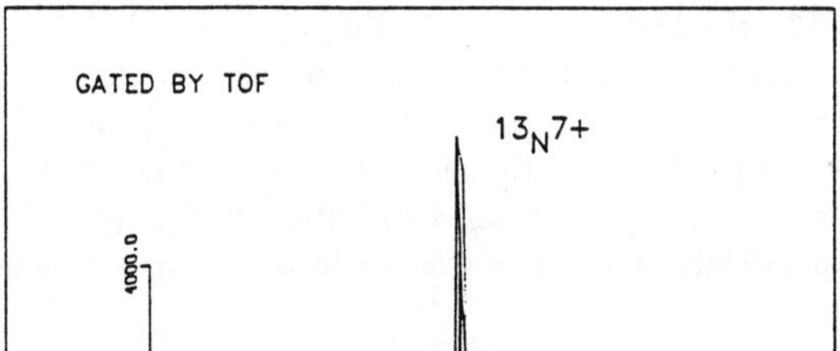


**Figure 3:** Bidimensional plot used for $^{13}$N identification after transport by the Quadrupole Sextuplet at the Lawrence Livermore National Laboratory [10].

### 2.2.2 Production of high energy fragment beams

Coming back to high energy primary beams, since the early experiments performed at Berkeley, for exotic nucleus production [6] and medical application purposes [12], several machines have produced secondary fragment beams. One must mention the CERN Synchrocyclotron [13], the GANIL French National Accelerator [14,15,16], the RIKEN facility [17]; the MSU Superconducting Cyclotron K.1200 [18], the SATURNE Synchrotron [19] and the DARMSTADT SIS Synchrotron [20].

By interaction of the primary beam with the target, a large number of fragments of various Z and A are produced, with velocities close to that of the projectile. At high energies ($\sim$ 1 GeV/u) the momentum distribution widths ($\sigma(P_{\|})$) and ($\sigma(P_{\perp})$) in the parallel and perpendicular directions relative to the beam are equal and given by the relation [21] :

$$\sigma^2 = \sigma_0^2 \frac{A_P(A_P - A_F)}{A_P - 1} \quad (1)$$

in which $\sigma_0 = 90$ MeV/c is a parameter,

in which $A_F$ and $A_P$ are the fragment and projectile masses, respectively.

When the incident energy is decreased, eq. (1) remains valid for calculating $\sigma(P_{\|})$ down to about 20 MeV/u [22] while a correction must be applied for calculating $\sigma(P_{\perp})$ below 200 MeV/u [23].

At the target exit, the momentum distribution of fragments is modified by the slowing down of projectiles and fragments in the target. The resulting energy and angular distributions of a given fragment may however be calculated with a satisfactory accuracy by taking into account the target stopping powers for projectile and fragment [24], and the associated energy and angular straggling [25,26]. Details about these calculations are given in ref. [27].

At high energy, the production yields may be predicted using the parametrization of fragmentation cross-sections elaborated by Sümmerer and Morrisey [28]. However such a prediction is more difficult at energies lower than 100 MeV/u, due to the lack of data.

### 2.2.3 Secondary beam selection

In order to select and transfer the exotic beam to its point of utilisation, a specific device must be used. This device may be considered either as a special beam transfer line or as a magnetic spectrometer. Several of these spectrometers, corresponding to various laboratories are shown in Figs. 4-9. Details about their characteristics will be found in the references indicated in each figure caption. Striking similarities can be observed between them. This is not surprising because the same principle of beam selection and purification is used in all of them :

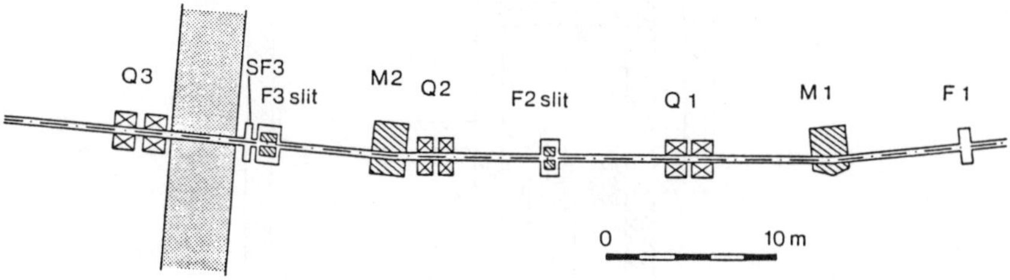


Figure 4: One of the secondary beam lines at the Berkeley Bevalac. The primary target is placed at F1 and the beam is travelling from the right to the left. M1 and M2 are dipoles. Q1 to Q3 are quadrupoles [5].

The production target being placed at a given point where the primary beam can be focused, the reaction products are selected by a first magnet (or set of magnets). The beam is thus dispersed in the focal plane of this magnet (intermediate focal plane of the spectrometer) where a slit is used to select the secondary beam according to its magnetic rigidity :

$$B\rho \;=\; k\frac{A}{Q}\left[\varepsilon(1+\frac{\varepsilon}{2m_u})\right]^{1/2} \quad (2)$$

In eq. (2),
B(T) is the spectrometer magnetic field
$\rho$(m) the curvature radius of the trajectory in the dipole magnet
Q the ionic charge (equal to the nuclear charge Z at high incident energy)
$m_u$ = 931.5 MeV the atomic mass unit
$k = (2m_u/c)^{1/2} = 0.1438$.

The second part of the spectrometer compensates the dispersion of the first one, and achieves an achromatic beam transport to the spectrometer focus. A set of quadrupoles, and sometimes one or several sextupoles are used to insure double achromatism and good optical qualities of the transmission.

It is clear from eq. (2) that, if the fragments have the same velocity (same $\varepsilon$ values), such a spectrometer insures a selection according to the A/Q ratio. At high energy, i.e. for fully stripped fragments, as Q = Z, the selection is made according to the A/Z ratio of the secondary beam. This is illustrated in Fig. 10 where the calculated relative intensities of various secondary beams produced through 800 MeV/u $^{12}C$ + $^{9}Be$ reactions are plotted versus the horizontal position in the intermediate focal plane after dispersion by the first spectrometer magnet (M1 in Fig. 4). It can be observed

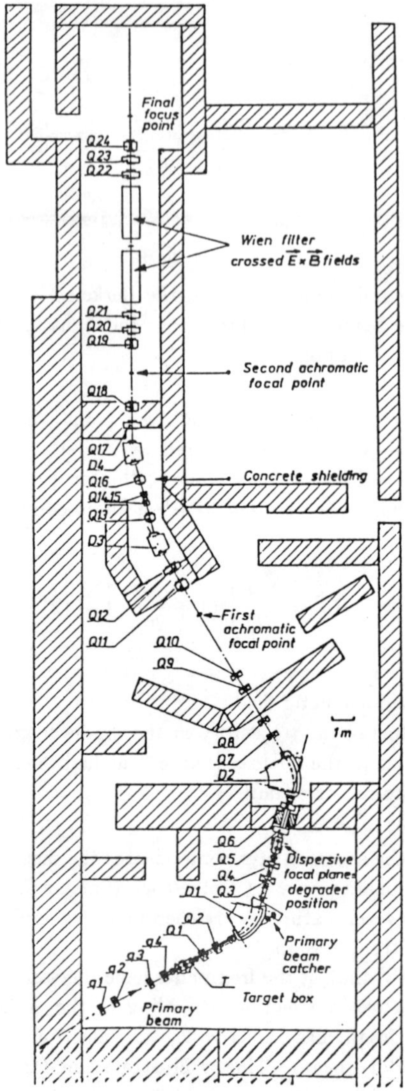


Figure 5: The LISE spectrometer at GANIL. Originally designed with 2 dipoles (D1 and D2) and 10 quadrupoles, the LISE spectrometer was recently modified in two ways : firstly a beam swinger placed at the spectrometer input was added in order to vary the incidence angle of the beam relative to the spectrometer axis. Secondly, the spectrometer was extended by an additional magnetic achromatic deviation, followed by a velocity filter [15,29].

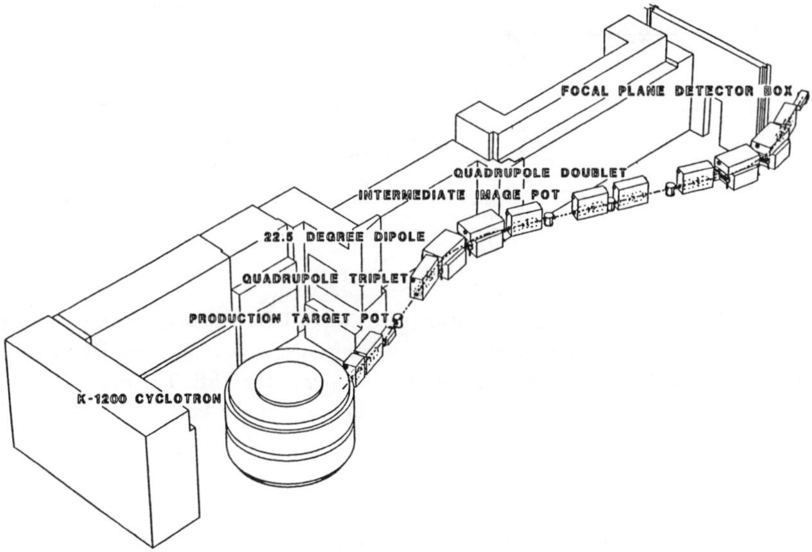


Figure 6: Magnetic spectrometer associated with the K-1200 superconducting cyclotron at Michigan State University for secondary beam selection and purification. Note that the position of this spectrometer at the exit of the machine makes it possible to distribute the secondary beams over the whole experimental area [18].

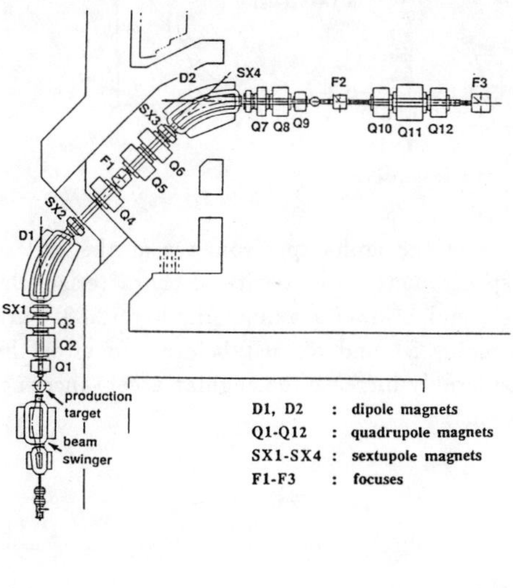


Figure 7: The RIKEN projectile fragment separator. Because of its very large angular and momentum acceptance ($\Omega = 5$ msr, $\Delta p/p = 6$ %), this spectrometer is equiped with four sextupoles magnets, in addition to the two dipoles and the nine quadrupoles preceding the first double achromatic focus F2. Note also the presence of a beam swinger at the entrance of the spectrometer to vary the incidence angle, and of an additional section F2-F3 which makes it possible to transport further the purified beam selected by a slit placed in F2 [17].

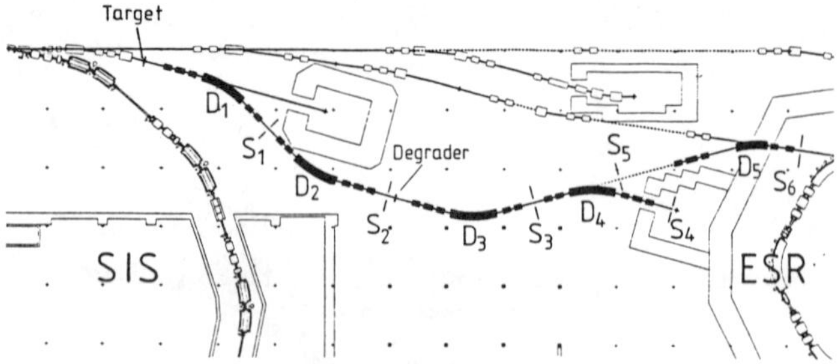


Figure 8: The fragment recoil separator (FRS) at GSI Darmstadt. Situated at the exit of the SIS synchrotron, this spectrometer is composed of four dipoles, D1 to D4 when operating by itself, and D1,D2,D3,D5 when injecting in the ESR. The radioactive beams from FRS can thus either be used directly or injected in the ESR for being cooled and stored [20].

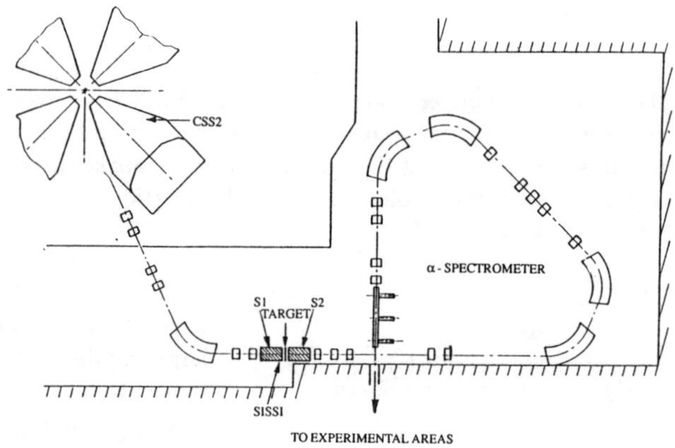


Figure 9: Selection of secondary beams using the alpha spectrometer at the exit of GANIL [16,58]. Relatively to the LISE spectrometer, this device is characterized by lower angular and momentum acceptances, and a lower maximal rigidity (2.88 Tm). The addition of two supraconducting solenoids S1 and S2 just before and after the production target (SISSI project) will considerably increase the angular acceptance, i.e. the intensity of transmitted radioactive beams.

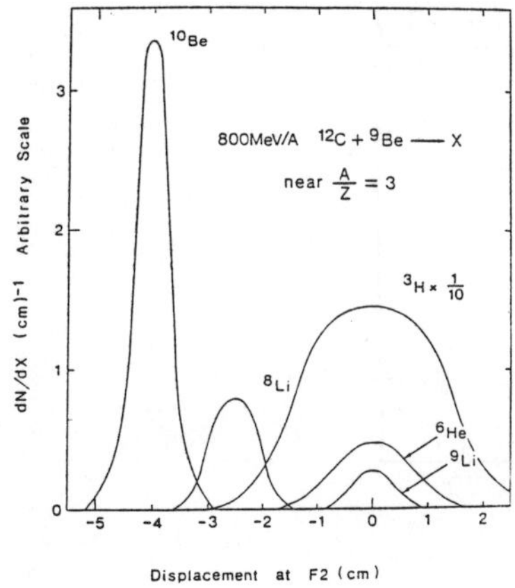


Figure 10: Example of the fragment distribution obtained at Bevalac energies in the F2 intermediate focal plane (see Fig. 4). The beam line is set for transmission of A/Z = 3 nuclei (from [5]).

that the $^{3}$H, $^{6}$He and $^{9}$Li beams, corresponding to the same A/Z ratio (A/Z = 3) are centred around the same position. On the other hand, the $^{8}$Li and $^{10}$Be beams, which correspond to different A/Z ratios are spatially separated from the previous ones. By tuning the slit in order to select the beams corresponding to a zero displacement (± 1 cm, for example), only the A/Z = 3 nuclei are transmitted through the spectrometer. The resulting identification matrix (experimental) is shown in Fig. 11.

At lower energies (several tens of MeV/u), the situation is less clear because the velocity spectrum of fragments is broadened by the reaction mechanism, and by beam target interaction. Under these conditions, several fragments with neighbouring values of A/Z (in fact A/Q) may be simultaneously transmitted through the spectrometer. This is illustrated in Fig. 12 which represents an identification matrix obtained at GANIL, the production reaction being $^{18}$O + Be at 65 MeV/u, and the LISE spectrometer being tuned for A/Q = 3 transmission. Together with the nuclei $^{6}$He to $^{15}$B, which do correspond to A/Q = 3, other isotopes of Li, Be and B are observed. For these nuclei, the ratio A/Q can be as low as 2.33. Note also the presence of the partially stripped $^{18}O^{6+}$.

### 2.2.4 Beam purification with an intermediate degrader

The magnetic selection of a secondary beam being generally not sufficient, a purification step based on another criterion should be used. A cheap and efficient technique consists in using a degrader placed in the intermediate focal plane. This technique has been

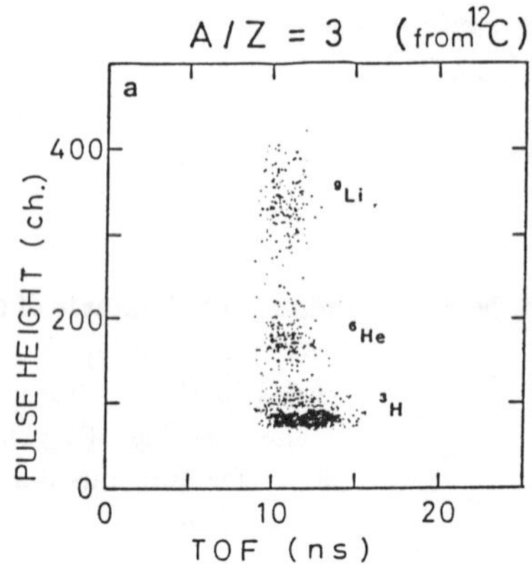


Figure 11: Bi-dimensional identification of A/Z = 3 secondary beams produced at the Bevalac (from [5]).

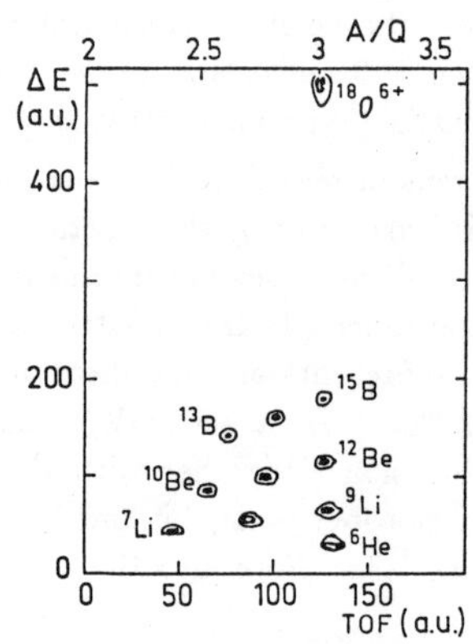


Figure 12: Bi-dimensional identification of secondary beams transmitted through the LISE spectrometer tuned for A/Q = 3 transmission. The reaction is 65 MeV/u $^{18}$O + Be. The ions are identified by their time of flight (TOF) and energy loss ($\Delta$E) in 1 mm thick Si detector (from [32]).

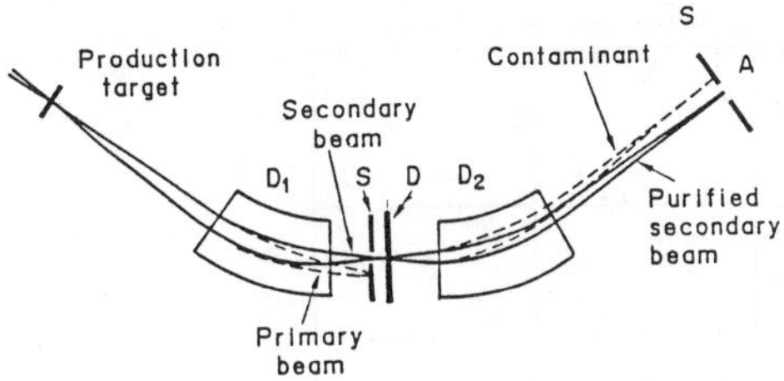


Figure 13: Principle of the secondary beam purification with an intermediate degrader.

widely described [14,15,20,30,31,32,33]. Its principle is shown in Fig. 13. The secondary beams of different nature (A,Z) but of the same magnetic rigidity fall onto a degrader of proper thickness. After its traversal, due to the differences in stopping powers, these secondary beams generally correspond to different magnetic rigidities. After proper tuning of the second part of the spectrometer only the purified beam is transmitted through a second slit placed at the spectrometer exit. Note that the use of a profiled degrader (wedge degrader) makes it possible to preserve the spectrometer achromatism. The splitting of a secondary beam into its components can be directly visualized with beam diagnostics, as in Fig. 14.

The quality of the purification depends on the number, abundance, and nature of the beam contaminants to eliminate (see Figs. 15-16).

### 2.2.5 Additional purification using a velocity filter

For further improvement of the secondary beam purification, it is necessary to use a new selection criterion. Electrostatic deflection provides an elegant solution when the energy is not too high. Thus, a velocity filter, composed of crossed electrostatic and magnetic fields was added to the LISE spectrometer at GANIL (see Fig. 6). These two fields act in opposite ways and their relative magnitude is tuned in such a way that an exact compensation is obtained for the nuclei with the selected velocity. The purification effect of such a device is shown in Fig. 17, in the absence of intermediate degrader.

If an intermediate degrader is also used, this additional selection leads to a pure secondary beam at the final focus.

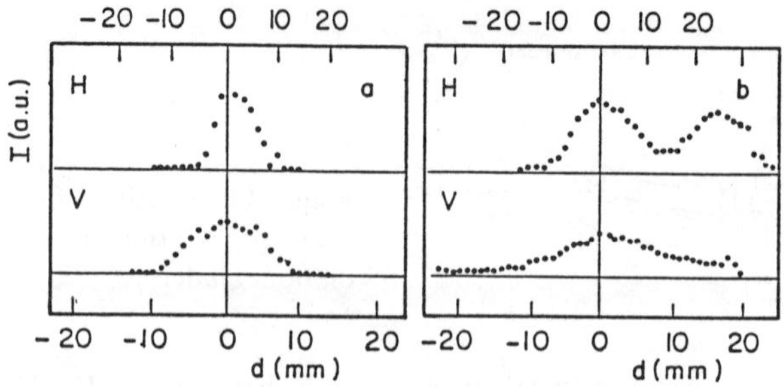


Figure 14: Secondary beam splitting at the achromatic focus after the traversal of a degrader. The horizontal (H) and vertical (V) profiles from wire chambers are displayed, the intensity collected on each wire being plotted versus the wire positions.
a) non purified beam, b) group of secondary beams dispersed after going through a wedge shaped Al degrader.

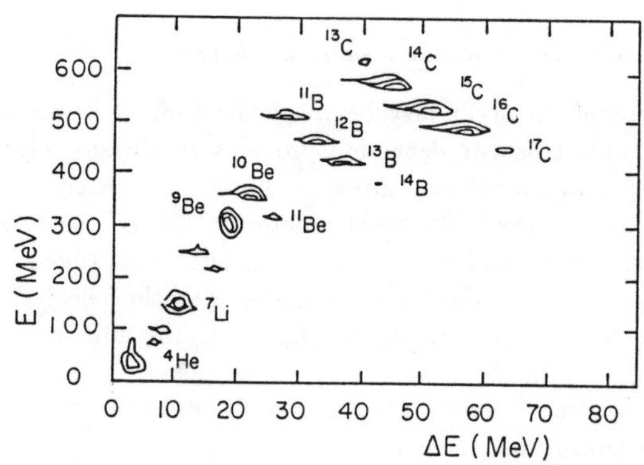


Figure 15: Bidimensional plot showing the isotopic composition of a secondary beam obtained in the following conditions : 65 MeV/u $^{18}O$ + 1036 mg/cm$^2$ Be. B$\rho$ = 2.248 Tm, $\Delta B\rho/B\rho = \pm$ 0.22 % [32].

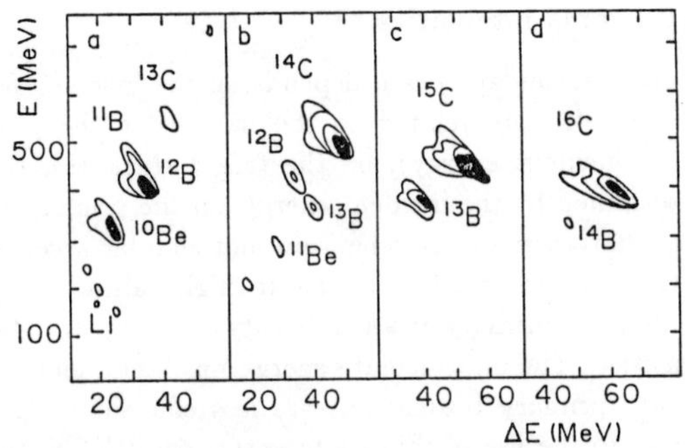


Figure 16: Isotopic composition of the beams obtained by purification of the beam analysed in Fig. 15 with an intermediate degrader. $B_1\rho = 2.248$ Tm, $\Delta B\rho/B\rho = \pm\ 2.78$ %.

a. $B_2 = 1.077$, b. $B_2 = 1.069$, c. $B_2 = 1.057$, d. $B_2 = 1.043$.

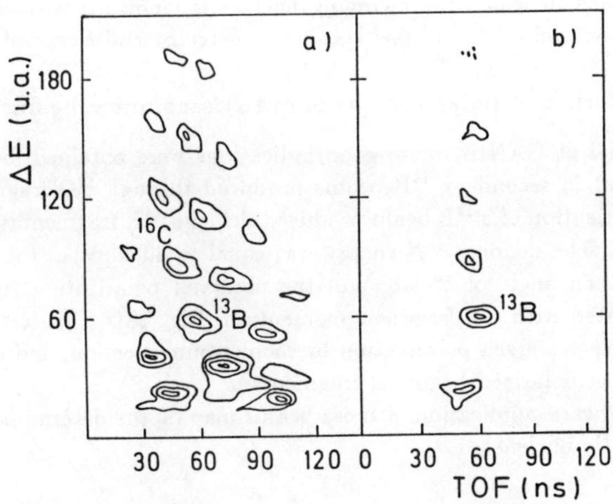


Figure 17: Secondary beam purification using the LISE 3 velocity filter : the isotopic composition of a secondary beam is displayed:
a. at the entrance of the filter, b. at the exit of the filter.

### 2.2.6 Secondary beams intensities

The intensity of a given secondary beam depends on the involved fragmentation cross-section (which is itself strongly related to the natures of the projectile, target and fragment, and to the incident energy), on the target thickness, the optimal value of which is mainly determined by the incident energy, on the primary beam intensity, and on the spectrometer characteristics (momentum and angular acceptances).

The high energy heavy ion machines lead to high values of the conversion rates (secondary beam intensity/primary beam intensity), but up to now, deliver very low primary beam intensities. The intermediate energy machines, and in particular GANIL produce high intensity primary beams, but are less efficient as far as the conversion rate is concerned (due to the use of thinner targets). At RIKEN, the high acceptance, and high maximum rigidity of the RIPS separator makes it possible to reach high transmission efficiencies, even for the most neutron rich nuclei.

Finally, the fragment beam intensities presently observed range in the interval from $10^2$-$10^3$ pps for very exotic beams to $10^8$ pps for nuclei not far from the stability, such as $^{17}$N, $^{14}$C [32,34]. These low intensities are the major inconvenient of secondary beams. However, this difficulty may be overcome by using high efficiency counters or multidetectors, a number of which are now available in the laboratories. Note also that, in all experiments in which the incoming nucleus is identified with a Si detector, the intensity must be limited to $10^3$ pps because of detector and electronics limitations.

### 2.2.7 Production of polarized (or aligned) secondary beams

After a first trial at GANIL where good indications were obtained for the presence of a spin alignment in secondary $^{14}$B beams produced through $^{18}$O fragmentation [35], a significant polarization of a $^{12}$B beam produced through $^{14}$N fragmentation was observed at RIKEN [36]. The incident $^{14}$N energy was equal to 40 MeV/u, the secondary beam was selected at an angle of 5° lab, and the observed polarization (up to 20 %) was strongly correlated with the fragment momentum (Fig. 18). This last property makes it possible to select a given polarization by momentum selection, and opens the way to the production of polarized beams of fragments.

A straightforward application of these beams may be the determination of unknown g factors of exotic nuclei.

# 3 PHYSICS WITH EXOTIC BEAMS

## 3.1 The "astrophysical" $^{13}$N(p,$\gamma$)$^{14}$O reaction

Among nuclear astrophysics studies, the determination of the cross section at very low energy for the reaction $^{13}$N(p,$\gamma$)$^{14}$O is a subject of high current interest, because of its

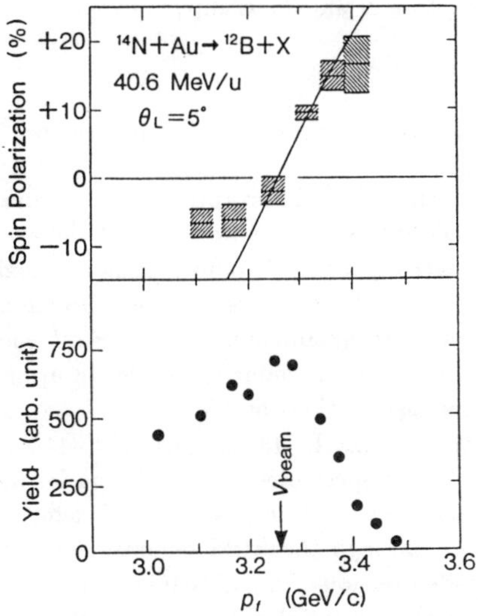


Figure 18: Spin polarization and yield of $^{12}$B nuclei produced through $^{14}$N + Au fragmentation, plotted versus the outgoing momentum. The solid curve represents the result to a calculation based on a simple fragmentation model [36].

importance in the so-called hot CNO cycle, which takes place, for example, in massive stars [37].

At such very low energies, the cross section reaction is dominated by a resonance at $E_R = 0.545$ MeV c.m., corresponding to the $1^-$, 5.17 MeV level in $^{14}$O, and may be well described by a single level Breit-Wigner expression, if the parameters $E_R$, $\Gamma$ (total width) and $\Gamma_\gamma$ (gamma width) are known [38]. The $E_R$ value [39] and total width [40] have been experimentally determined, and the recent efforts concentrated on the $\Gamma_\gamma$ determination. While the Louvain-la-Neuve equipment devoted to the production of a $^{13}$N beam was being elaborated, indirect measurements of $\Gamma_\gamma/\Gamma$ using the $^{12}$C($^3$He,n$\gamma$)$^{14}$O reaction were performed in France [41] and in USA [38]. They led respectively to an upper limit $\Gamma_\gamma \leq 7.6$ eV, and to the value $\Gamma_\gamma = 2.7 \pm 1.3$ eV not far from the theoretical estimations of this parameter which ranged in the 1-10 eV interval, with higher probability around 3 eV.

However, this experimental value needed confirmation and experiments based on a completely different approach, the Coulomb break-up of $^{14}$O at high energy, were elaborated, in France and Japan. The basic idea [42] was to get information on the reaction $^{13}$N(p,$\gamma$)$^{14}$O at very low energy through the break-up of $^{14}$O in the Coulomb field of a heavy nucleus ($^{208}$Pb), according to the reaction

$$^{14}\text{O} \longrightarrow \text{p} + {}^{13}\text{N},$$

which is interpreted as being due to the capture of an equivalent photon

$$^{14}\text{O} + \gamma \longrightarrow \text{p} + {}^{13}\text{N}.$$

This last reaction is obviously inverse of the radiative capture of a proton by $^{13}$N, i.e. of the reaction of interest here.

Therefore, the experiment consists in producing a $^{14}$O beam through $^{16}$O fragmentation at intermediate energy, and to detect the proton and $^{13}$N from the $^{14}$O break-up induced by a $^{208}$Pb target. Accurate determinations of the $^{13}$N and p emission angles and energies are needed in order to determine their relative energy in the $^{14}$O system.

The equivalent gamma flux is computed from the Coulomb break-up theory. A check of the predominance of Coulomb over nuclear excitation in inducing the break-up may be performed by measuring the relative angular distributions of $^{13}$N and p.

Such experiments have been realised both at GANIL [43] and RIKEN [44]. At GANIL, the $^{14}$O beam was produced in the ALPHA spectrometer, the $^{13}$N nuclei were detected with the SPEG spectrometer and the protons with a set of CsI detectors. At RIKEN, $^{14}$O was produced and transmitted through the RIPS separator, and the reaction products were detected using an upgraded version of the EMRIC multidetector, built at Grenoble [45]. Both experiments have been successful. The data reduction is in progress and should lead to new $\Gamma_\gamma$ determinations in the near future (see Figs. 19-20).

Meanwhile, the first results from Louvain-la-Neuve have come out. A preliminary value of $\Gamma_\gamma = 3.8 \pm 1.2$ eV has been obtained [9].

This $^{13}$N story is a good illustration of the fast progress of techniques involving radioactive beams and of their efficiency in solving nuclear astrophysics problems.

## 3.2 Reaction cross-sections, nuclear radii and neutron halos

### 3.2.1 Reaction cross-sections and nuclear radii

The first experiments in which the total reaction cross-sections were measured for exotic beams were performed at Berkeley by a Japanese group [46,47]. Secondary beams of stable and radioactive isotopes of He, Li, Be, B, C were produced around 800 MeV/u, and the transmission method was used to measure the reaction cross sections from beam intensity attenuation through the target. From these experiments, the root-mean square radius of nuclear matter distribution was calculated for each projectile. Extremely large radii were found for the neutron drip line nuclei $^{11}$Li and $^{14}$Be.

These experiments were extended to lower energies (25-50 MeV/u) by two GANIL programs. In the first one, the $\gamma$-rays emitted from radioactive beam interaction with a thick target were measured with a high efficiency, and the total reaction cross-sections radii derived from these measurements. In their first publication [48], the authors reported a surprising linear dependance of the projectile reduced strong absorption

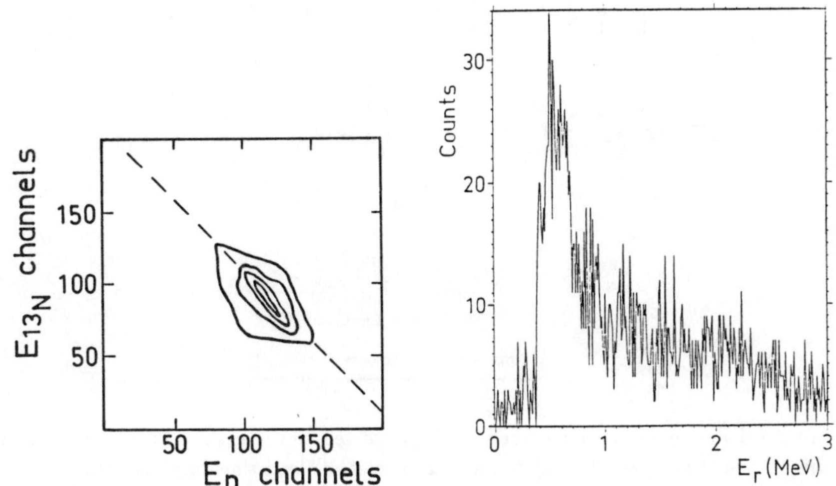


Figure 19: Coulomb break-up of $^{14}$O : experimental results from [43]. a. bidimensional plot showing the correlation between proton and $^{13}$N energies. The dotted line represents the expected correlation from $^{14}$O break-up. b. Relative energy spectrum. The $1^-$ state at 5.17 MeV corresponds to the peak at about 500 keV.

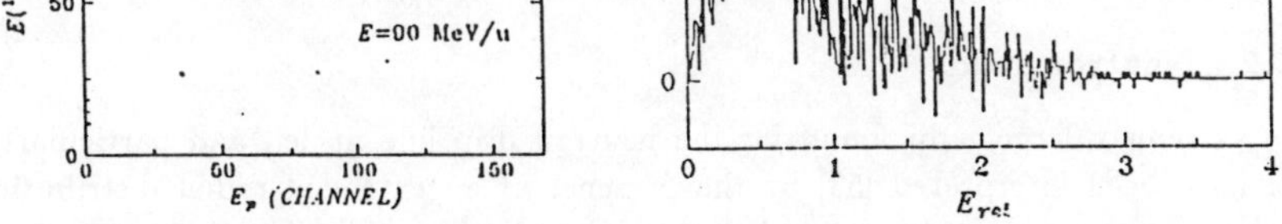


Figure 20: Same as Fig. 19. Results from [44].

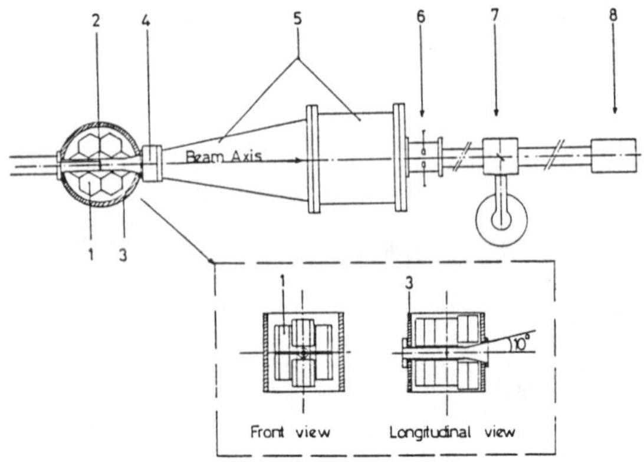


Figure 21: Experimental device used in the total reaction cross-section measurements by the $4\pi$-$\gamma$ method [50]. 1) NaI counter, 2) target, 3) lead shielding, 4) beam control, 5) conical extension of the reaction chamber, 6) Si detectors for monitoring the beam axis, 7) X detection monitoring, 8) Faraday cup. Note the presence of a conical aperture of $10^o$ without detectors at foward angles.

radius $r_0$ versus the projectile neutron excess. However, after further investigation (new experiments and re-analysing the data), the same authors concluded to the presence of a much smaller effect [49].

Meanwhile, another experiment had been performed using thin targets and the $4\pi$-$\gamma$ method [50]. The corresponding experimental device is shown in Fig. 21. This experiment confirmed the presence of very large radii for the nuclei $^{11}$Be, $^{14}$Be, $^{17}$B, but showed no systematic variation of $r_0$ [50,51] versus the neutron excess (see Fig. 22). It is also important to note that no significant reaction cross-section enhancement was observed for $^{11}$Li projectiles. This absence of effect will be interpreted in the next section.

Finally, in a recent RIKEN experiment [52], an enhanced total reaction cross-section was also observed for $^{11}$Be at 33 MeV/u.

### 3.2.2 Neutron halos

The extremely large radii found for the neutron drip line nuclei, and particularly for $^{11}$Li have been interpreted [53] by the presence of an extended radial distribution of the loosely bound neutrons around the nucleus core (i.e. $^{9}$Li). This interpretation was confirmed by the narrow momentum distribution (Fig. 23) observed for the $^{9}$Li nuclei from $^{11}$Li fragmentation at 800 MeV/u [54] and by the large cross sections for Coulomb

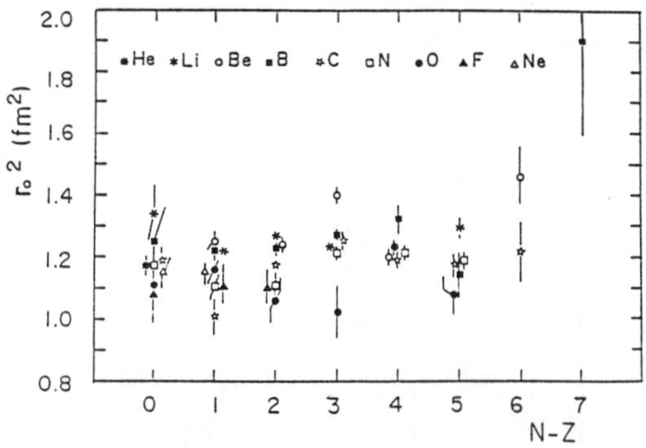


Figure 22: The reduced strong absorption radius $r_0^2$ as a function of the neutron excess N-Z for all the isotopes measured in [50].

dissociation observed at the same energy with lead targets [55].

In order to obtain more information about the spatial extent of the neutron halo, an exclusive experiment was performed at GANIL to measure the angular distribution of forward neutrons from the break-up of the $^{11}$Li halo [56]. The experimental device is shown in Fig. 24. The incoming $^{11}$Li beam, selected through the LISE spectrometer, was identified and monitored using a transmission silicon detector $\Delta E_1$, immediately before the target. The $^{9}$Li nuclei resulting from $^{11}$Li break-up were identified with a $\Delta E_2$-$E_3$ telescope, and the associated neutrons detected in a 23 neutron hodoscope covering an angular range from zero to 50 degrees. In order to investigate the relative influence of Coulomb vs nuclear break-up, targets of Be, Ni and Au were used. A comparative run was performed with a $^{9}$Li beam of the same velocity as the $^{11}$Li one. The inclusive neutron angular distributions obtained for $^{11}$Li and $^{9}$Li beams on a Be target are shown in Fig. 25. Obviously the neutrons from $^{11}$Li dissociation are more strongly peaked forward than those from $^{9}$Li fragmentation. The opening angle for the former distribution is $\theta_{1/2} = 2.9 \pm 0.4^o$ and corresponds to a characteristic spread in transverse momentum of 24 $\pm$ 4 MeV/c, well below typical nucleon momenta in a nucleus. This is a strong evidence for the existence of a halo of loosely bound neutrons.

The very forward peaking of the neutron angular distribution provides an explanation for the fact that no enhancement was found for the $^{11}$Li induced total reaction cross-section measured in [50] by the $4\pi$-$\gamma$ method (see sect. 3.2.1). Indeed, most of the reactions involving the neutron halo lead to $^{9}$Li and 2n emission, probably without any photon emission. As the two neutrons are emitted at very small forward angle, they could not be detected in the counter used in this experiment (see Fig. 21).

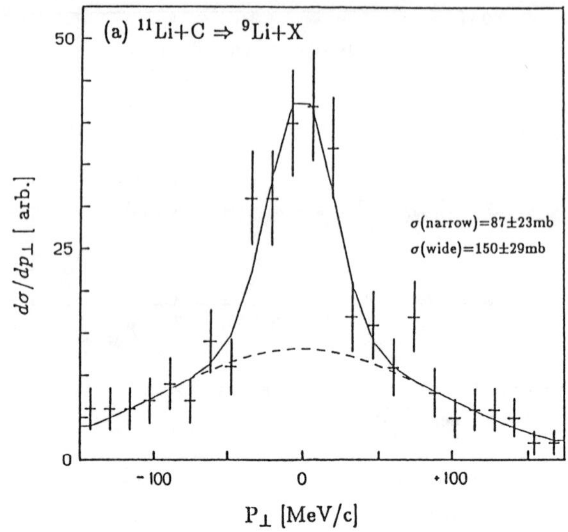


Figure 23: Distribution of transverse momenta for $^{9}$Li nuclei from $^{11}$Li fragmentation [54].

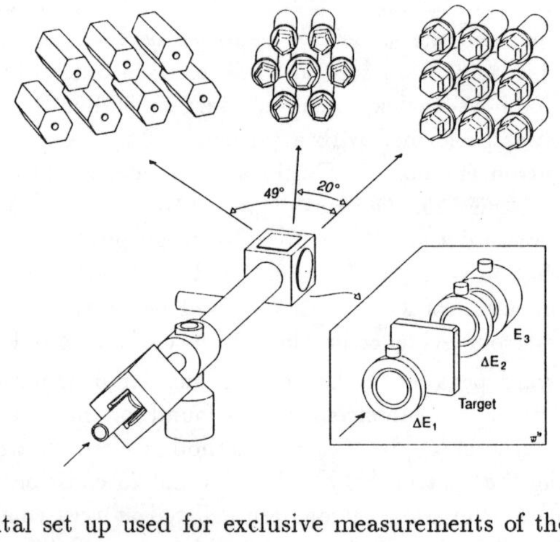


Figure 24: Experimental set up used for exclusive measurements of the angular distribution of neutrons from $^{11}$Li break-up [56].

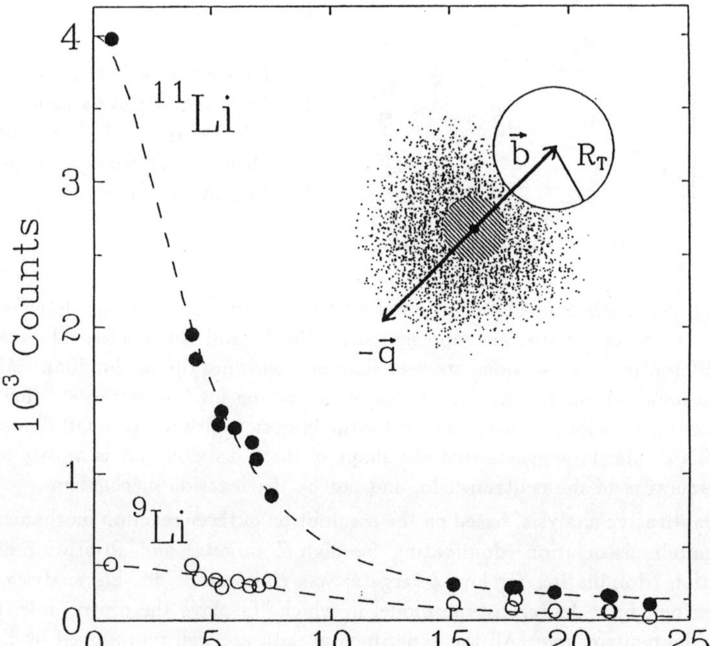


Figure 25: Inclusive angular distribution of neutrons from secondary beams of $^9$Li and $^{11}$Li at the same velocity incident on a Be target. The data were normalised to the same number of incident ions. The insert shows a schematic model for $^{11}$Li induced reaction [56].

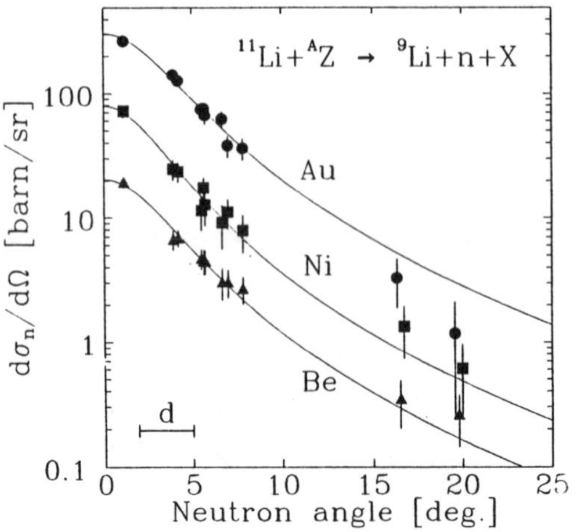


Figure 26: Differential cross-sections for observation of a neutron in coincidence with a $^{9}$Li fragment for a 29 MeV/u $^{11}$Li beam incident on Be, Ni and Au targets [56].

In Fig. 26, the angular distribution of neutrons from $^{11}$Li break-up, detected in coincidence with a $^{9}$Li residue, are represented for Be, Ni and Au targets. The magnitudes of the differential cross-sections are very different, showing the predominant influence of Coulomb break-up for the Au target, and its vanishing for the Be target. However, the three distributions look similar, except for the largest angles where small differences appear. This similarity suggests that the shape of these distributions is mostly influenced by the structure of the neutron halo, and not by the reaction mechanism.

A quantitative analysis, based on the assumption of three reaction mechanisms which are Coulomb dissociation (dominating for high Z targets) and absorption-diffraction dissociation (dominating for low Z targets) was elaborated. In this analysis, the $^{11}$Li was described by a deuteron-type model in which $^{9}$Li plays the proton role, and a di-neutron the neutron role. All the experimental data are well reproduced by this model which leads to a radius of about 12 fm for the neutron halo.

# 4 CONCLUSION

The physics of radioactive beams is a field of high current interest in rapid evolution. Several machines have just started to produce such beams, and a number of projects have been elaborated over the world to improve the existing devices or build new facilities. In Europe, significant improvements are expected in the near future from :

- an increase of the primary SIS beam intensity at GSI,

- the starting of the SISSI project at GANIL which consists in increasing significantly the angular acceptance of the ALPHA spectrometer, by adding two supraconducting solenoidal lenses near its object point. The consequence will be the possibility of transmitting intense radioactive beams over the whole experimental area. After completing the so-called O.A.I. modifications which will lead to a significant intensity increase of the GANIL primary beams, this machine will thus be able to produce a wide variety of radioactive fragment beams of significant intensities ($\geq 10^6$ pps), and to use the full experimental potential available, including the very efficient multidetectors which exist on the site.

For a further future, several exciting projects are anticipated. One must mention the creation of an Isospin Laboratory in the United States, the Radioactive Beam Arena in the Japanese Hadron Facility, and several projects in Europe, concerning the production of sub-barrier and barrier energy radioactive beams. Among these last ones, let us mention i) the Louvain-la-Neuve ARENAS$^3$ project [57] which combines the use of the two existing cyclotron (see Fig. 1), of the isotope separator LISOL, and of a linear post-accelerator, and ii) the recent GANIL project which involves:

- radioactive nucleus production in a primary target using the high intensity light ion beams ($^{12}$C, $^{16}$O) at 100 MeV/u,

- efficient ionization of these radioactive nuclei using ECR sources,

- isotope separation,

- secondary beam acceleration to 1-15 MeV/u with a linear accelerator.

Finally, this review has shown that Exotic Beam Physics is now a well established field of Nuclear Physics. The number and quality of existing facilities and projects all over the world is a guarantee of success and long life for this new born Research Area.

This paper has been prepared with the valuable help of several friends and colleagues. I thank R. Anne, K. Asahi, D. Bazin, D. Guillemaud-Mueller, T. Kubo, T. Motobayashi, A.C. Mueller and J. Vervier who provided fresh documents and information. I am particularly grateful to P. Aguer and Y. Kiener for illuminating discussions, and to E. Davanture for typing and reducing the entropy of the handwritten draft.

## References


[1] Proceedings of the Workshop on Radioactive Ion Beams and small cross-section measurements, Burr Oak, Ohio (1981), IEEE Trans. Nucl. Sci. ns 30 (1983).

[2] Proceedings of the Workshop on prospects for research with radioactive beams from heavy ion accelerators, April 1984, LBL Report 18187 (1984).

[3] Proceedings of the Accelerated Radioactive Beams Workshop, Parksville, Canada, Sept. 1985, Triumf Report TRI-85-1 (1985).

[4] Proceedings of the First International Conference on Radioactive Nuclear Beams, Berkeley (USA), Oct. 1989, Ed. by W.D.Myers, JM.Nitschke and E.B.Norman (World Scientific, Singapore, 1990).

[5] I.Tanihata, RIKEN Report AF-NP-58 (1987) in Treatise on Heavy Ion Science, vol. 8 (D.A.Bromley edr).

[6] Y.P.Viyogi et al, Phys. Rev. Lett. 42 (1979) 33.
G.D.Westfall et al, Phys. Rev. Lett. 43 (1979) 1858.

[7] See Ref. 4, p. 3-72.

[8] D.Darquennes et al, Phys. Rev. C42 (1990) R804.

[9] J.Vervier, private communication (feb. 1991).

[10] R.C.Haight et al, Nucl. Instr. and Meth. 212 (1983) 245.
G.J.Mathews et al, Ref. 2, p. 21.

[11] R.N.Boyd et al, Ref. 4, p. 311.

[12] J.R.Alson et al, IEEE ns 26 (1979) 3003.

[13] A.Fleury et al, Nucl. Instr. and Meth. B10/11 (1985) 369.

[14] R.Bimbot et al, Z. Phys. A322 (1985) 443.

[15] R.Anne et al, Nucl. Instr. and Meth. A257 (1987) 215.

[16] A.Gillibert et al, Phys. Lett. B176 (1986) 317.

[17] T.Kubo et al, Ref. 4, p. 563.

[18] B.M.Sherrill et al, Ref. 4, p. 72.

[19] J.P.Dufour et al, Ref. 4, p. 321.

[20] H.Geissel et al, A proposal for the SIS-ESR Experimental Programme (1987). E.Roeckl, this meeting.

[21] A.S.Goldhaber and H.H.Heckmann, Ann. Rev. Nucl. Sci. 28 (1978) 161.

[22] C.K.Gelbke et al, Phys. Rev. Lett. 37 (1976) 1191.

[23] K.Van Ribber et al, Phys. Rev. Lett. 43 (1979) 840.

[24] F.Hubert et al, At. Data Nucl. Data Tables 46 (1990) 1.

[25] J.Hérault, Thesis Orsay 1988.

[26] R.Anne et al, Nucl. Instr. and Meth.

[27] R.Bimbot, Proceedings EULIMA Workshop on the potential value of Light Ion Beam Therapy, Nice (1988), P.Chauvel and A.Wambersie Edrs, Report EUR-12165-EN, p. 443.

[28] K.Summerer and D.J. Morrisson, Ref. 4, p. 122.

[29] A.C.Mueller and R.Anne, Report IPNO-DRE-90-17 (1990).

[30] J.P.Dufour et al, Nucl. Instr. and Meth. A248 (1986) 267.

[31] S.Hoath, Nucl. Instr. and Meth. A248 (1986) 247.

[32] R.Bimbot, Report IPNO-DRE-87-35, Orsay (1987).

[33] K.H.Schmidt et al, Nucl. Instr. and Meth. A260 (1987) 287.

[34] T.Kubo and M.Ishihara, Proc. $2^{nd}$ IN2P3-RIKEN Symposium on Heavy Ion Collisions, Obernai, April 1990, B.Heusch and M.Ishihara Edrs (World Scientific, Singapore), p. 40 (1990).

[35] K.Asahi et al, Phys. Rev. C43 (1991) 456.

[36] K.Asahi et al, Report RIKEN-AF-NP-87 (1990).

[37] W.A.Fowler, Rev. Mod. Phys. 56 (1984) 149.

[38] P.B.Fernandez et al, Phys. Rev. C40 (1989) 1887.

[39] F.Ajzenberg-Selove, Nucl. Phys. A449 (1986) 1.

[40] T.E.Chupp et al, Phys. Rev. C31 (1985) 1023.

[41] P.Aguer et al, Proc. Int. Symposium on Heavy Ion Physics and Nuclear Astrophysics Problems, Tokyo, Japan (1988), Edited by S.Kubono, M.Ishihara and T.Nomura (World Scientific, Singapore) (1989) 107.

[42] G.Baur et al, Nucl. Phys. A458 (1986) 188.

[43] P.Aguer et al, Ganil proposal 124E (1989).
J.Kiener et al, Private communication (1991).

[44] T.Motobayashi et al, Private communication (1991).

[45] F.Merchez et al, Nucl. Instr. and Meth. A275 (1989) 133.

[46] I.Tanihata et al, Phys. Rev. Lett. 55 (1985) 2676.

[47] I.Tanihata et al, Phys. Lett. B206 (1988) 592.

[48] W.Mittig et al, Phys. Rev. Lett. 59 (1987) 1889.

[49] A.C.C.Villari et al, Proc. XXVII Int. Winter Meeting on Nuclear Physics, Bormio (1989).

[50] M.G.Saint-Laurent et al, Z. Phys. A332 (1989) 457.

[51] E.Liatard et al, Europhys. Lett. 13 (1990) 401.

[52] M.Fukuda et al, preprint RIKEN (1991).

[53] P.G.Hansen and B.Jonson, Europhys. Lett. 4 (1987) 409.

[54] T.Kobayashi et al, Phys. Rev. Lett. 60 (1988) 2599.

[55] T.Kobayashi et al, Phys. Lett. 232B (1989) 51.

[56] R.Anne et al, Phys. Lett. 250B (1990) 19.

[57] M.Arnould et al, The ARENAS$^3$ project, Unpublished Report (1990).

[58] W.Mittig, Nucl. Phys. News 1 (1990) 30.

# DECAY STUDIES OF NUCLEI FAR FROM STABILITY: FROM MeV/u TO GeV/u HEAVY-ION BEAMS

E. ROECKL
GSI Darmstadt
Postfach 110 552
D-6100 Darmstadt

## Abstract

Some recent decay studies of very proton-rich nuclei are presented as examples for nuclear-structure investigations far from the line of beta stability. Most of these investigations are based on heavy-ion beams from the accelerators UNILAC (5 to 20 MeV/u) or GANIL (< 85 MeV/u) in combination with an on-line mass separator or a zero-degree spectrometer, respectively. The prospects for the latter type of instrument on-line to SIS (< 2 GeV/u) are also described.

## 1. Introduction

Studies of nuclei far from the beta stability line offer the possibility (i) to extend nuclear structure data in a systematical way (e.g. as a function of proton number, neutron number or isospin) and (ii) to encounter nuclear phenomena (e.g. new shell closures, new decay modes) that are substantially different from those close to the stability line. Thus a more stringent test of nuclear-model predictions become possible, including astrophysically relevant ones.

In an attempt to give examples of recent studies of very proton rich nuclei, this report deals mainly with nuclear ground-states far from stability, with decay properties of these nuclei, in particular for beta decay, and with excited levels of daughter nuclei populated in the decay. For discussing these topics, the nucleon drip lines are of special interest. They are defined as the limits, beyond which a single nucleon becomes unbound in the respective nuclear ground-state. This line has been reached in experiments, admittedly in a somewhat tentative way, for proton-rich (odd-Z) nuclei up to bromine and for neutron-rich nuclei up to neon. For heavier nuclei, even a few cases of ground-states beyond the proton drip-line, hence proton-radioactive, have been identified. Well before reaching these "ground-state drip lines", beta-delayed particle emissions sets in. This process, in which the nucleon drip-line (or threshold) is passed for excited states, is one of the decay processes typical for off-stability nuclei - and typical for a disintegration mode whose study yields rich information.

This report is organized in the following way. After a sketch of the relevant experimental techniques (section 2), decay-spectroscopic studies of proton-rich nuclei are presented that use heavy-ion beams from UNILAC or from GANIL (section 3). Related studies from other accelerators are also taken into account. In section 4, experiments in progress at the new SIS/ESR facility are described. This paper is meant as a short summary of some recent developments. For more details and a more extended list of references the reader is referred to a forthcoming review article (Roeckl 1991). Separate contributions to this workshop deal with the production of secondary beams of unstable nuclei (Bimbot 1991) and with the investigation of neutron-rich isotopes of astrophysical interest (Kratz 1991), and these topics will therefore be discussed here only very briefly.

## 2. Experimental Techniques

### 2.1 ISOL systems

The isotope separation on-line (ISOL) methods are based on stopping reaction products in a solid-state or gas medium inside of or connected to an ion source, extracting them as single-charged ions and mass-separateng them as beams of typically 60 keV energy by means of a magnetic field. This method has been pioneered by the CERN-ISOLDE group, using high-enery proton beams. However, the operation with 4 to 15 MeV/u heavy-ion beams, pursued at the GSI On-Line Mass Separator, offers special advantages related to the thin target-catcher technology and to the selective production pattern. Examples of fusion-evaporation reactions used for the production of the very proton-rich nuclei $^{48}$Mn and $^{104}$Sn are $^{12}$C($^{40}$Ca,p3n) and $^{50}$Cr($^{58}$Ni,2p2n) , respectively (see section 3). For producing neutron-rich isotopes, neutron-rich beams between $^{76}$Ge and $^{238}$U were used for multinucleon-transfer reactions with tantalum or tungsten targets placed inside the ion source. Many new isotopes were identified in the regions chromium-to-copper and holmium-to-lutetium. The source strengths of the mass-separated samples were high enough to enable determination of beta-decay halflives and to deduce excited-level energies in the daughter nuclei from $\gamma$ singles and $\gamma$-$\gamma$ coincidence data. These data allow to test theoretical halflife predictions, which are also of astrophysical interest (Kratz 1991) and of other nuclear properties calculated on the basis of a macroscopic-microscopic model. As an example of the latter studies we mention the investigation of allowed-unhindered beta decays in the region of neutron-rich holmium-to-lutetium isotopes, which allowed to assign Nilsson configurations to low-lying states and to trace the influence of quadrupole and hexadecapole deformation (Rykaczewski et al 1989, Becker et al 1991). This investigation is of interest as this region with $N \approx 108$ is homologous to the region of the heaviest known elements with Z = 107 to 109, which are interpreted as being stabilized by a deformed shell-closure (Patyn and Sobiczewski 1991).

In times, where excitement about "radioactive beams" and "phase-space cooling" is common place, it may be appropriate to recall the attractive feature of the 60 keV radioactive ISOL beams, namely their small energy spread of $\lesssim$ 5 eV and their small beam emittance of $\lesssim 10\pi$mm mrad: Compared to the original phase-apace distribution of the reaction products, this cooling effect is quite impressive and leads to ideal conditions for collinear laser spectroscopy, for capture of such ions in a trap, and for decay studies on thin point-like sources.

### 2.2 Zero-degree magnetic spectrometer

Due to the forward peaking of fragments from peripheral heavy-ion reactions at relativistic (and intermediate) energies, zero-degree magnetic spectrometers offer an effective way, especially if combined with energy degraders, for isotope separation (Bimbot 1991). With respect to studies of exotic nuclei, zero-degree spectrometers offer the following advantages over ISOL systems: (i) short separation time; (ii) conceptually simple isotope identification (e.g. by determining energy loss and time of flight) that is independent upon chemical properties and has a high signal-to-background ratio; (iii) straightforward determination of decay branching-ratios on the basis of the known number of implanted nuclei; (iv) high efficiency for the detection of decay radiation by deep implantation of the nuclei of interest into suitable detectors.

Based on the pioneering work at the BEVALAC, performed about a decade ago, this method has been used very successfully at GANIL in recent years and is now being implemented at several other accelerators. Experimental results from the LISE facility (GANIL) and from the A1200 spectrometer (MSU) will be presented in section 3, and first measurements performed with the projectile-fragment separator (FRS) at GSI will be described in section 4 together with plans for future FRS studies.

## 3. Properties of Very Proton-Rich Nuclei

### 3.1 Nuclei near the proton drip-line

The proton drip-line represents a borderline of nuclear stability that, due to progress in experimental techniques, has attracted considerable interest recently. Experimentally, the particle instability of very proton-rich nuclei can be deduced from the mere observation that they do not arrive at the final focus of the spectrometer, hence that their half-lives are larger (or at least not much shorter) than the time-of-flight from the target to the detector (typically 100ns).

The procedure of "isotope hunting" and "mapping of the proton drip-line" is illustrated by means of two recent experiments on $^{36}$Ar and $^{78}$Kr fragmentations, performed by means of the spectrometer LISE (Saint-Laurent et al 1987) and A1200 (Mohar et al 1990), respectively. The first one succeeded to identify the new isotope $^{22}$Si which has an isospin-projection quantum number $T_z$ of -3 and is thus the most proton-rich nucleus known to date. Together with $^{22}$Si, $^{20}$Mg was transmitted as an isotonic contaminant to the focus of LISE, while $^{21}$Al, known to be proton unbound, was not observed. The second experiment led to the identification of several new isotopes in the gallium-to-strontium region, which may partly lie beyond the proton drip-line (see also section 3.3).

The fundamentally simple nuclear disintegration mode of ground-state or isomeric proton decay ("proton radioactivity") yields, if only the lifetime and the proton energy are measured, already nuclear-structure information on, e.g., the proton separation energy and the spectroscopic factor for proton emission. So far, the attempts to observe proton or two-proton radioactivity for light ($A < 109$) nuclei failed, and $^{53m}$Co is thus still the only known case of direct proton decay in this region. As far as the heavier proton-radioactivity nuclei between $^{109}$I and $^{151}$Lu are concerned, qualitative agreement was generally found between measured proton decay half-life and single-particle estimates based on a spherical shell model; for the two lightest nuclei within this series $^{109}$I and $^{113}$Cs, however, the measured proton decay half-lives is two orders of magnitude longer than such theoretical estimates. Bugrov and Kadmensky (1991) have recently performed calculations within the framework of a deformed shell model that yield better agreement with experiments also for $^{109}$I and $^{113}$Cs, assuming Nilsson configurations $1/2^+[420]$ and $3/2^+[421]$ with deformation parameters $\beta_2$ of 0.05 to 0.10 and 0.10 to 0.15, respectively. A rather puzzling result, namely the tentative assignment of an alpha line to the decay of the presumably proton-unbound nucleus $^{108}$I, claimed by Woods et al (1990), awaits further confirmation.

It is interesting to note that the masses of some heavier nuclei close to the proton drip-line are rather accurately known now from experiment: Combining measurements of $Q_{EC}$ values for $^{104}$In (Hagberg et al 1990, Reusen et al 1990) and $^{104}$Sn (see chapter 3.2) with alpha and proton decay data, the masses of the nuclei $^{109}$I and $^{113}$Cs , situated beyond the proton drip line, were determined with 140 keV accuracy. More stringent tests of mass formulae can thus be performed for this region. Similar "decay-link" considerations, performed from the $^{146}$Gd region on towards heavier nuclei, yield detailed information on the structure of the proton drip-line in the rhenium-to-tantalum region (Keller 1991c).

3.2 Beta decay of very proton-rich nuclei

Beta decay of very proton-rich ($T_Z<0$) nuclei is governed by Fermi ($\Delta T=\Delta I=0$; $\Delta\pi=$ no) and Gamow-Teller ($\Delta T=\Delta I=0,1$; $\Delta\pi=$ no) transitions. The isobaric analogue state (IAS) in the daughter nucleus is strongly fed by Fermi decay, and can thus be identified rather easily by measuring the energy spectra of beta-delayed charged particles (see Fig. 1). So far, beta-delayed emission of protons, two-protons (or $^{2}He$) and alpha particles has been detected, but in addition beta-delayed $^{3}He$ emission would be observable close to the proton drip line: Energy windows of 3 MeV or more for this process can be found in a series of light nuclei between $^{22}Si$ ($T_z=-3$) and $^{43}Cr$ ($T_z=-5/2$).

The interest for investigating Fermi β decay close to the proton drip-line can be seen from the recent LISE study (Pougheon et al 1989) of the $^{28}S$ ($T_z=-2$) decay (see Fig. 2) that yielded the mass of the IAS in $^{28}P$ and completed thereby the A=28, T=2 isospin quintet. A fit of the isobaric multiplet mass equation (IMME) to these experimental masses sets upper limits of 7 keV and 4 keV to the third and fourth order IMME term, respectively. This confirms the insignificance of charge-dependent mix-

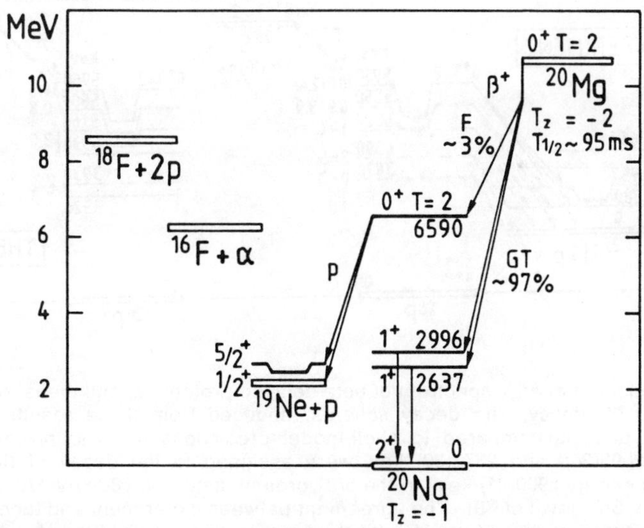


**Fig. 1:** Fermi (F) and Gamow-Teller (GT) β decay of $^{20}Mg$. So far, only beta-delayed protons (3.95 MeV, 4.16 MeV) were observed (Moltz et al 1979) from this decay, while beta-delayed γ-rays remain to be studied, e.g. in an experiment proposed for LISE. Out of the many excited $^{20}Na$ levels known from reaction work, the indicated $1^+$ state at 2637 keV is of special interest, since it may lead to an s-wave resonance in p+ $^{19}Ne$ reactions at stellar energies (see section 3.3). The theoretically predicted $1^+$ level at 2996 keV is also shown, wheras some known low-lying $^{20}Na$ states are omitted.

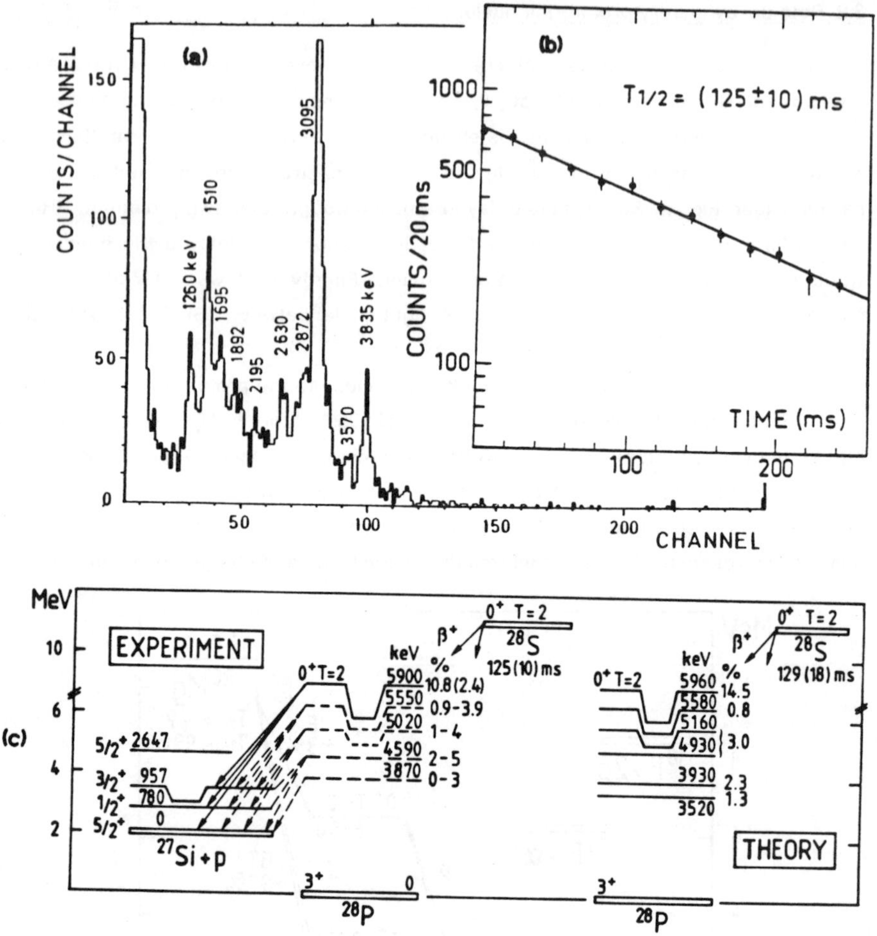


**Fig. 2:** Measured energy spectrum of beta-delayed protons (a) and measured half-life (b) for the $^{28}$S decay. The decay scheme, deduced from these results and from $^{28}$Si(p,n)$^{28}$P data, is compared to shell-model predictions (c). The proton lines at 3835(20), 3095(20), and 2872(30) keV were assigned to the decay of the $^{28}$P IAS (excitation energy 5900(21) keV) to the $5/2^+$ ground-state, the 780-keV $1/2^+$ level, and the 957-keV $3/2^+$ level of $^{27}$Si. The agreement between experiment and theory is good for the excitation energy of the IAS, for the beta-decay feeding of the IAS, and for the half-life of $^{28}$S. Satisfactory agreement is also found for the other beta-decay branching ratios, in spite of experimental ambiguities in assigning measured proton energies and determining excited-level energies. Disagreement exists, however, between measured and calculated proton widths for the isospin-forbidden IAS decay (Pougheon et al 1989).

ing, shown in corresponding IMME analyses at A = 20, 24, and 36. In view of the A = 8 discrepancy, it would be interesting to extend such considerations by measuring new masses of multiplet members near the proton drip line (and/or by determining known multiplet masses more accurately): Decay studies of, e.g., $^{44}$V, $^{44}$Cr, $^{46}$Mn, and $^{48}$Fe would allow to determine the IAS masses of $^{44}$Ti, $^{44}$V, $^{46}$Cr, and $^{48}$Mn , respectively. We also note that the complete knowledge of these and other isobaric mass multiplets requires accurate mass measurements for the lowest $T_z$ members of such multiplets, which might become feasible using time-of-flight methods at GANIL or at the ESR (see section 4).

In spite of the success of identifying the IAS from beta-delayed proton studies, performed for the decays of $^{28}$Si (Pougheon et al 1989), $^{27}$S and $^{31}$Ar (Borrel et al 1991), it is clear, that more refined decay measurements, including detection of beta-delayed γ-rays and p-γ coincidences, are mandatory for yielding the following results for the beta decay under consideration: (i) unambiguous identification of the (isospin forbidden) proton emission from the IAS (such ambiguities seriously hamper the comparison with shell-model predictions for the cases of $^{28}$S (see Fig. 2), $^{27}$S, and $^{31}$Ar (Borrel et al 1991)); (ii) unambiguous and/or more accurate determination of the IAS mass; (iii) Fermi β-decay branching ratios and corresponding reduced transition probabilities. The accurate measurement of Fermi transition probabilities is of fundamental interest, since it allows to determine the vector coupling constant $g_V$ for heavy nuclei (taking Gamow-Teller contributions as well as radiative and charge corrections into account) and to probe, e.g., a possible dependence of $g_V$ on the nuclear spin. At present, with the longstanding 3% discrepancy for $^{35}$Ar being removed by a recent experiment, all four $g_V$ values deduced for non $0^+ \rightarrow 0^+$ decays ($^{19}$Ne, $^{20}$Na, $^{24}$Al, $^{35}$Ar ) agree with the very precise result from $0^+ \rightarrow 0^+$ studies and with the $g_V$ value for the neutron decay (Clifford et al 1989, Hardy et al 1990). It is a challenge to reach the accuracies, required for $g_V$-relevant half-life and branching-ratio measurements, in decay studies at the FRS, and also to consider related mass measurements.

Detailed decay spectroscopy, discussed so far for the investigation of Fermi transitions, is also necessary for studying Gamow-Teller (GT) decays. For the latter case, the motivation is to test nuclear models by comparing calculated GT transition probabilities with values obtained from βp, βγ, pγ, and γγ measurements. Some recent examples of such studies will be discussed in the following.

For sd-shell nuclei, the measured GT strength is "quenched" globally by a factor of 0.58(2) compared to predictions from large-basis shell model calculations using the free-nucleon $g_A$ value (Brown and Wildenthal 1985); however, the most recent ISOLDE data for $^{32-35}$Ar show quenching factors between 0.66 and 1.19 based on the same model calculation (Borge et al 1989), indicating some deficiency of the shell-model at least in this "corner" of the sd shell.

In a decay measurement of the fp-shell nucleus $^{48}$Mn ($T_Z$ = -1) at the GSI On-Line Mass Separator, a more accurate halflife value of 158.1(22) ms was obtained, the $\gamma$ singles, $\gamma\gamma$ coincidence, and proton single data were improved, and proton-$\gamma$ coincidences were measured for the first time (Szerypo et al 1991). A comparison between the experimentally deduced GT strength distribution and predictions from shell-model calculations, shown in Fig. 3, yields GT quenching factors of 0.29 ± 0.02 and 0.48 ± 0.08 for the $^{48}$Cr excitation-energy intervals of 0-5.75 and 8.75-11.75 MeV, respectively. The preliminary conclusion is that these factors tend to be somewhat below the global quenching of 0.58(5) for the middle of the sd shell and also below the average quenching of 0.560(54) for the fp shell (Miyatake et al 1987).

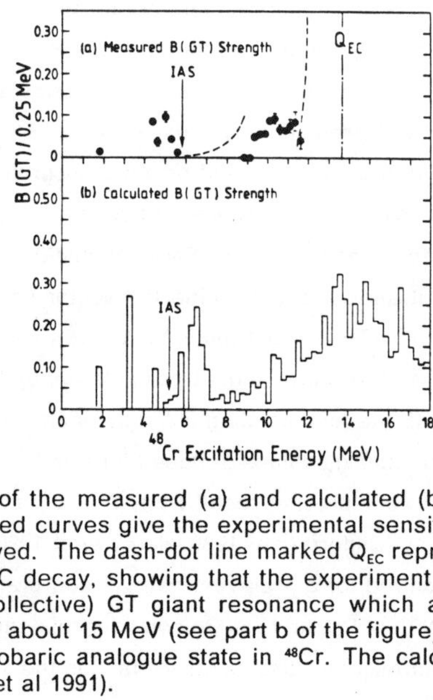


**Fig. 3:** Comparison of the measured (a) and calculated (b) B(GT) strength for the $^{48}$Mn decay. The dashed curves give the experimental sensitivity limits, below which decay was not observed. The dash-dot line marked $Q_{EC}$ represents the upper energy limit of the $^{48}$Mn $\beta^+$/EC decay, showing that the experiment can only probe the low-energy tail of the (collective) GT giant resonance which apparently is situated at excitation energies of about 15 MeV (see part b of the figure). The label IAS indicates the position of the isobaric analogue state in $^{48}$Cr. The calculation is based on 1ph excitations (Szerypo et al 1991).

For decay studies near $^{100}$Sn and $^{146}$Gd , a novel method has been applied to measure the endpoint energy of positron spectra, using a Si(Li)-BGO-Ge spectrometer at the GSI On-Line Mass Separator (Keller et al 1991a). Positrons were measured in a Si(Li) detector in coincidence with a characteristic $\gamma$ transition observed in a germanium detector; summing of positron annihilation radiation in the Si(Li) detector was suppressed by detecting both 511 keV quanta in opposite segments of a BGO ring (see Fig. 4). In order to deduce accurate endpoint energies and

$Q_{EC}$ values, the measured positron spectra are compared to theoretical ones, taking level-scheme details from $\gamma$-ray and conversion-electron spectroscopy into account. As an example, the summation-free $\beta^+$ spectrum of $^{102}$Cd, displayed as part (c) of Fig. 4, is compared in Fig. 5 to a curve obtained by folding the theoretical shape of a $\beta^+$ spectrum with the experimental response function. An endpoint energy of 1075(8) keV was obtained for this $\beta^+$ decay branch of $^{102}$Cd, corresponding to a $Q_{EC}$ value of 2587(8) keV. The remaining background above the endpoint, which amounts to only 0.3% of the total number of 25 000 events in the summation-free $^{102}$Cd spectrum, yields a negligible contribution to the uncertainty of the endpoint determination.

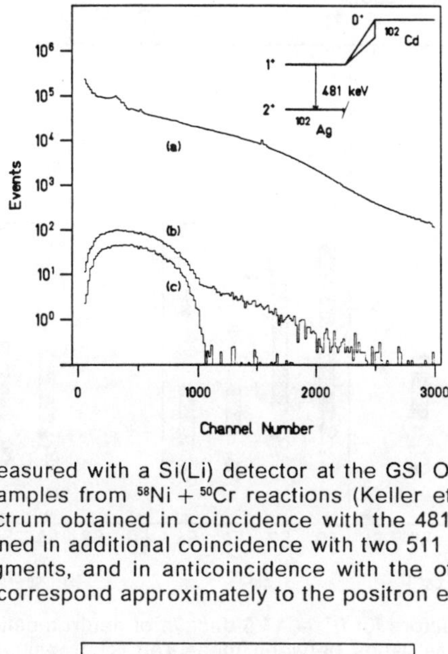


**Fig. 4:** Spectra measured with a Si(Li) detector at the GSI On-Line Mass Separator, using mass-102 samples from $^{58}$Ni + $^{50}$Cr reactions (Keller et al 1991b): (a) singles spectrum, (b) spectrum obtained in coincidence with the 481 keV transition in $^{102}$Ag, (c) spectrum obtained in additional coincidence with two 511 keV quanta, detected in opposite BGO segments, and in anticoincidence with the other six segments. The channel numbers correspond approximately to the positron energy in keV.

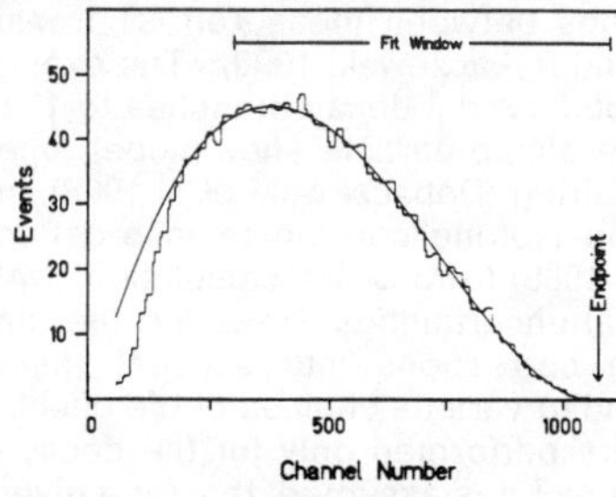


**Fig. 5:** The summation-free measured $\beta^+$ spectrum of $^{102}$Cd (histogram) compared to the fitted theoretical shape for a single-component $\beta^+$ spectrum (solid curve). The assumption of a single component is made on the basis of detailed $\gamma\gamma$ coincidence measurements. The fit extends over the whole energy-range of the spectra except for a low-energy threshold of 300 keV, which is related to the electronic discriminator setting for the Si(Li) detector (Keller et al 1991b).

Such accurate $Q_{EC}$ values, obtained for several nuclei near $^{100}$Sn and $^{146}$Gd, are important both as "decay links" (see section 3.1) and, together with measured beta halflives and beta branching ratios, for deducing the experimental GT strength. The resulting quenching factors for $(\pi g_{9/2})_{0^+} \rightarrow (\pi g_{9/2}\, \nu g_{7/2})_{1^+}$ GT transitions around $^{100}$Sn differ substantially from those for $(\pi h_{11/2})_{0^+} \rightarrow (\pi h_{11/2}\, \nu g_{9/2})_{1^+}$ GT transitions around $^{146}$Gd, if measurements are compared to the extreme single-particle model. Attempts have been made to improve this model by taking pairing correlation, core polarization and/or particle-particle interaction into account (see Fig. 6). A more detailed discussion of GT quenching near $^{100}$Sn and $^{146}$Gd was recently presented by Klepper and Rykaczewski (1990).

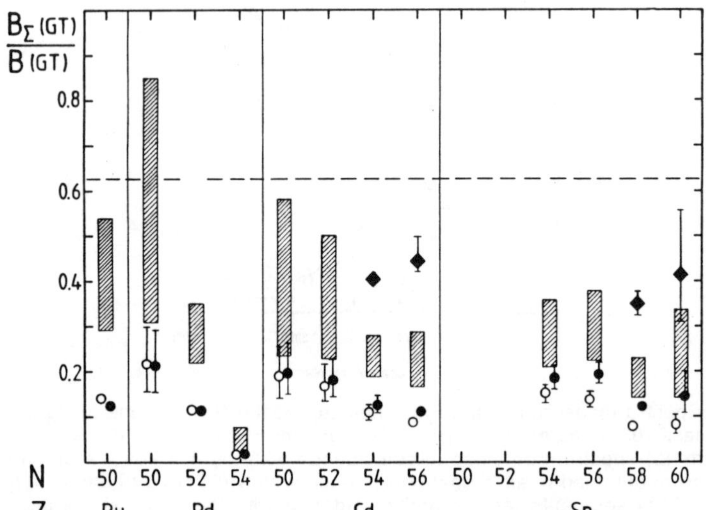


**Fig. 6:** GT quenching factors for $0^+ \rightarrow 1^+$ $\beta$ decays of neutron-deficient ruthenium-to-tin isotopes, obtained as ratios between measured GT transition probabilities and model predictions (Klepper and Rykaczewski 1990). The experimental values $B_\Sigma$ (GT) represent summing over all observed $\beta$ decay branches to $1^+$ states. The predictions B(GT) result from an extreme single-particle shell-model (open circles) and from a correction of this model for pairing (Dobaczewski et al 1988) (circles). The circles are drawn slightly aside from the isotone coordinate in order to avoid overlap. The squares present results (Cha 1983) from QRPA calculations. All error bars in the figure indicate only experimental uncertainties. However, the range for quenching factors taking core-polarization corrections into account (hatched bars) is due to experimental uncertainties and to various choices of the effective interaction (Towner 1985). These calculations were performed only for the decay of N = 50 isotones between $^{94}$Ru and $^{100}$Sn. In this plot it is assumed, that for a given proton number these factors stay constant independent upon neutron number. The horizontal dashed line indicates how much quenching one can compensate by renormalizing $|g_A/g_V|$ from its free-nucleon value 1.263 to a nuclear matter value (Towner 1985) of about one.

### 3.3. Break-out from the hot CNO cycle and termination of the rp process: An example for the astrophysical relevance of nuclear structure data far from stability

Studies of very proton-rich nuclides are also of astrophysical relevance. The examples of the $^{20}$Mg → $^{20}$Na decay and of the proton stability of nuclei near $^{65}$As may serve to prove this claim.

For the hydrogen-rich stellar sites with very high temperatures ($>3\times10^{8}$ K), a rapid proton (rp) process was suggested (Wallace and Woosley 1981) to give a chance for a leak (or break-out) from the hot CNO cycle and for subsequent heavy-element synthesis. The rp process, characterized by a flow path near the proton drip line, can produce considerably higher energies at early times than the β-limited CNO cycle. Nuclear structure information on excited states of $^{20}$Na is crucial for determining the break-out. Kubono et al (1988) succeeded recently in identifying a 2637(15)-keV, $1^{+}$ level in $^{20}$Na from ($^{3}$He,$^{3}$H) and (p,n) studies. This state can lead to a s-wave resonance of large amplitude in p + $^{19}$Ne reactions at stellar energies, which increases the stellar reaction rate of the $^{19}$Ne(p,γ)$^{20}$Na process by more than two orders of magnitude compared to previous theoretical predictions (Kubono et al 1989). This interesting hypothesis could be verified by studies of the $^{20}$Mg decay (see Fig. 1). By measuring beta-delayed γ-rays, $^{20}$Na levels can be identified with much better energy resolution than in reaction experiments, and unambigious spin/parity information can be obtained: The 2637-keV, $1^{+}$ level in $^{20}$Na is expected to be fed by a GT transition with a log ft value of the order of 4.5, corresponding to a β-decay branching ratio of 10%; other $1^{+}$ levels above the p + $^{19}$Ne threshold, such as the theoretically predicted one at 2996 keV, are expected to be also strongly fed in GT decay. The $^{20}$Mg rate, available at LISE from $^{36}$Ar fragmentation, is estimated to be of the order of $10^{4}$ $s^{-1}$ , which should suffice for improved investigation of these $1^{+}$ levels in $^{20}$Na.

The rp process can lead to production of heavy elements up to and beyond the iron group, and was applied in attempts to explain the observation of neon and heavier elements in novae (see Kubono et al 1989). This process is expected to be terminated at $^{64}$Ge if $^{65}$As is p unbound. In this context, the recent identification of $^{65}$As and $^{69}$Br as being relatively long-lived and of $^{73}$Rb as being most probably proton-unbound (see Fig. 7) means that the rp process is significantly slowed down around $^{72}$Kr (Mohar et al 1990). More quantitative conclusions can only be made if experimental halflives and proton binding-energies (eventually from observation of proton radioactivity) in this region become available.

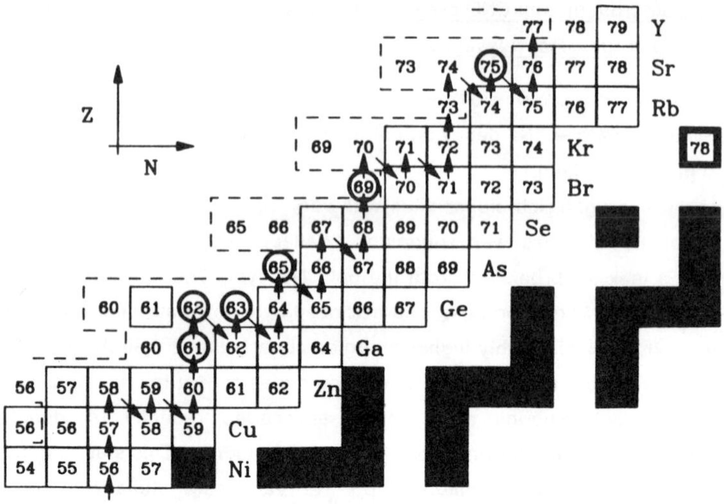

**Fig. 7:** Section of the chart of the nuclides around $^{65}$As. Stable nuclei are indicated by filled squares and the projectile $^{78}$Kr is specially noted. Open squares indicate nuclei that have been previously identified, and those circled were identified at MSU. Nuclei to the left of the dashed line are predicted to be unstable by the mass model of Jänecke and Masson (1988). The arrows indicate a possible flow for the rp process (Mohar et al 1990).

## 4. First Experiments with the Projectile-Fragment Separator at GSI

The projectile-fragment separator FRS (Geissel et al 1987) is an energy-loss achromat similar to LISE and is installed at GSI in the beam transfer hall between the heavy-ion synchrotron SIS and the experimental storage ring ESR (see Fig. 8). The instrument has been completed in October last year. Using SIS beams between neon and gold with energies up to 1.65 GeV/u, various detector tests and ion-optical measurements have been performed (Geissel and Sümmerer 1991). The results show good agreement with performance expectations. Furthermore, first experiments have been carried out, concerning (i) ionic charge distributions of 600 MeV/u $^{197}$Au (Stöhlker and Geissel 1990), (ii) separation of fragments from 800 MeV/u $^{136}$Xe induced reactions and determination of the production cross sections, (iii) measurement of nuclear charge-changing cross sections for N=Z nuclei between $^{14}$N and $^{36}$Ar, which were produced by 820 MeV/u $^{40}$Au fragmentation. Only preliminary results are available from these experiments so far.

From the long list of experiments planned with secondary beams from the FRS, we cite a few here due to their relation with the topics of this report: (i) study of excited states of the doubly magic nucleus $^{56}$Ni by inverse (p,p′) reactions; (ii) Coulomb excitation of $^{132}$Sn; (iii) measurement of neutron-neutron correlations from the break-up of [11]Li; (iv) mass measurements in the ESR. The latter two experiments are waiting for the transfer line between FRS and ESR which will be installed soon. However, the experiment, that is most directly related to the physics questions raised in this report, is the one on fragmentation of $^{78}$Kr and $^{86}$Kr beams. A first beam-time is scheduled for the second half of April, and the FRS group is looking forward with optimism to finding copious production of nuclei far from stability - and to making experimental observations that include surprises.

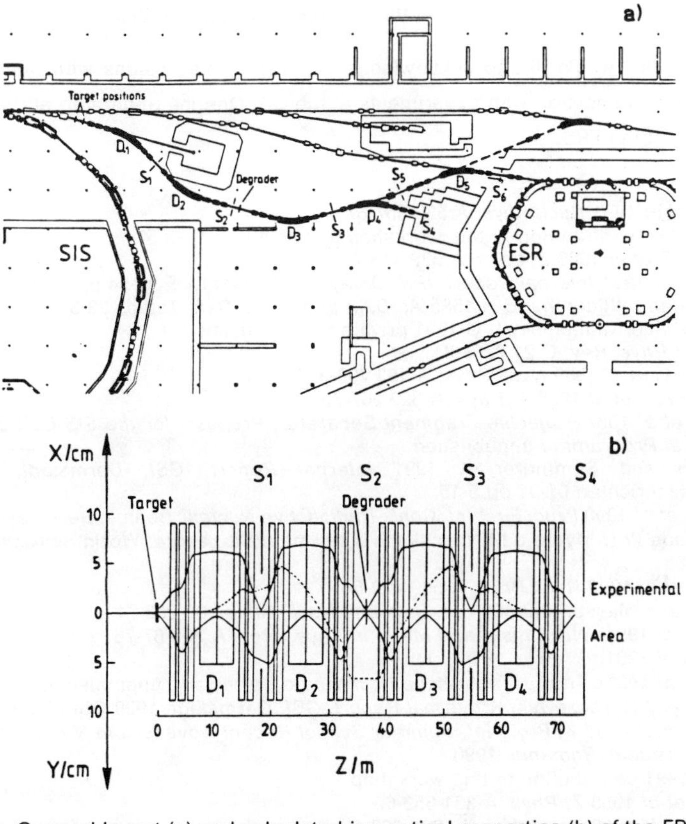


**Fig. 8:** General layout (a) and calculated ion-optical properties (b) of the FRS. The 30° dipole magnets and the slits at the different focal planes are indicated by $D_1$ to $D_5$ and $S_1$ to $S_6$, respectively. Calculated beam envelopes are shown for the target position characterized by an angular acceptance of 0.7 m sr. The dashed area, given together with the beam envelopes, represent the dispersion for $\Delta p/p = 1\%$.

## 5. Concluding Remarks

A good deal of our knowledge on nuclear structure far from stability has come from measurements that were perfomed with heavy-ion beams around Coulomb-barrier energies. While such experiments are being continued successfully, the past decade has already witnessed an increasing contribution from work done with intermediate-energy and relativistic heavy ions. It is hoped that the application of new tools, such as the application of the energy-loss achromat method and the storage ring at relativistic energies, will meet the present optimistic expectations.

The author would like to acknowledge interesting discussions with numerous colleagues in connection with experiments at the GSI On-Line Separator, at the LISE facility, and at the FRS.

Becker K *et al* 1991 *Nucl. Phys.* A 522 557-67
Bimbot R 1991 contribution to this workshop 51 527-40
Borge M J G *et al* 1989 *Z. Phys.* A 332 413-7
Borrel V *et al* 1991 *Internal Report, IPN, Orsay,* 1991 IPNO DRE 91-04 pp 1-11
Brown B A and Wildenthal B H 1985 *At. Data and Nucl. Data Tables* 33 347-404
Bugrov V P and Kadmensky S G 1991 private communication
Cha D 1983 *Phys. Rev.* C 27 2269-81
Clifford E T H *et al* 1989 *Nucl. Phys.* A 493 293-322
Dobaczewski J *et al* 1988 *Z. Phys.* A 329 267-73
Geissel H *et al* 1987 *Projectile Fragment Separator, Proposal for the SIS-ESR Experimental Programme* unpublished
Geissel H and Sümmerer K 1991 *Internal Report, GSI, Darmstadt,* 1991, GSI-Nachrichten 01-91 pp 5-15
Hagberg E *et al* 1990 *Proc. First Int. Conf. Radioactive Nuclear Beams, Berkeley, USA,* 1989 eds W D Myers J M Nitschke E B Norman (Singapore: World Scientific) pp 489-98
Hardy J C *et al* 1990 *Nucl. Phys.* A 509 429-60
Jänecke J and Masson P J 1988 *At. Data and Nucl. Tables* 39 265-71
Keller H *et al* 1991a *Nucl. Instr. and Meth. in Phys. Res.* A 300 67-76
Keller H *et al* 1991b *Z. Phys.* A in print
Keller H *et al* 1991c *Thesis, Technische Hochschule Darmstadt*, unpublished
Klepper O and Rykaczewski K *Internal Report, GSI, Darmstadt,* 1990 GSI-90-63 p 1-15 to be published in *Proc. Int. Summer School Recent Advances in Nuclear Structure, Predeal, Romania*, 1990
Kratz K L 1991 contribution to this workshop
Kubono S *et al* 1988 *Z. Phys.* A 331 359-60
Kubono S *et al* 1989 *Astrophys. J.* 301 629
Miyatake H *et al* 1987 *Nucl. Phys.* A 470 328-38
Mohar M F *et al* 1990 private communication
Moltz D *et al* 1979 *Phys. Rev. Lett.* 42 43-6
Patyn Z and Sobiczewski A 1990 *Internal Report, GSI, Darmstadt* 1990 GSI-90-53 pp 1-46
Pougheon F *et al* 1989 *Nucl. Phys.* A 500 287-300
Reusen G *et al* 1990 *Z. Phys.*A 336 381-6
Roeckl E 1991 *Rep. Prog. Phys.* to be published
Rykaczewski K *et al* 1989 *Nucl. Phys.* A 499 529-45
Saint-Laurent M G *et al* 1987 *Phys. Rev. Lett.* 59 33-5

Stöhlker T and Geissel H 1990 *Internal Report, GSI, Darmstadt,* 1990 GSI-Nachrichten 11-90 pp 3-7

Szerypo J *et al* 1991 *Nucl. Phys.* A in print

Towner I S 1985 *Nucl. Phys.*A 444 402-18

Wallace R and Woosley S E 1981 *Astrophys. J. Suppl.* 45 389

Woods P J *et al* 1990 *Proc. First Int. Conf. Radioactive Nuclear Beams, Berkeley, USA,* 1989 eds W D Myers J M Nitschke E B Norman (Singapore: World Scientific) pp 380-90

# SUPERDEFORMATION IN THE MASS 190 REGION

*C.W. Beausang*
*Oliver Lodge Laboratory*
*University of Liverpool*

## ABSTRACT

Over the past two years some fourteen nuclei with twenty eight superdeformed bands have been reported in the mass 190 region and have revealed a wealth of new information regarding nuclear properties at large deformations and high spin. Questions regarding identical bands, pseudospin-symmetry and octupole correlations in the second minimum are discussed. A review is given of the experimental status of superdeformation in the mass 190 region using the $^{193}$Hg nucleus as an example.

# INTRODUCTION
## A BRIEF HISTORY OF SUPERDEFORMATION

The experimental study of superdeformed states dates back almost thirty years to the discovery of the first fission isomers in the actinides by Polikanov in 1962 [1]. These isomers, with abnormally short half-lives against fission, were subsequently explained as prolate shape isomers, with an axis ratio of 2:1, existing in a second minimum in the nuclear potential energy surface [2].

The existence of the second minimum is caused by a combination of macroscopic (liquid drop) and microscopic (shell) effects. Put simply, for these heavy nuclei (large $Z$) a near balance between the Coulomb energy (which favours large deformations) and the surface energy (which favours spherical shapes) leads to a relatively flat potential energy surface with deformation. This relative flatness allows the possibility of shell effects stabilizing a second minimum in the potential energy surface for certain favoured particle numbers and at certain favoured deformations. In other words, deformed shell gaps add stability to the nucleus just as is the case for the more familiar spherical magic numbers.

That such deformed shell gaps exist is illustrated in Fig. 1 for a harmonic oscillator potential. For this relatively simple potential gaps occur at axis ratios of 2:1 and 3:1. Furthermore, the gaps appear at particle numbers which are related to the particle numbers of the spherical shell gaps. Thus, for the 2:1 axis ratio every second deformed shell-gap corresponds to twice the particle number of a spherical gap (for example $80 = 40 + 40$) while, for the remaining 2:1 gaps, the particle numbers correspond to the sum of the particle numbers in consecutive spherical gaps ($110 = 70 + 40$). This simple symmetry relationship has been pointed out by W. Nazarewicz [3] and has implications regarding the existence of octupole correlations in superdeformed states. Finally, each deformed shell gap is characterised by the occupation of a deformation-driving, high-$j$ intruder orbital originating from two oscillator shells above. It is the occupation of such deformation-driving, high-$j$ orbitals which is believed to be chiefly responsible for the final nuclear deformation in superdeformed bands and for the value and variation (with rotational frequency) in the nuclear moments of inertia. When a more realistic nuclear potential, such as a Woods-Saxon, is used the principal features discussed above remain valid although the precise details differ slightly.

It was soon realized that the centrifugal force, from nuclear rotation, acts to complement the Coulomb repulsion by favouring large deformations and so the hunt was on for high-spin (but lower $Z$) superdeformed bands. It took until the mid 1980's for the advances in detector arrays to catch up with the theoretical predictions. The first indications of high-spin superdeformation were observed by Nyako in 1984 in the $^{152}$Dy continuum spectrum [4]. Two years later discrete superdeformed bands were observed, almost simultaneously, by two groups working at Liverpool. The first, in $^{132}$Ce, was observed by P.J. Nolan's group [5] and the second in, $^{152}$Dy, by P.J. Twin's [6].

The $^{152}$Dy band, shown in Fig. 2, consists of $\sim$ 19 evenly spaced transitions extending in energy from around 650 to 1450 keV and in spin from around 26 to 60 $\hbar$. The spacing (the increase in $\gamma$-ray energy between successive transitions) is very constant along the band and is around 47 keV. Following on these initial discoveries many other superdeformed bands have been observed in three mass regions; the mass 130 region, centred around $^{132}$Ce, the mass 150 region, centred around $^{152}$Dy and most recently, the mass 190 region centred, around $^{192}$Hg.

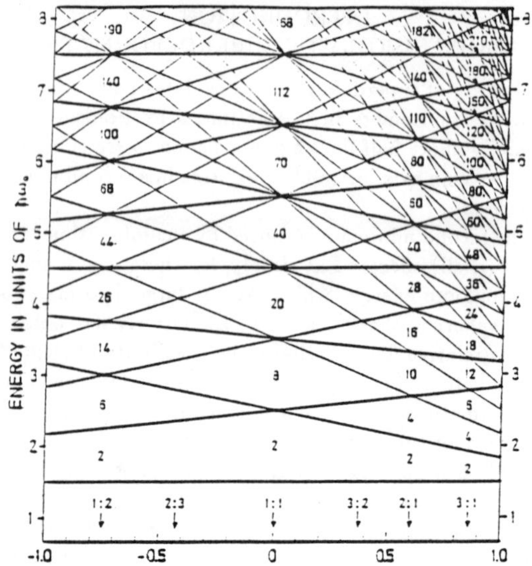


Figure 1: Single particle levels of the axially symmetric harmonic oscillator as a function of deformation. The shell gaps are labelled by their particle numbers.

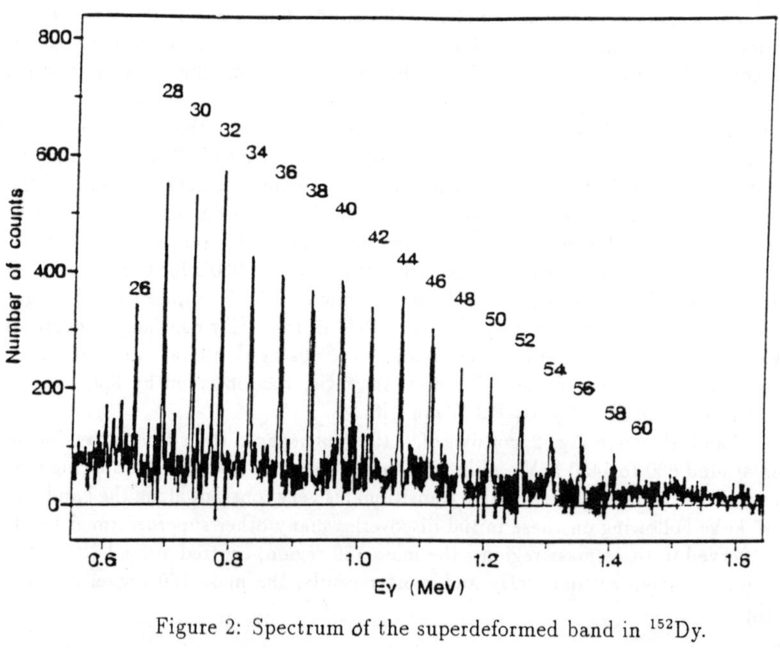


Figure 2: Spectrum of the superdeformed band in $^{152}$Dy.

A reasonable question to ask is how one knows that a band is superdeformed? The initial evidence comes from the band moment of inertia. Two moments of inertia are usually defined when discussing rotating nuclei, the kinematic moment of inertia $\mathcal{I}^{(1)} = I/\omega$ and the dynamic moment of inertia $\mathcal{I}^{(2)} = dI/d\omega \sim 4/\Delta E_\gamma$. For superdeformed bands the precise spin values are usually not known and, therefore, one generally relies on the $\mathcal{I}^{(2)}$ values. However, the moment of inertia, and particularly the $\mathcal{I}^{(2)}$ moment of inertia, depends not only on the nuclear deformation but is also very sensitive to changes in the nuclear pairing and to single particle alignments and, therefore, may not always be a reliable measure of the deformation.

A better measure is the nuclear quadrupole moment $Q_0$ which depends mainly on the deformation and shape. Under certain, fairly reasonable, assumptions $Q_0$ can be obtained from the collective $B(E2)$ of the states, which in turn is obtained from lifetime measurements. Such measurements on superdeformed bands, where they have been carried out, have confirmed the deformation inferred from the moment of inertia values. In particular, for $^{152}$Dy the lifetime measurement implies that $Q_0 \sim 18$ eb corresponding to a deformation of $\beta \sim 0.6$ or to an axis ratio of 2:1.

## SUPERDEFORMATION IN THE MASS 190 REGION
## INITIAL OBSERVATIONS

The first observation of a superdeformed band in the mass 190 region was in $^{191}$Hg by a group working at Argonne [7], thereby confirming many theoretical predictions [3, 11, 8] that a region of superdeformation should exist around $Z = 80$. The band consisted of twelve transitions spanning the energy range 350 to 750 keV. The energy spacing between adjacent $\gamma$-ray transitions was found to decrease from $\sim 40$ keV to $\sim 30$ keV with increasing excitation energy. The average energy spacing is 37 keV implying an average $\mathcal{I}^{(2)}$ moment of inertia of $\sim 110\hbar^2$/MeV and a deformation $\beta \sim 0.5$. A lifetime measurement, using the doppler shift attenuation method, yielded an average quadrupole moment of $Q_0 \sim 18$ eb for the band, confirming the large deformation inferred from the moment of inertia value.

The observation of a second superdeformed band in this region, in $^{192}$Hg, was announced almost immediately by the Livermore/Berkeley and by the Argonne groups working independently [9, 10]. The $\gamma$-ray spectrum of this band is plotted in Fig. 3. The $^{192}$Hg band, consisting of $\sim 17$ transitions, extends over a wider energy range than the $^{191}$Hg band, with $\gamma$-ray transition energies extending from 215 to 790 keV. However, the energy spacing varies in a very similar manner to the $^{191}$Hg band, decreasing from $\sim 40$ keV to $\sim 30$ keV with increasing $\gamma$-ray energy, with the average spacing again being around 37 keV. This similarity implies that the $\mathcal{I}^{(2)}$ moments of inertia ($\mathcal{I}^{(2)} = 4/\Delta E_\gamma$) for the two bands are also very similar. These are plotted in Fig. 4 and indeed the variation with rotational frequency is very similar, although the moment of inertia of the $^{191}$Hg band is consistently smaller than that for the $^{192}$Hg band.

The observation of superdeformation in $^{191}$Hg and $^{192}$Hg opened up a whole new region of superdeformed nuclei to explore and raised many new and exciting questions. For example,

- What is the extent of this new region of superdeformation?
- Why do the $\mathcal{I}^{(2)}$ moments of inertia show such a smooth increase with rotational

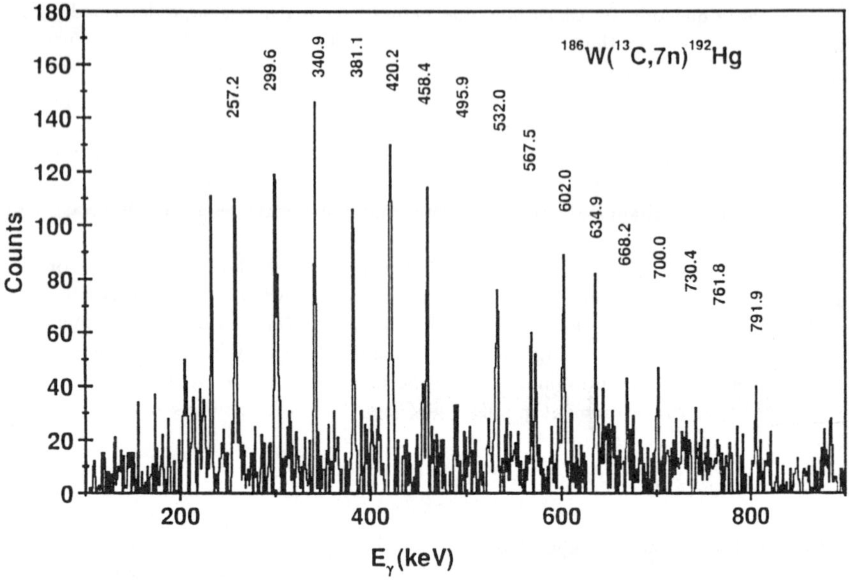


Figure 3: The superdeformed band in $^{192}$Hg populated following the $^{13}C+^{186}W \to ^{192}Hg+7n$ reaction.

frequency? Is the increase due to centrifugal stretching, to single particle alignments or to changes in pairing? Will other superdeformed nuclei in this mass region have moments of inertia which behave in a similar manner?

- How are the superdeformed bands populated and how do they de-excite into the normal-deformed states? The bands extend to very low frequencies , $\omega \sim 0.1 - 0.15$ MeV/$\hbar$ (compared to the mass 150 region where the bands terminate around $\omega \sim 0.3$ MeV/$\hbar$). Does this imply that the bands also extend to low spins? Can we determine the bandhead spins or possibly even observe the transitions out of the bands into the normal deformed yrast states?

Over the last two years, and particularly the first months following these initial discoveries, there has been an explosion of activity among the spectroscopy community with many new superdeformed bands being reported in the mass 190 region. Over this time the discovery rate of new bands has averaged around one band per month!

## THE CURRENT SITUATION

The present situation in the mass 190 region is summarized in Fig. 5. The contour lines are from a calculation by Chasman [11] and show the predicted superdeformed well depth at $I = 40\hbar$. The thin line shows the 3.5 MeV contour while the thick lines indicate the 2.0 MeV contour. The nuclei where superdeformed bands have been observed are indicated by the number of observed bands. The results summarized in Fig. 5 are due to the work of groups working at LBL/LLNL, Argonne, Liverpool, Bonn/Berlin and ORNL.

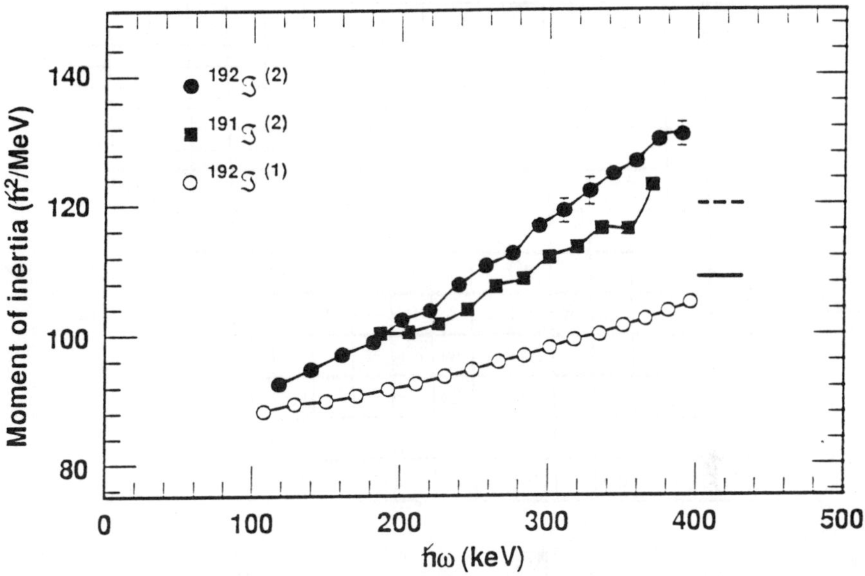


Figure 4: Variation in the moments of inertia of the superdeformed bands in $^{191}$Hg and $^{192}$Hg with rotational frequency.

Fourteen superdeformed nuclei have been found, to date, in the mass 190 region, with a total of twenty eight bands between them. In general, more than one band has been observed per nucleus, the record being six bands in $^{194}$Tl [12]. With one exception, $^{192}$Pb, the superdeformed nuclei all lie inside the 2.0 MeV contour calculated by Chasman. In addition, the more strongly populated bands lie closer to the 3.5 MeV well-depth contour.

The mass numbers of nuclei where unsuccessful searches for superdeformation have been made are indicated by italic numbers in Fig. 5. My knowledge of these nuclei is certainly incomplete, as generally one only hears about successful experiments! However, in general, these nuclei lie on or outside the border of the 2.0 MeV contour, that is outside the region predicted by Chasman to be favourable for superdeformation. The most notable exception is $^{195}$Hg. However, in this case, the angular momentum in the reaction was probably insufficient to populate the superdeformed structure, as will be discussed in more detail later. Production of heavier Hg, Au and Pt nuclei by conventional fusion-evaporation reactions is very difficult, or impossible, because of the lack of suitable (stable) beam-target combinations and may have to await the availability of an exotic beams facility.

I have also indicated, by a question mark, in Fig. 5 several Rn and Bi nuclei where we have recently carried out superdeformation searches in Liverpool. The competition from fission in the light Rn nuclei is very severe and further experiments there must wait for a more sensitive spectrometer such as Eurogam coupled to a mass selection device (recoil mass spectrometer). The analysis of the Bi data is currently under way, and indications of a superdeformed ridge-valley structure have been observed in the two-dimensional $E_\gamma \times E_\gamma$

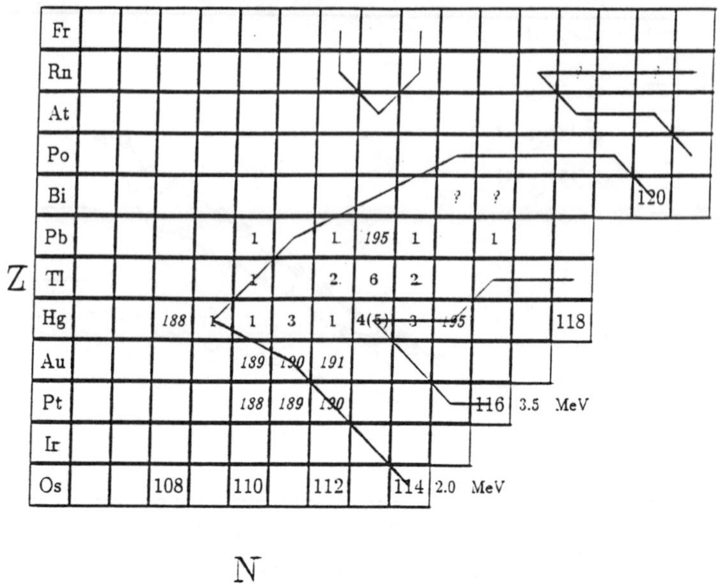


Figure 5: The superdeformed mass 190 region. The contour lines shown the calculated superdeformed well depth at spin $I = 40\hbar$ [11]. Nuclei where unsuccessful searches for superdeformation have been made are indicated by their (italicised) mass numbers.

matrix.

The superdeformed nuclei in the mass 190 region exhibit several interesting features.

- Single particle calculations indicate that $^{192}$Hg is a doubly magic superdeformed nucleus, having closed shells for both protons and neutrons which persist as a function of rotational frequency (see Fig. 6). Experimental evidence for this lies in the fact that only one superdeformed band has been observed in $^{192}$Hg. Both calculations [3] and lifetime measurements [13] for $^{192}$Hg indicate that the deformation is essentially constant along the band. The rise in the $\mathcal{I}^{(2)}$ moment of inertia is then due to a combination of alignment effects and to a reduction in pairing with increasing frequency. Single particle proton and neutron energies relevant for $^{192}$Hg are plotted in Fig. 6 as a function of rotational frequency.

- The bands generally extend to low frequencies, around 0.1-0.15 MeV/$\hbar$. This has allowed the LLNL/LBL group [14] to develop a method for determining the spins of the levels based on a fit to the $\mathcal{I}^{(2)}$ moment of inertia and a subsequent integration to give the spins (remember that $\mathcal{I}^{(2)} = dI/d\omega$). The problem with the method (and a cause of controversy) is what value to assume for the constant of integration (essentially the aligned angular momentum at zero frequency). Knowledge of the spins of the levels, together with the existence of many identical bands has lead to speculation regarding the observation of pseudospin symmetry in nuclei and of pseudospin alignment [16].

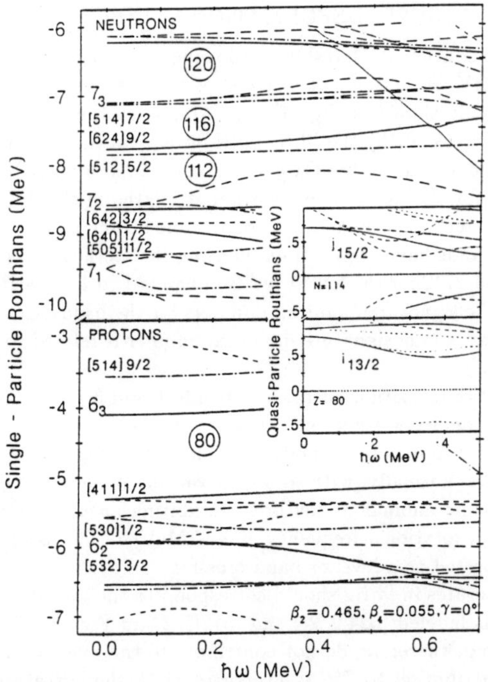


Figure 6: Calculated single particle proton and neutron energy levels for superdeformed $^{192}$Hg. Notice the shell gaps for $Z = 80$ and $N = 112$ which persist to high rotational frequencies

- With only two exceptions, the moments of inertia of all bands in the mass 190 region behave in a very similar fashion. In addition, and most surprisingly, several bands exist where the $\gamma$-ray transition energies are identical, within 1 keV to those in another nucleus. This close identity of rotational bands is one of the most surprising results in nuclear physics in recent times.
- Generally there are multiple superdeformed bands per nucleus
- A superdeformed band crossing has been found in $^{193}$Hg.
- Two superdeformed bands in $^{193}$Hg appear to 'see' each other. That is, gates placed on transitions in one band bring back transitions in both bands.
- These last few points have been taken as evidence for the occurrence of octupole correlations in superdeformed states.

The superdeformed bands in $^{193}$Hg illustrate most of the interesting features of the mass region and I shall use their properties to discuss in more detail some of the fascinating features of the superdeformed mass 190 region.

## SUPERDEFORMATION IN $^{193}$Hg

The nucleus was produced at Daresbury following the reaction $^{150}$Nd($^{48}$Ca, $5n$) at a beam energy of 213 MeV. A subsequent analysis of the data revealed the four (possibly five) superdeformed bands shown in Fig. 7 [17]. Almost at the same time, two of these bands were observed in Berkeley by the LLNL/LBL group via the $^{176}$Yb($^{22}$Ne, $5n$) reaction at $E_{beam} \sim 116$ MeV and also using the Ca plus Nd reaction at a lower beam energy of 205 MeV [18]. The reason for the different population strengths in the different reactions is, presumably, due to differences in the input angular momentum, as will be discussed later.

The spins in these bands have been estimated using the method described above, yielding 15/2, 21/2, 19/2 and 27/2 for the lowest levels in bands 1 to 4, respectively. From these estimated spin values the signature quantum numbers for the bands are -1/2 for bands 1,2 and 3 and +1/2 for band 4 (the signature quantum number is related to the time reversal symmetry of the system).

At low frequencies the transition energies in bands 1 and 3 are almost identical. The $\gamma$-ray energies deviate from each other only above $\sim 400$ keV.

The experimental $\mathcal{I}^{(2)}$ moments of inertia are plotted in Fig. 8 for the $^{193}$Hg bands. Two of the bands behave 'normally' with smoothly varying moments of inertia. The other two (bands 1 and 4) show anomalous features, exhibiting an upbend and a downbend, respectively, centred at a rotational frequency of 0.27 MeV/$\hbar$. This behaviour is obviously correlated and is interpreted as a level or band crossing.

Superdeformed structures in $^{193}$Hg should correspond to the valence neutron occupying the two high-$K$ orbitals labeled [624]9/2$^+$ and [512]5/2$^-$ in Fig. 6. These orbitals have a flat slope in Fig. 6 and, therefore, do not contribute to the dynamic moment of inertia (the single particle contribution to $\mathcal{I}^{(2)}$ is proportional to the curvature of the orbital). Inclusion of pairing in the calculation moves the $j_{15/2}$ high-j intruder orbital closer to the Fermi surface, as shown in the insert to Fig. 6. This orbital is affected by the nuclear rotation and is expected to contribute to the dynamic moment of inertia.

The interaction observed between band 1 and 4 is interpreted as a crossing between the favoured signature ($\alpha = -1/2$) of the negative parity $N = 5$ [512]5/2$^-$ orbital and the favoured signature ($\alpha = -1/2$) of the $N = 7$ $j_{15/2}$ orbital. Therefore, at low frequencies, band 1 is assigned the [512]5/2 ($\pi = -, \alpha = -1/2$) structure and band 4 the $j_{15/2}$ $(-, -1/2)$ structure, while at high frequencies ($\omega \geq 0.3$ MeV/$\hbar$) the characters of the bands are interchanged.

Band 3, which has the same assigned signature as band 1, is totally unaffected by the crossing and, therefore, must have the opposite parity. Thus, band 3 is assigned the [624]9/2 $(+, -1/2)$ configuration.

The signature splitting (energy difference between the bands based on opposite signatures of the same orbital) for both the high-$K$ [624]9/2 and [512]5/2 orbitals is expected to be very small, as can be seen in Fig. 6. In such situations one normally populates bands built on both signature components. On the other hand the signature splitting of the $j_{15/2}$ orbital is quite large and, hence, one expects to observe only the band based on the favoured signature component of this orbital.

Band 2, with $\alpha = +1/2$, and $\gamma$-ray energies lying almost exactly between those of band 3 is thus assigned as the signature partner band of band 3, i.e. the [624]9/2 $(+, +1/2)$ configuration.

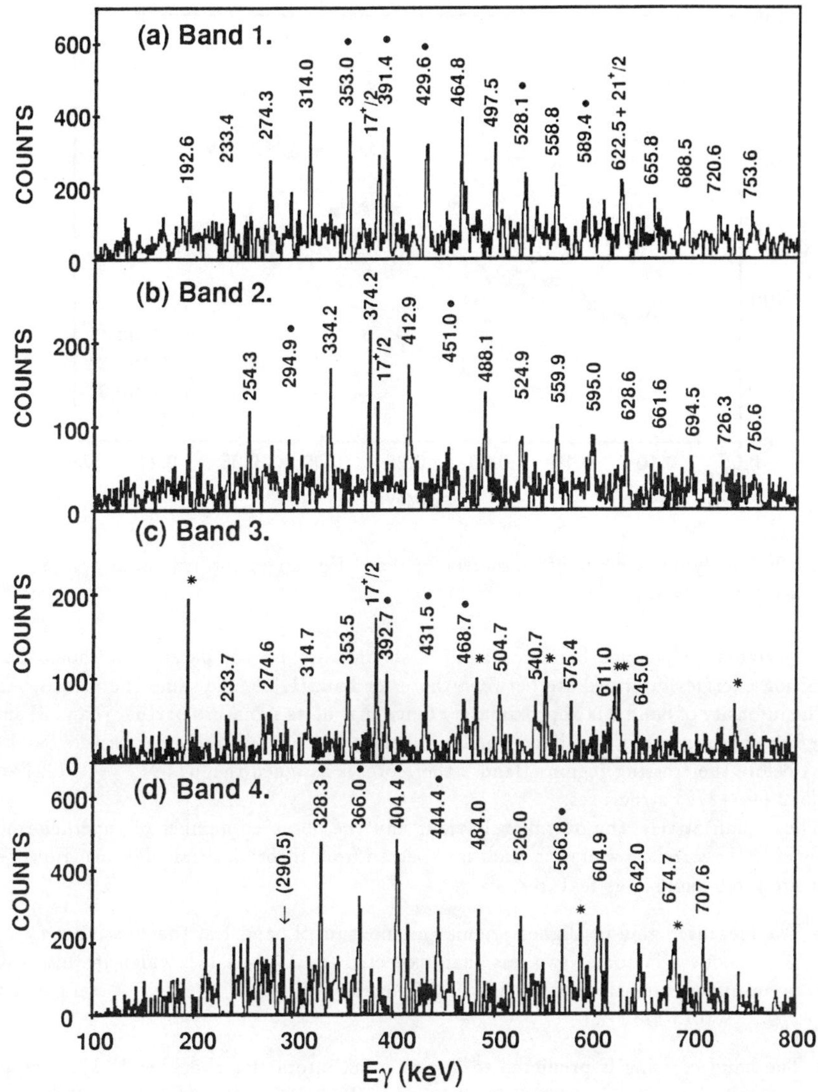


Figure 7: Spectra of the four (possibly five) superdeformed bands in $^{193}$Hg. Gating transitions are marked by a dot and normal deformation yrast transitions are labelled by their spins

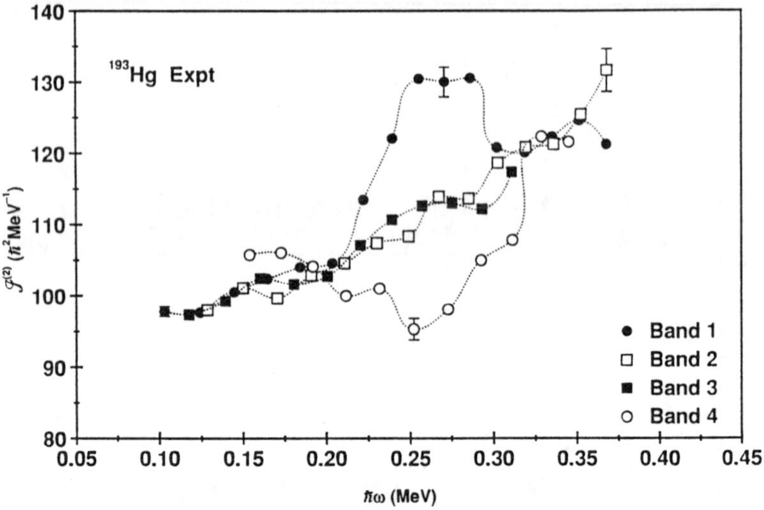


Figure 8: The dynamic moments of inertia for the $^{193}$Hg superdeformed bands as a function of rotational frequency.

The signature partner band for band 1 should also be populated and should have transition energies intermediate between those for band 1, i.e. the same energies as band 2. The intensity of band 2 is approximately twice that of its signature partner (band 3) and, therefore, it is suggested that band 2 may in fact be two 'identical' bands, as were bands 1 and 3 before the crossing region. Band 2 therefore, is assigned the [624]9/2 (+, +1/2) and [512]5/2 (−, +1/2) structures.

Thus, qualitatively, the overall behaviour and the observed number of superdeformed bands in $^{193}$Hg correlate well with what is expected from theoretical calculations. However, there are some anomalous features, viz.

- The measured gain in aligned angular momentum of band 1 at the crossing is $i_{expt} \sim 1.3\hbar$, about a factor of two less than expected ($i_{theory} \sim 3.0\hbar$), which in turn leads to an underestimation of the crossing frequency ($\omega_{expt} \sim 0.27$ MeV/$\hbar$ compared to $\omega_{theory} \sim 0.13$ MeV/$\hbar$).

- The band crossing is predicted to have a weak interaction (backbend) while experimentally a strong interaction is observed with an estimated interaction strength of $V_{int} \sim 26$ keV.

- The observation of cross-talk between two of the superdeformed bands is a very surprising and unexpected feature.

The cross-talk is illustrated in Fig. 9 which shows the low energy portion of the spectrum resulting when two gates are placed on transitions in band 1. As expected, other transitions in band 1 appear in the spectrum. However, low-energy $\gamma$-rays belonging to

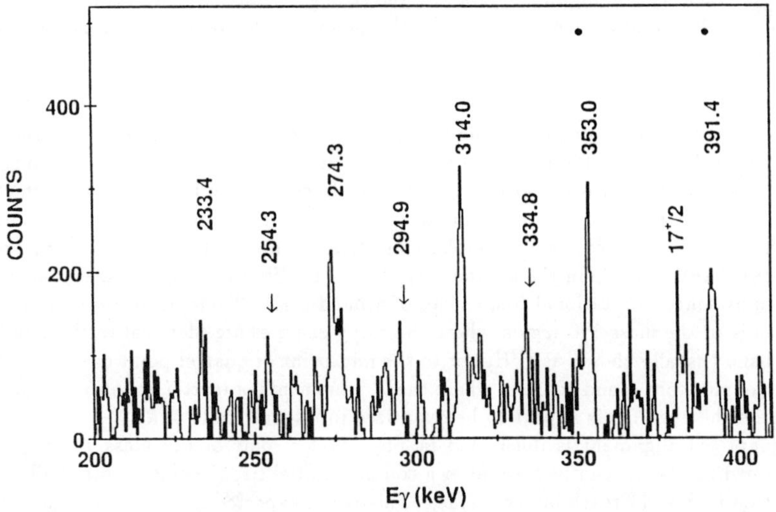


Figure 9: Spectrum illustrating the cross-talk between two superdeformed bands in $^{193}$Hg. The gates were set on the 353 and 391 keV transitions belonging to band 1, but the coincidence spectrum also shows transitions (254, 295 and 335 keV) belonging to band 2.

band 2 (254, 295, and 335 keV) also appear! This spectrum indicates that up to 30% of the intensity of band 1 crosses from the $\alpha = -1/2$ to the $\alpha = +1/2$ signature!

Unfortunately, the linking (dipole?) transitions were not observed. Since superdeformed bands of both parity exist in $^{193}$Hg these dipole transitions may be either magnetic or electric in nature. An estimate of the strength of the dipole transitions needed to compete with the in-band superdeformed E2 transitions gives about 2.0 Wu if M1 and 1.5 $\times 10^{-2}$ Wu if E1. The M1 strength seems much too strong and hence E1 transitions seem more plausible. Additionally, any energy separation between the superdeformed bands of opposite parity would increase the E1 transition energies and hence reduce the E1 strength required to compete. Other evidence for E1 transitions lies in the fact that the cross talk seems to go only one way, from band 1 to band 2 and not vice versa.

All of these unexpected and anomalous features could be explained by the existence of octupole correlations in the second minimum. Indeed calculations [3] show that the $^{192}$Hg core is very soft with respect to octupole deformation. That such correlations may be expected in superdeformed nuclei can be seen from a consideration of the particle numbers for deformed shell gaps in the harmonic oscillator potential (see Fig. 1). For superdeformed shapes, protons and neutrons fill adjacent shells ($\sim 80$ and $\sim 110$ for the mass 190 region). The particle numbers for these gaps are related to those for the spherical shell gaps [3]. Thus the protons, $Z \sim 80 = 40 + 40$, should generally favour reflection symmetric shapes, while for the neutrons $N \sim 110 = 40 + 70$, being the sum of two different 'size' spherical gaps favour reflection asymetric shapes, i.e. octupoles. Calculations based on realistic mean field potentials confirm this general trend, i.e. regions of particle numbers that favour shapes with $\beta_3 = 0$ and $\beta_3 \geq 0$ alternate [3]. The observation of cross-talk in $^{193}$Hg, together with the reduced alignment gain and the strong interaction at the band crossing is the first experimental evidence to support the prediction for octupole correlations in superdeformed nuclei.

## IDENTICAL BANDS

One of the outstanding problems in understanding the properties of the superdeformed mass 190 region (and mass 150 region) is the behaviour of the so called 'identical bands.' These are pairs of bands, or whole groups of bands, in different superdeformed nuclei which have transition energies identical to within an average of 1 keV – much closer than could possibly have been expected. For example, transitions in an excited band in $^{194}$Hg are within about 1 keV of those in the ground state band in $^{192}$Hg over a spin range of about 20 $\hbar$, an equivalence in rotational frequencies of around 0.1%. There are many examples of such bands in the mass 190 region where the $\gamma$-ray energies are identical to those in a reference band (usually chosen as $^{192}$Hg) or to the mid-point or quarter point energies. All of these examples correspond to cases where the $\mathcal{I}^{(2)}$ moments of inertia are also the same and to where the alignments appear to be quantized in integral units of $\hbar$.

The quantized alignment is illustrated in Fig. 11 for some of the mass 190 region superdeformed bands. As can be seen, after a common initial rise the alignments of all the bands plotted in Fig. 11 reach integer values. This is not expected, a priori, as alignment is not quantized. The initial rise is presumably due to the onset of alignment with rotation or to a decrease in the pairing correlations or both. In general, these results support the occurrence of pseudospin alignment.

It is well known that the nuclear shell structure is affected in a major way by the strong spin-orbit coupling. This modifies each major shell, characterised by the principal quantum number $N$, by the removal of the state with the largest $j = N + 1/2$ and by the addition of the corresponding state (of opposite parity) from the $N + 1$ shell above. These are the so called 'intruder orbitals' and, since they possess opposite parity, they do not mix with the remainder of the $N$-shell states.

One may consider that the remaining normal parity states form a pseudo $N-1$ shell and they may be relabelled using pseudo quantum numbers [15]. These states are mixed due to deformation effects and are also split due to the spin-orbit interaction and by deviations from the harmonic oscillator potential. By what appears to be a coincidence, these effects lead to the occurrence of close lying doublets of states with parallel and antiparallel spin-orbit coupling, e.g. $h_{9/2}$ and $f_{7/2}$, which can be relabeled as $\tilde{g}_{7/2}$ and $\tilde{g}_{9/2}$ in the pseudospin representation, see Fig. 10. These pseudospin doublets have a weak residual coupling of the pseudo intrinsic spin to the pseudo intrinsic orbital angular momentum so that the pseudo intrinsic spins can be more easily aligned by the Coriolis force (implying quantized alignments in $1/2\hbar$ units).

This fascinating proposal that the half integer and integer alignments, observed in these superdeformed bands is due to the alignments of the intrinsic pseudospins while the pseudo orbital angular momentum remains strongly coupled to the rotation axis has been put forward by the Berkeley group [16]. To confirm this proposal one needs additional experimental information plus a more certain knowledge of the spins of the states.

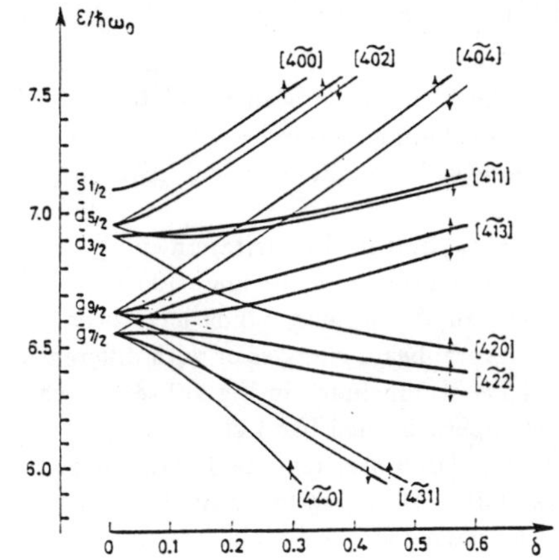


Figure 10: Single particle energy levels for the neutron shell $82 \leq N \leq 126$ labelled by their pseudospin quantum numbers. Note the occurrence of close lying doublets of states with parallel and antiparallel couplings of the pseudo intrinsic spin.

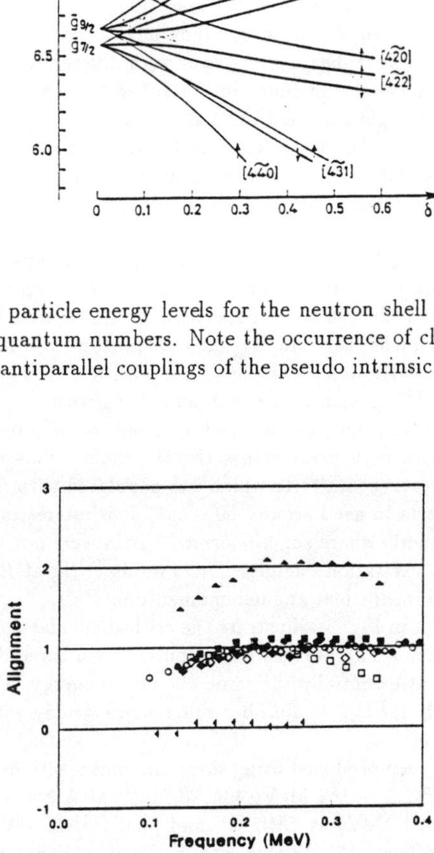


Figure 11: Total aligned angular momentum for eight bands in the mass 190 region as a function of rotational frequency. The labels are [nucleus (signature)]; □, $^{191}$Hg (+); ■, $^{191}$Hg (−); ◇, $^{193}$Hg (+); ◆, $^{193}$Hg (-); •, $^{194}$Hg (1); ○, $^{194}$Hg (0); ◁, $^{194}$Tl and △, $^{194}$Pb. The figure is taken from reference [16].

## PRODUCTION OF SUPERDEFORMED NUCLEI

At present the mass 190 region consists of $\sim 14$ nuclei with around 28 superdeformed bands between them. All of the bands have been populated following heavy-ion fusion-evaporation reactions. In this concluding section I want to briefly examine the systematics of the band production for trends which might act as guides in the production of additional superdeformed nuclei.

Figure 12 is a plot of the entry point in the excitation energy versus maximum angular momentum plane for superdeformed nuclei in the mass 190 region. The symbols indicate the locations where the known superdeformed bands have been populated. Where the same band was populated at several beam energies or with different reactions the symbols are connected with a line. I have also included in Fig. 12 several cases where experiments have been carried out but no superdeformed band observed.

The individual bands are populated over reasonably large regions in $(E^*_{Res}\ l_{max})$ space, typically over 5 – 10 MeV and 10 – 20 $\hbar$ (see for example, $^{194}$Hg in Fig. 12). Therefore, one should be cautious about drawing strong conclusions from Fig. 12, however, it is fun to speculate.

The first point to take from Fig. 12 is that, on average, all the nuclei were produced with approximately the same excitation energy, $E^*_{Res} \sim 26$ MeV. Secondly, there appears to be evidence for an odd-even effect in the angular momentum required for superdeformed band production in the Hg nuclei, the odd Hg nuclei requiring $10 - 15\hbar$ more $l_{max}$ than the even. For example, the $^{190}$Hg band was populated with $l_{max} \sim 41\hbar$, while $^{191}$Hg was populated with $l_{max} \sim 54\hbar$, $^{192}$Hg with $l_{max} \sim 45\hbar$ and $^{193}$Hg with $l_{max} \sim 60\hbar$ and finally $^{194}$Hg with $l_{max} \sim 50\hbar$. In addition, the angular momentum required to populate the superdeformed band increases with mass across the Hg chain. This is presumably the reason why no bands were observed in $^{195}$Hg which was populated with $l_{max}$ of only $\sim 36\hbar$, while from Fig. 12 one expects to need around $60 - 65\hbar$. It is interesting to note that the entry points for the experiments where superdeformed bands were not observed generally lie at much higher, or lower excitation energies, for example $^{193}$Hg at $E^*_{res} \sim 35$ MeV, or, as is the case for $^{195}$Hg with insufficient angular momentum.

The data for $^{193}$Hg, shown in Fig. 12 illustrate the critical angular momentum required to produced superdeformation. The nucleus was populated via three different reactions, each at a different $l_{max}$ but at essentially the same excitation energy. The superdeformed bands were observed strongly for $l_{max} \sim 60\hbar$, but much more weakly with $l_{max} \sim 46\hbar$ and hardly at all with $l_{max} \sim 43\hbar$.

The $^{192}$Hg band has also been produced using three different reactions, at three different laboratories, $^{36}$S+$^{160}$Gd at $E_{beam} \sim 162$ MeV $(43\hbar, 26$ MeV) at Argonne [10], $^{22}$Ne+$^{176}$Yb at $E_{beam} \sim 122$ MeV $(48\hbar, 23$ MeV) at LBL [9], and finally $^{13}$C+$^{186}$W at $E_{beam} \sim 105$ MeV $(46\hbar, 27$ MeV) at Daresbury [19]. In this case, however, each reaction populates the nucleus with essentially the same excitation energy and angular momentum. Although it is difficult to compare the absolute intensities of the band in the three reactions the relative intensities, plotted in Fig. 13 show that the feeding pattern for the superdeformed band is identical for the three reactions even though the projectile mass has changed by almost a factor of three.

Figure 12 also plots the entry points for the known Tl and Pb superdeformed nuclei. In this case the scatter on the points is greater than for the Hg nuclei but the average entry

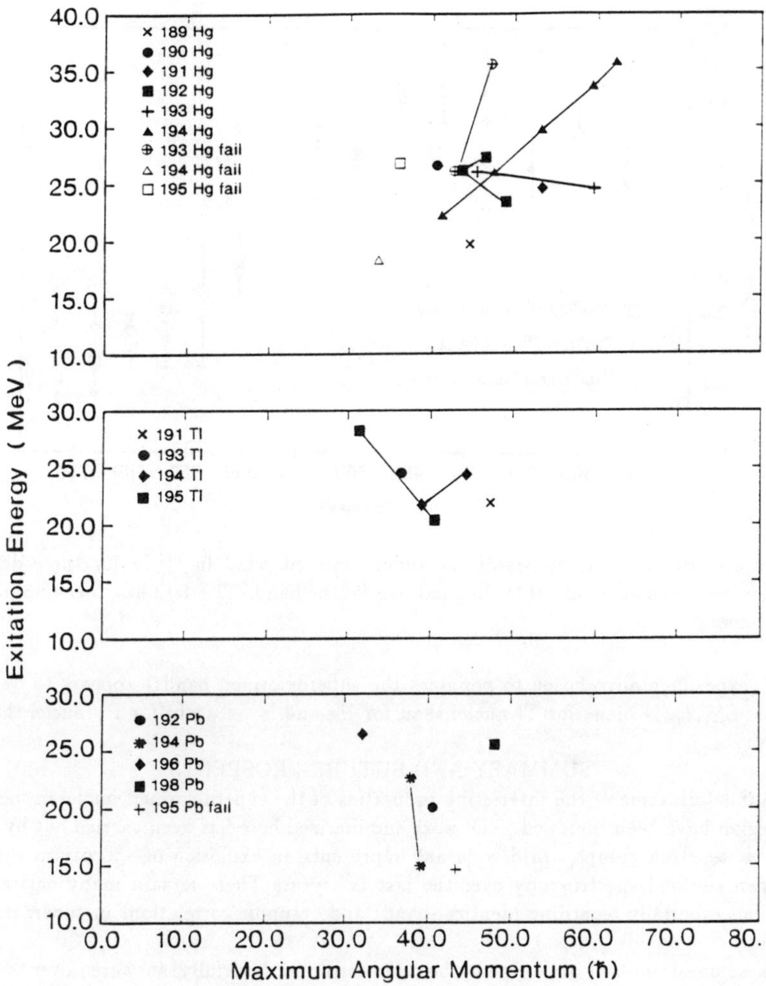


Figure 12: Estimated entry points for superdeformed bands in the mass 190 region plotted in the excitation energy angular momentum plane.

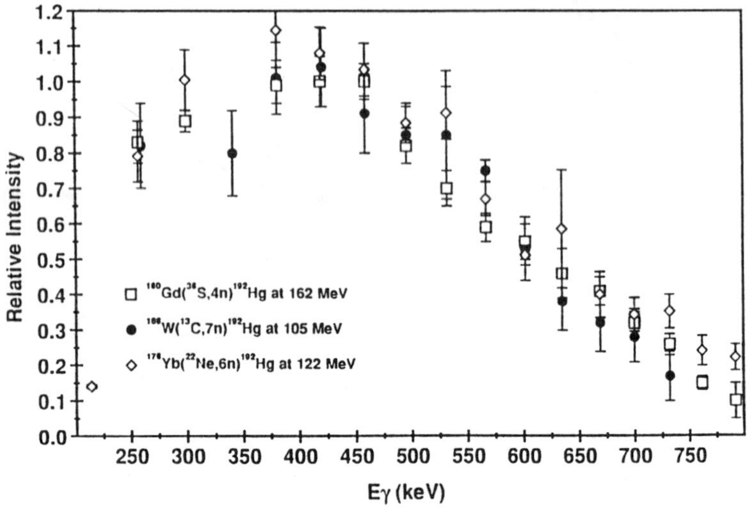


Figure 13: Relative intensities of the superdeformed band in $^{192}$Hg for three different reactions showing an identical feeding pattern for the band. The data have been normalised at low energies.

point (experimentally chosen to populate the superdeformed bands) appears to be lower in the $(E^*_{res}, l_{max})$ plane for Tl nuclei than for Hg and lower again for Pb nuclei than for Tl.

## SUMMARY AND FUTURE PROSPECTS

In this talk some of the interesting properties of the superdeformed nuclei in the mass 190 region have been reviewed. The work summarized here has been carried out by many different research groups world wide and represents an explosion of activity in the field of superdeformed spectroscopy over the last two years There remain many outstanding questions especially regarding identical bands and octupole correlations in superdeformed nuclei.

These questions will undoubtedly be addressed and, hopefully, answered over the next few years as experiment and theory continue to push back the frontiers of our knowledge of superdeformation. The next generation of $\gamma$-ray arrays, the Eurogam and Gammasphere arrays, should become operational over the next year and will give a huge increase in sensitivity, in efficiency and in selectivity over existing arrays. This increase in sensitivity should allow the observation of even more exotic nuclear shapes, for example the hyperdeformed 3:1 axis ratio shape.

The remaining question is what to do first with such powerful instruments? Two outstanding experiments in the mass 190 region, which could be addressed by the Eurogam array, are to search for linking transitions between; a) superdeformed and normal states, $^{192}$Hg is probably the best candidate to search for these transitions, and b) the linking transitions between the two superdeformed bands in $^{193}$Hg in order to establish the relative parities of the two communicating bands.

This presentation represents the work of many groups in many laboratories. Particular thanks are due to Dave Cullen in Liverpool. Helpful discussions with P. Fallon and E. Paul are greatfuly acknowledged. Finally, I would like to thank the organisers for inviting me to give this presentation and for their hospitality during my stay in Megeve.

# References

[1] S.M. Polikanov, *et al*, Zh. Eksp. Teor. Fiz. **42**, 1464, (1962).

[2] V. Strutinski, Nuc. Phys. **A95**, 420, (1967).

[3] W. Nazarewicz, private communication, and W. Satula, *et al*, submitted to Nuc. Phys. **A** (1990).

[4] B.M. Nyako, *et al*, Phys. Rev. Lett. **52**, 507 (1984).

[5] P.J. Nolan, *et al*, J. Phys. **G11**, 217 (1985).

[6] P.J. Twin, *et al*, Phys. Rev. Lett. **57**, 811, (1986).

[7] E.F. Moore, *et al*, Phys. Rev. Lett **63**, 360 (1989).

[8] S. Aberg, Phys. Scr. **25**, 23, (1982), and M. Cuilliau *et al* Phys. Lett **B46**, 11, (1973).

[9] J.A. Becker, *et al*, Phys. Rev. **C41**, R9, (1990).

[10] D. Ye, *et al*, Phys. Rev. **C41**, R13, (1990).

[11] R.R. Chasman, Phys. Lett. **B219**, 227, (1989).

[12] F. Azaiez, *et al*, Phys. Rev. Lett. **66**, 1030, (1991).

[13] E.F. Moore, *et al*, Phys Rev. Lett. **64**, 3127, (1990).

[14] J.E. Draper, *et al*, Phys Rev. C**42**, R1791, (1990), and J.A. Becker, *et al*, submitted to Nucl. Phys. (1990).

[15] A. Bohr, *et al*, Phys. Scripta, **26**, 267, (1982).

[16] F. S. Stephens, *et al*, Phys Rev. Lett. **64**, 2623, (1990), and F. S. Stephens, *et al*, Phys Rev. Lett. **65**, 301, (1990).

[17] D.M. Cullen, *et al*, Phys Rev. Lett. **65**, 1547, (1990).

[18] E.A. Henry *et al*, Z. Phys. **A335**, 361, (1990).

[19] D.M. Cullen, *et al*, Institute of Physics Conference, Liverpool, April 1991, and to be published.

# $\pi$ AND K MESON PRODUCTION BELOW THE NUCLEON-NUCLEON THRESHOLD.

G. ERICSSON and B. JAKOBSSON

*Div. of Cosmic and Subatomic Physics, University of Lund, Sölvegatan 14, S-223 62 Lund, Sweden*

Meson production in heavy ion reactions at an energy per nucleon below the free nucleon-nucleon threshold could be a signature for either cooperative effects or high momentum tails in the nuclei. The present experimental situation on $\pi$ meson emission is reviewed and possible production mechanisms are discussed. The advantage of measuring K mesons is explored as well as recent results and experimental programs in progress.

A meson(M) can be produced in any reaction A + B $\rightarrow$ M + .... when the available energy($E_{av}$) exceeds the meson mass(m). The relation between $E_{av}$ and the beam energy per nucleon($\epsilon_A$) is given by,

$$(E_{av} + \sum_f m_f)^2 = (\gamma_A m_A + \gamma_B m_B)^2 - (\gamma_A \beta_A m_A + \gamma_B \beta_B m_B)^2 \qquad (1)$$

where $\epsilon_A = (\gamma_A - 1)m_A$ and f stands for all final products. This implies,

$$E_{av} = -\sum_f m_f + \sqrt{m_A^2 + m_B^2 + 2m_A m_B[\sqrt{\gamma_A^2 - 1}\sqrt{\gamma_B^2 - 1} + \gamma_A\gamma_B]} \qquad (2)$$

This relation is plotted in fig. 1 for a number of meson producing scenarios in heavy ion collisions, namely:

1) A free nucleon-nucleon(NN) scattering; $m_A = m_B = m_f = m_N$, f=2 and $\beta_B$=0.

2) The most favourable doubly Fermi boosted NN collision (see fig. 3a). This means that $\beta_A = (\beta_{beam} + \beta_{Fermi})/\beta_{beam}\beta_{Fermi}$ and $\beta_B = \beta_{Fermi}$, where the Fermi velocity depends on which nucleus it relates to. In the presented curve (number 2 in fig. 1) $\beta_{Fermi} = 0.26$ for both nuclei ($p_F$=253 MeV/c).

3,4,5) The cooperative interaction between clusters (nuclei) of A = B = 4, 16, 40 nucleons, $m_A = m_B = Am_N$, f=2 and $\beta_B = 0$.

The available energy necessary for pion production($\sim$140 MeV), $\Delta$ production (297 MeV if $m_\Delta$=1236 MeV), $K^+$ production (669 MeV for NN $\rightarrow$N$\Lambda$K) and $K^-$ production (986 MeV for NN $\rightarrow$NN$K^+K^-$) is shown as the horizontal lines and the crossing points between these lines and the $E_{av}$ curves mark the production thresholds($\epsilon_{th}$).

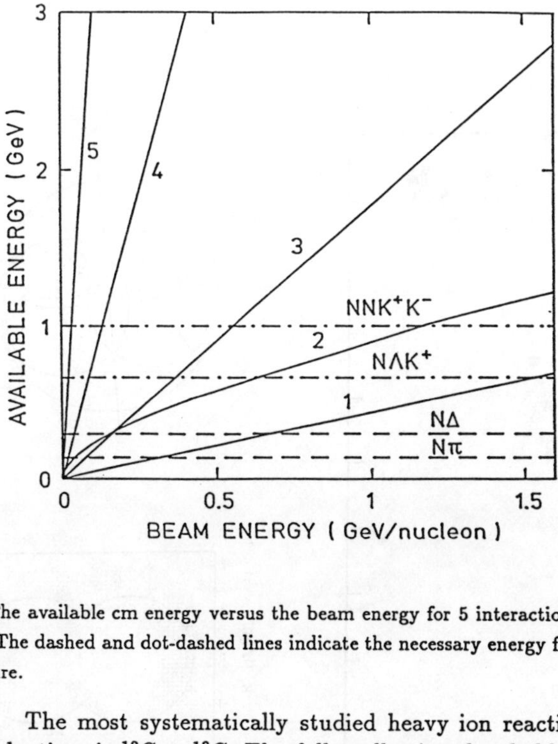


FIG. 1. The available cm energy versus the beam energy for 5 interaction scenarios described in the text. The dashed and dot-dashed lines indicate the necessary energy for the processes labelled in the figure.

The most systematically studied heavy ion reaction, with respect to pion production, is $^{12}C + ^{12}C$. The fully collective threshold is here $\sim$23A MeV. When the binding energies and the $\pi$ mass are properly introduced this threshold is even lower, $\sim$10A MeV for the $^{12}C + ^{12}C \rightarrow ^{24}Mg$(isospin,T=0) + $\pi^o$ reaction. The Fermi boosted threshold is $\sim$43A MeV and the free NN threshold as always $\sim$290A MeV.

Fig. 2 shows the total production cross section[1-6] from a number of experiments. Since the interacting system has T = 0 there should be no differences in $\sigma_{tot}$ for the three charge states of the pion. The conclusion is obvious, - cooperative processes or very high momentum tails are required to account for $\sigma_{tot}$ far below the NN threshold.

Unfortunately, many models, quite different in their spirit, do explain most of the $\sigma_{tot}(\epsilon_A)$ behaviour. Only in the lowest energy region can we find clear selectivity. Thus, an independent NN scattering model[7, 8], where nucleons are placed in sharp (temperature,$\tau$=0) Fermi momentum distributions, does not explain $\sigma_{tot}$ below ~100A MeV[3,9]. More extended momentum tails are needed, which means that one must recognize the problem of the off-shell kinematics, that makes the proper energy-momentum relation less favourable for pion production[10].

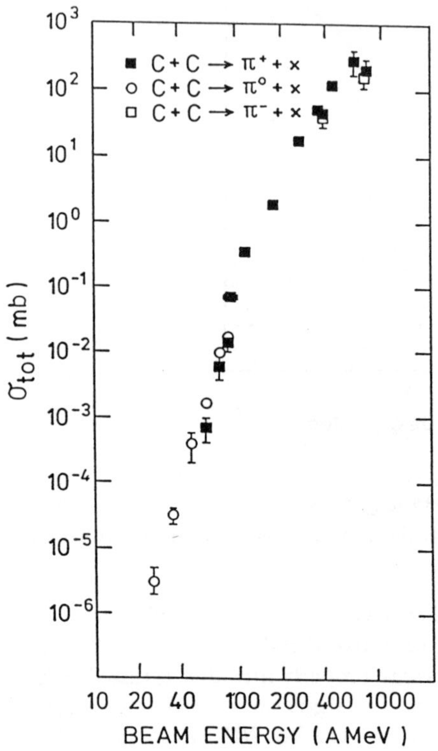


FIG.2. The total pion production cross section in C + C reactions[1-6].

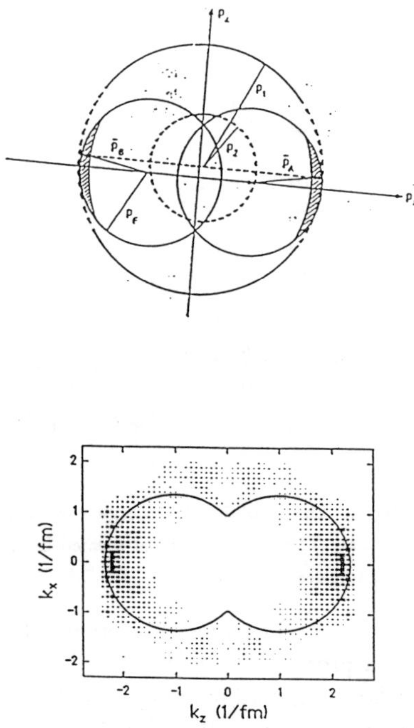


FIG.3. The phase space utilized for pion production in a sharp Fermi sphere and a VUU prescription.

A more complete description of NN interactions in the nuclear medium requires that both many-body dynamics and the elementary scattering process are considered. Based on the Vlasov-Uehling-Uhlenbeck(VUU) equation for the one-body phase-space distribution and a non-perturbative inclusion of NN scattering, Cassing and coworkers[11] find the energetic pions in a perturbative prescription. In this way the well defined part of phase-space(fig. 3a) corresponding to pion production above threshold 2 in fig. 1. is smoothed out and diffuse high momentum tails are introduced(fig. 3b). This favours the production of subthreshold pions and not only $\sigma_{tot}$ but also the spectral shape($d^2\sigma/d\Omega dE$) of pions even from extreme subthreshold reactions have been successfully described.

However also cooperative models, based on interactions between clusters of nucleons[10] or coherent bremsstrahlung models[12,13] can explain the inclusive data. The most extreme degree of collectivity is utilized in the pionic fusion description[14,15] where a thermally equilibrated emission source is introduced. This raises the question about pions as probes for hot and dense subvolumes of nuclear matter. In order to explore this possibility a systematic investigation of several parameters must be performed in the experiments.

The isospin conditions in the production region can be studied via measurements of the $\pi^o/\pi^-/\pi^+$ ratios in T$\neq$0 systems. If pions are produced solely in NN collisions these ratios should follow from the weights for the isospin decomposition branches of the NN $\rightarrow NN\pi$ channel:

$$\begin{aligned}
\sigma_{\pi^+} &= \frac{Z_A Z_B}{A_A A_B}[\sigma_{10}(d)+\sigma_{10}(pn)+\sigma_{11}] + \frac{1}{2}(\frac{N_A Z_B + Z_A N_B}{A_A A_B})[\sigma_{11}+\sigma_{01}] \\
\sigma_{\pi^o} &= \frac{Z_A Z_B + N_A N_B}{A_A A_B}\sigma_{11} + \frac{1}{2}(\frac{N_A Z_B + Z_A N_B}{A_A A_B})[\sigma_{10}(d)+\sigma_{10}(pn)+\sigma_{01}] \\
\sigma_{\pi^-} &= \frac{N_A N_B}{A_A A_B}[\sigma_{10}(d)+\sigma_{10}(pn)+\sigma_{11}] + \frac{1}{2}(\frac{N_A Z_B + Z_A N_B}{A_A A_B})[\sigma_{11}+\sigma_{01}]
\end{aligned} \tag{3}$$

A few experiments have reported on simultaneous measurements of $\pi^+$ and $\pi^-$. In [3], energy spectra at 90° were presented in the 30 - 70 MeV region for $^{12}$C induced reactions. The $^{12}$C + $^{12}$C spectra confirms a $\pi^-/\pi^+$ ratio of unity, but only after a proper Coulomb shift ($\sim 2 \cdot 4$ MeV) has been introduced. On the other hand, the overall $\pi^-/\pi^+$ ratio is found to be very high ($\sim$ 7) for the $^{12}C +^{208} Pb$ reaction after the inclusion of a similar (compound nucleus) Coulomb shift. At this beam energy the isospin weights are dominated by the first terms in (3) which therefore gives a ratio $\pi^-/\pi^+ \approx (A_B - Z_B)/Z_B = 1.54$ . Other reports [16] on very strong $\pi^-$ excesses have recently come from GANIL experiments.

It should be noticed that these experiments, as well as the subthreshold $\pi^o$ experiments, suffer from important efficiency corrections due to indirect pion registration methods and from extrapolations of the yields into non-covered parts of momentum-space. More systematic measurements of the $\pi^o/\pi^-/\pi^+$ ratios are thus highly recommended at the dedicated medium-energy heavy ion accelerators.

However, if we accept the strong $\pi^-$ excess as real, we must consider the possibility that the effective Coulomb source for pions is strongly reduced or that the strength of the reapsorption of $\pi^-$ and $\pi^+$ is very different. Without having performed serious calculations on these effects, we hardly believe they are strong enough to account for the $\pi^-$ excess. If so, the remaining possibility should be that the pions are produced in extremely neutron rich parts of the interacting system.

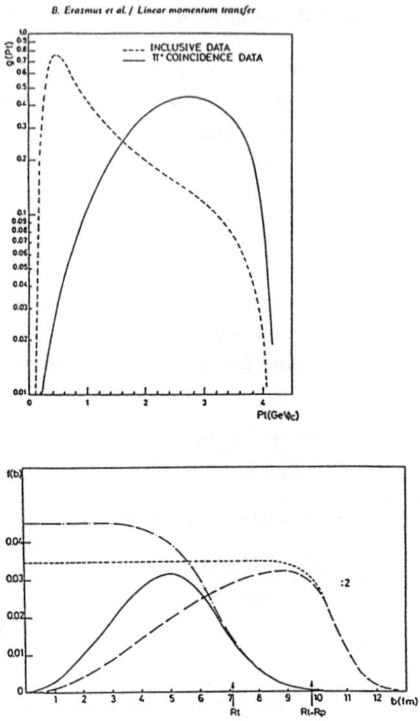


Fig.4 The linear momentum transfer(upper) and impact parameter distributions for inclusive(dotted) and pionic(solid resp. dot-dashed) 95A MeV O + Th reactions. Fission events are dashed(inclusive) or solid(pionic) in the lower figure[20].

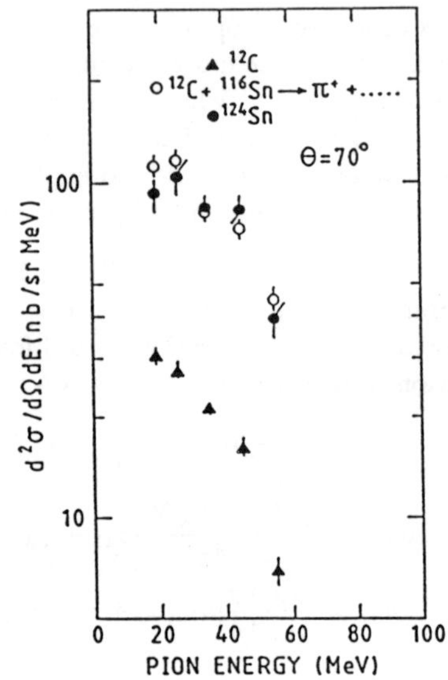


Fig.5 Doubly differential $\pi^+$ cross sections in 85A MeV reactions[17].

The strongest neutron excess is normally found in the surface region of a nucleus which means that either very peripheral reactions or a very early reaction stage must dominate the pion events. Some experiments, in the 100A MeV region, all agree about the conclusion that peripheral pion producing events are quite rare. This conclusion is drawn in ref. [17] from a strongly reduced abundance of heavy projectile-like fragments in pion events, in ref. [18] from the enhanced energy transfer and in ref [19] from the enhanced light particle multiplicity. Erazmus et al.,[20] determine the momentum transfer in pion events from a measurement of the folding angle between two fission-like fragments in the very fissile reaction $^{16}O+^{232}Th$ at 95A MeV. Fig. 4 summarizes their results. The upper part confirms the strong momentum (and energy) transfer. The lower part shows an attempt to construct impact parameter distributions based on the assumption that the momentum transfer is correctly described in VUU calculations. Again, it can be stated that the pion production events are fairly central, with a b-distribution that peaks at b$\simeq$ $1/2(R_A+R_B)$, but with a rather large width.

Thus, the possibility remains that pions are produced in the surface region in early phases of central collisions. In order to investigate this possibility and also to avoid effects from efficiency corrections and extrapolations of the differential cross section our last experiment at the CERN SC [17], before the shutdown of this accelerator, contained a measurement of the $\pi^-/\pi^+$ ratio in $^{12}C+^{116}Sn$ and $^{12}C+^{124}Sn$ reactions at 85A MeV. As a reference a $^{12}C+^{12}C$ measurement was also included. Fig. 5 shows an example of the $d\sigma/d\Omega dE$ distributions for $\Theta_{lab} = 70^o$. The $\pi^+$ results, based on the most efficient registration method, indicates that the yield is the same for both Sn targets(0.96±0.09 ). If indeed pions are produced in the neutron skin, one would expect an excess of $\pi^-$ in the $^{124}$Sn reaction. Such an effect was not observed. In fact $(\pi^-/\pi^+)_{124}/(\pi^-/\pi^+)_{116} = 1.2\pm0.3$ and this ratio is expected to be 1.12 from the NN isospin formulae (3). The prediction of the $\pi^-/\pi^+$ ratio from a straightforward fusion calculation [21],

$$R = \sigma_-^{abs}/\sigma_+^{abs} \cdot exp[-2(\mu_p - \mu_n)/\tau] \qquad (4)$$

where $\sigma^{abs}$ stands for absorption cross sections and $\mu$ for the chemical potentials, agrees better with the data.

A careful conclusion is that pion collisions below 100A MeV seem to be rather - but not perfectly - central. It seems likely that with decreasing energy the degree of centrality gets stronger. The question if pions come from the early phases of the reaction or, at least at the lowest energies, rather from typical fusion times is still not clarified. Another problem, which is related to this, is now discussed.

The question is how good pions actually are as probes for the conditions in the production region. This question arises naturally in view of the strong rescattering and reabsorption effects experienced by the pion on its passage through the nuclear medium. Both in $\pi^o$ [22] and in $\pi^+$[3] experiments it has been made clear, that the absorption plays an essential role for the degree of assymmetry and anisotropy in the final pion spectra. Fig. 6 gives one such example[22] where the deviation from symmetry in the assumed production system (here the NN system) implicates the importance of the reabsorption. In recent calculations[23] we have predicted the effects of the absorption processes for pions as well as for kaons. These calculations, which are based on energy dependent mean free paths from optical calculations and meson paths determined in Saxon-Woods density– and potential distributions, are still preliminary. The goal is to predict final emission spectra, once the production scenario is given. This initial distribution of the produced mesons is still more safely predicted for proton-nucleus collisions substantially above the collective threshold. Here, the production point in space can be assumed to follow the impact parameter weight (∼b) for the coordinates perpendicular to the incident direction and the NN mean free path for the depth into the target nucleus. In fig. 7 we show preliminary results for escape factors of $\pi^+$ and $K^+$ in p + $^{208}$Pb collisions. Obviously the total amount, as well as the asymmetry of the absorption is substantially smaller, but still not negligible, for kaons. Anyway, the kaon is much less affected than the pion on its way from the production point to the detector and this makes it an interesting probe.

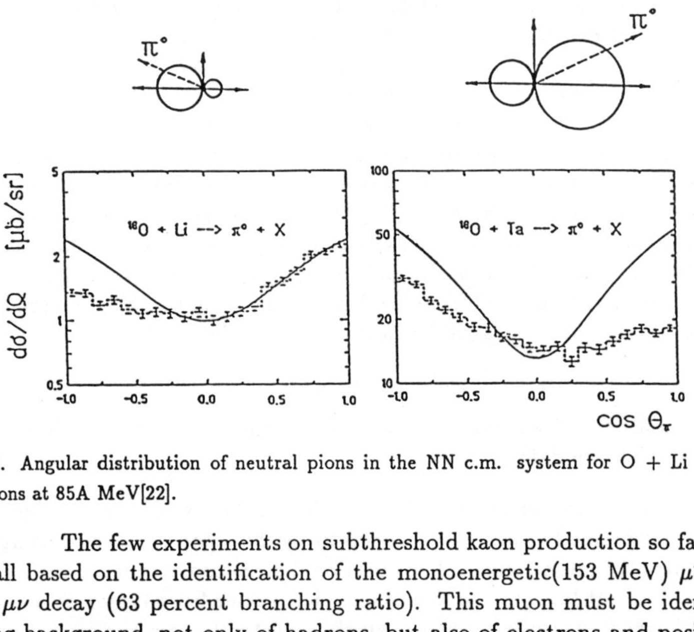


FIG.6. Angular distribution of neutral pions in the NN c.m. system for O + Li and O + Ta reactions at 85A MeV[22].

The few experiments on subthreshold kaon production so far reported, are all based on the identification of the monoenergetic(153 MeV) $\mu^+$ from the K→ $\mu\nu$ decay (63 percent branching ratio). This muon must be identified in a strong background, not only of hadrons, but also of electrons and positrons from pion decay - from the $\pi\mu e$ decay chain (very frequent but with a long decay time) as well as from the $\pi \rightarrow e\nu$ decay ($10^{-4}$ branching ratio of the pion decay) with a lifetime in the same region as the $K^+$ decay(26 ns as compared to 12 ns). This is a delicate technical task. We present here one detector(fig. 8), which we have just tested in beam at GSI(SIS). The $K^+$ is stopped in an active plastic range telescope which gives $\Delta E$ as well as time information. In addition, the decay muon is identified in one Cerenkov detector and two (x and y) arrays of scintillators, arranged in a set of shields surrounding the range telescope. In fact, the most recent tests indicate that four scintillator planes may be nessecary for a clean identification of the 153 MeV muon.

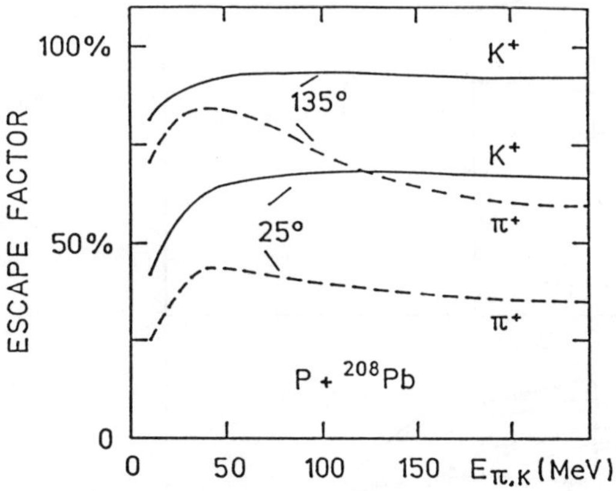


FIG.7. Fractions of escaping pions and kaons in p + Pb reactions[23].

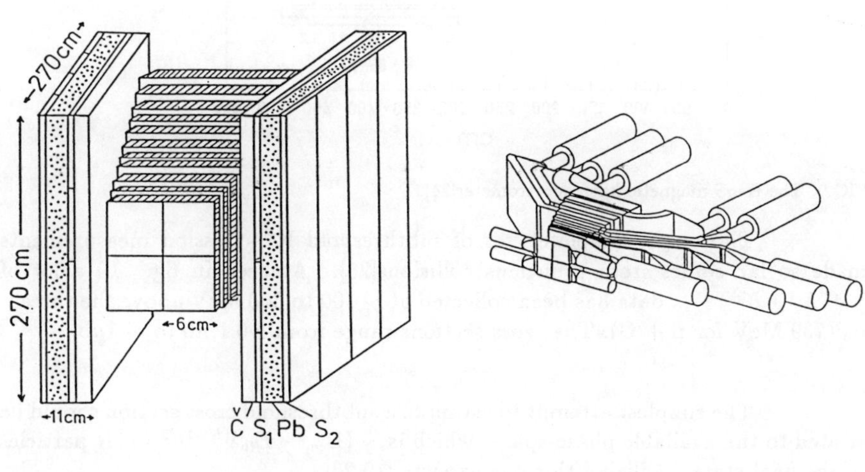


FIG.8. The first test version of the CHIC $K^+$ detector.

A number of such telescopes, covering a kaon energy range of 10 - 130 MeV, is planned to be set up as a complement to the double focusing QD spectrometer (fig. 9) which has been built by the KaoS group at GSI[24]. In its normal setting the spectrometer covers kaon energies above 200 MeV, with a possible lower field setting for $\sim$ 100 - 200 MeV kaons. The $K^+$ telescopes are of course particularly important to introduce for subthreshold experiments.

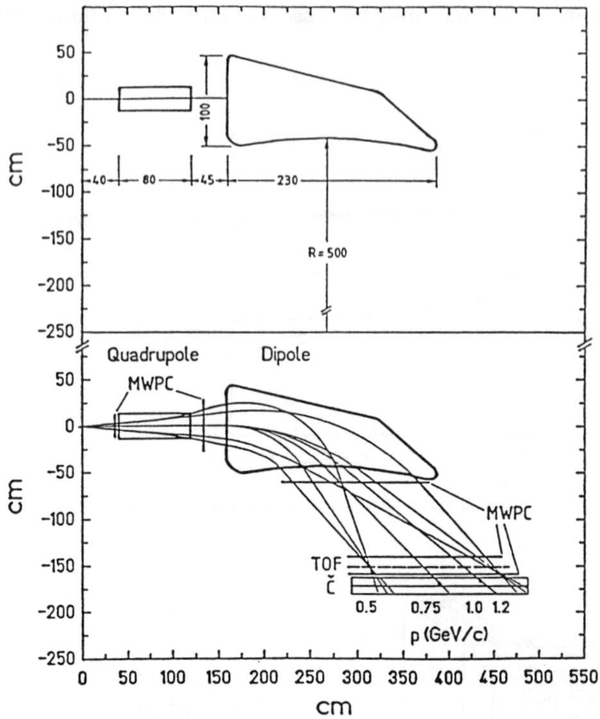


FIG.9. The KaoS magnetic kaon spectrometer[24].

The only "complete" set of subthreshold $K^+$ emission measurements made so far comes from p-nucleus collisions[25]. As seen in fig. 10 a lot of p + A $\rightarrow K^+ + ...$ data has been collected at $\sim$ 100 to 300 MeV above the threshold(739 MeV for p + C). The cross sections range from $\sim$0.1 nb to $\sim 1\mu$b.

The simplest attempt to estimate a subthreshold cross section should be related to the available phase-space, which is $\sim (E_{cm} - E_{th})^{(N-1)/2}$ for N particles in the final state. Utilizing this one expects [10,26],

$$\sigma^K_{pC}(E) = \alpha^K_{pC}\sqrt{E_{cm}}[E_{cm}/(m_\Lambda - m_p) - m_K]^2. \quad (5)$$

However, the assumption that the energy independent(threshold) parameter is $\alpha^K_{pC} = \frac{\alpha^\pi_{pC}}{\alpha^\pi_{pp}}\alpha^K_{pp}$ ,where the right side parameters are empirically determined, gives much too large cross sections as compared to data. Therefore, the proper $\alpha^K_{pC}$ parameter must be related to the momentum transfer(q) in the process via the form factor, F(q). In fact $\sigma^K_{pC} \sim \mid F(q) \mid^2$ is the natural approach, which we will come back to at the end of this talk.

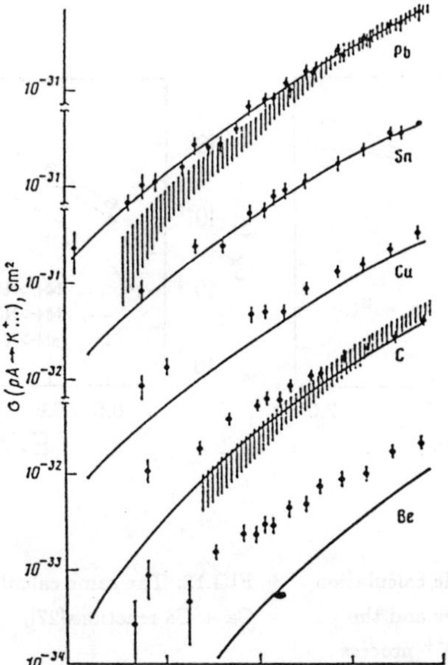


FIG.10. K+ production cross section in p-nucleus reactions[25].

Detailed calculations of the pN→ $K^+\Lambda$N process, where the NN collisions benefit from extended tails of the internal momenta, have been performed within the VUU prescription[27] as well as within a cascade calculation using Hartree-Fock groundstate momentum distributions and the optical (Glauber) approach to calculate the number of NN collisions[28]. These calculations underestimate the K+ production(fig. 11) which raises a very important question: "Is it actually the two-step NN→ $NN\pi, \pi N \to \Lambda K^+$ which dominates the K+ production?" From kinematics point of view the two-step mechanism is favoured since e.g. the Fermi boosted threshold (corresponding to curve 2 in fig. 1) is only 0.7 GeV as compared to ~ 1.1 GeV for the direct NN→ $N\Lambda K$ process. Fig. 11 shows the comparison between the predicted cross sections for the direct and the two-step process in p + C reactions. The data clearly favour the secondary production mechanism which dominates the scene up to ~1.5A GeV from where the direct mechanism takes over.

How do the two processes compare in the doubly boosted nucleus-nucleus collision? Batko et al.[27] present the VUU calculation also for $^{40}Ar+^{40}Ar$

collisions(fig. 12) which states the opposite result in the subthreshold region as compared to the p + C reactions. In the more dense surrounding the direct

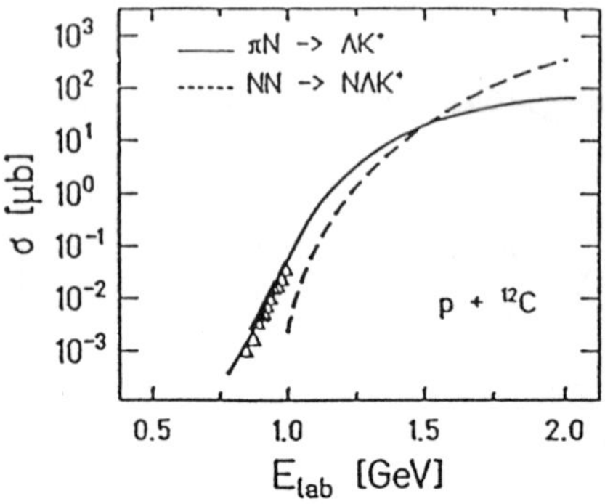


FIG.11. The VUU based cascade calculation of the direct NN→ $N\Lambda K$ process and the two-step NN→ $NN\pi, \pi N \rightarrow \Lambda K^+$ process in p + Creactions[27].

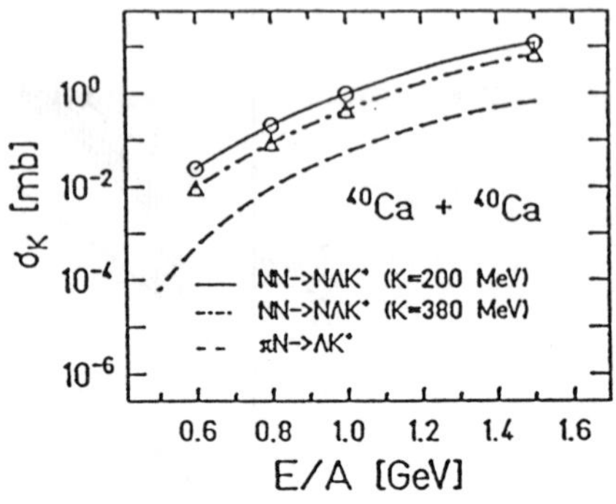


FIG.12. The same calculation as in fig. 11 for Ca + Ca reactions[27].

NN→NΛK channel is favoured both in production and from the enhanced pion absorption which reduces the secondary channel.

Let us now come to the essential question: "Can we determine the nuclear equation of state(eos) from K measurements?" The degree of sensitivity to the compressibility is not impressive(fig. 12) if we believe the calculations in ref. [27]. Stock et al.[29] once started the debate about the possibility to probe the nuclear eos by their observation of a substantially lower pion production in the Ar + KCl reaction at 1A GeV as compared to a cascade calculation(fig. 13). Their original interpretation of a strong energy loss to the compression channel had to be modified due to e.g. the great importance of the pion rescattering processes. There is now the hope that the kaon will prove to be a much better probe for the nuclear eos.

Finally we come back to the region very close to the K$^+$ threshold. In [26] a prediction is made for the cross section of K$^+$ production in nucleus-nucleus collisions. This is based on (5) with an estimate of the $\alpha_{CC}^{K}$ from the form factor, F(q). Even under the favourable condition that the momentum transfer is uniformly distributed among all nucleons in the nucleus, extremely low, in practise experimentally unavailable, cross sections are obtained since $\alpha_{pp}^{K} \simeq 6 \cdot 10^{-30} mbGeV^{-3/2}$. From a preliminary data analysis of $K^+$ emission from

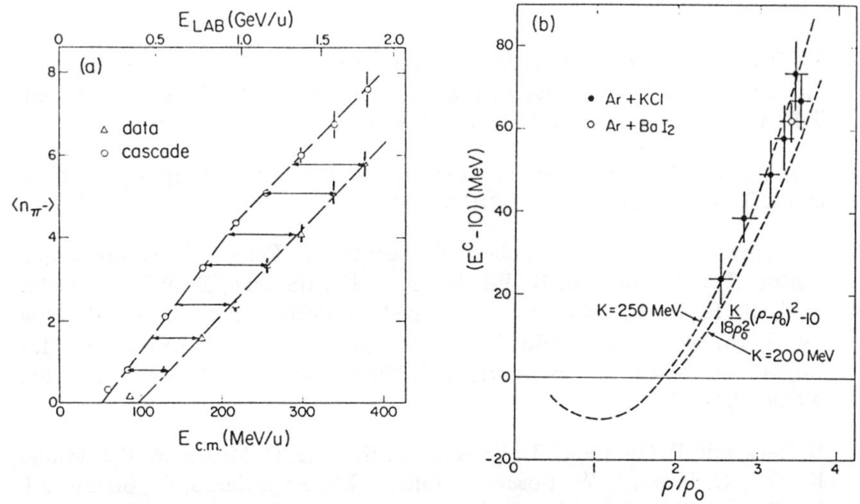


FIG.13. The mean $\pi_-$ multiplicity in Ar + KCl reactions as compared to a cascade calculation. The right side part shows the attemprt to derive information about the compressibility(K)[29].

80A MeV heavy ion collisions at GANIL[16] it is however reported measurable cross sections (10 -100 pb), though the statistics of identified 153 MeV muons is low (10 - 20 candidates). Full scale experiments along this line, which in fact are in progress, appear to be one of the most important tasks for heavy ion physics in the near future.

# Bibliography

[1] W. Benenson, G. Bertsch, G.M. Crawley, E. Kashy, J.A. Nolen, H. Bowman, J.C. Ingersoll, J.O. Rasmussen, J. Sullivan, M. Koike, M. Sasao, J. Peter and T.E. Ward, Phys. Rev. Lett. 43(1979)683 and J. Sullivan, Private comm.

[2] S. Nagamiya, M-C. Lemaire, C. Moeller, S. Schnetzer, G. Shapiro, H. Steiner and I. Tanihata, Phys. Rev. C24(1981)971.

[3] V. Bernard, J. Girard, J. Julien, R. Legrain, J. Poitou, A. Oskarsson, L. Carlin, H-A. Gustafsson, B. Jakobsson, P. Kristiansson, B. Norin, I. Otterlund, H. Ryde, T. Johansson, G. Tibell, R Bertholet, C. Guet, M. Maurel, H. Nifenecker, P. Perrin, F. Schussler,M. Buenerd, D. Lebrun, P. Martin, G. Lovhoiden, J.P. Bondorf, O-B. Nielsen and A. Palmeri, Nucl. Phys. A423(1984)511.

[4] H. Heckwolf, E. Grosse, H. Dabrowski, O. Klepper, C. Michel, W.F.J. Muller, H. Noll, C. Brendel, W. Rosch, J. Julien, G.S. Pappalardo, G. Bizard, J.L. Laville, A.C. Mueller and J. Peter, Z. Physik A315(1984)243.

[5] H. Noll, E Grosse, P. Braun-Muntzinger, H. Dabrowski, H. Heckwolf, O. Klepper, C. Michel, W.F.J. Muller, and H. Stelzer, Phys. Rev. Lett. 52(1984)1284.

[6] G.R. Young, F.E. Obenshain, F. Plasil, P. Braun-Muntzinger, R. Freifelder, P. Paul and J. Stachel, Phys. Rev. C33(1986)742.

[7] G.F. Bertsch, Phys. Rev. C15(1977)713.

[8] B. Jakobsson, J.P. Bondorf and G. Fai, Phys. Lett. 82B(1979)35.

[9] C. Guet and M. Prakash, Nucl. Phys. A428(1984)119c.

[10] R. Shyam and J. Knoll, Nucl. Phys. A426(1984)606.

[11] W. Cassing, V. Metag, U. Mosel and K. Niita, Phys. Reports 188(1990)363.

[12] D. Vasak, B. Muller and W. Greiner, Phys. Scr. 22(1980)25.

[13] D. Vasak, H. Stocker, B. Muller and W. Greiner, Phys. Lett. B93(1980)243.

[14] K. Klingenbeck, M. Dillig and M.G. Huber, Phys. rev. Lett. 47(1981)1654.

[15] J Aichelin and G.F. Bertsch, Phys. Lett. B138(1984)350.

[16] J. Julien, Private comm., February 1991.

[17] B. Noren, B. Jakobsson, A. Kristiansson, L. Karlsson, A. Oskarsson, M. Westenius, J, Julien, V. Bellini, R. Fonte, T. Johansson, J. Thun and C. Guet, Nucl. Phys. A489(1988)763.

[18] S. Aiello et al., Europhys. Lett. 6(1988)25.

[19] V. Bizard et al., LPC Caen Preprint LPCC 90-10(1990).

[20] B. Erazmus, C. Guet, R. McGrath, M.G. Saint-Laurent, Y. Schutz, J.L. Ciffre, M. Bolori, Y Cassagnou, H. Dabrowski, M. Hisleur, J. Julien, R. Legrain, C. le Brun, J.F. Lecolley and M. Louvel, Nucl. Phys. A481(1988)821.

[21] A. Bonasera and G.F. Bertsch, Phys. Lett. B195(1987)521.

[22] E. Grosse, Proc. Int. School of Physics "Enrico Fermi", Varenna, Italy (1987).

[23] G. Ericsson and B. Jakobsson, to be published.

[24] E. Grosse et al., A Proposal for a QD Magnetic Spectrometer at SIS, GSI (1987).

[25] V.P Koptev, S.M. Mikirtychyants, M.M. Nesterov, N.A. Tarasov , G.V. Shcherbakov, I.K. Abrosimov, V.A. Volchenkov, A.B. Gridnev, V.A. Eliseev, E.M. Ivanov, S.P. Kruglov, Yu.A. Malov and G.A. Ryabov, Sov. Physics JETP 67(1988)2177.

[26] S. Mrowczynski, Phys. Lett. B220(1989)337.

[27] G. Batko, W Cassing, K Niita and Gy Wolf, GSI Sci Report(1989), p.192.

[28] W. Cassing, G. Batko, U Mosel, K. Niita, O Schult and Gy Wolf, Phys. Lett. B238(1990)25.

[29] R. Stock, R Bock, R. Brockman, J.W. Harris, A. Sandoval, H. A. Dacal and M.E. Ortiz, Phys. Rev. Lett. 49(1982)1236.

# ANALOGY BETWEEN GROUND-STATE BANDS IN NEUTRON-RICH A≃100 ISOTOPES AND SUPER DEFORMED BANDS IN HEAVY NUCLEI *

G. LHERSONNEAU[1], H. GABELMANN[1], B. PFEIFFER[1], K.-L. KRATZ[1], THE OSTIS[2] AND ISOLDE[3] COLLABORATIONS

[1]Institut für Kerchemie, Universität Mainz, Germany
[2]Institut Laue-Langevin, Grenoble, France
[3]CERN, Geneva, Switzerland

**Abstract**

*Very neutron-rich isotopes at A≃100 have been investigated at the OSTIS and ISOLDE facilities by means of $\gamma$-spectroscopy. Detailed studies of the ground state bands in these strongly deformed isotopes reveal that their moments of inertia only depend on the magnitude of quadrupole deformation but not on the mass. This indicates that the degenerate transition energies observed in super-deformed bands of heavy nuclei at high spin and high energy may be a more general phenomenon than expected originally.*

The neutron-rich isotopes of $_{38}$Sr, $_{39}$Y and $_{40}$Zr show a dramatic change of their ground-state shapes at neutron number N=60. Isotopes with N≤58 have spherical shape; shell structure near the doubly closed $^{96}$Zr nucleus (E($2^+$)=1751 keV) has, for example, been observed up to high excitation energy and high spin ($I^\pi = 27/2^-$) in $^{97}_{39}Y_{58}$ [1]. In contrast, the N≥60 isotopes have strongly deformed ground states with prolate shape. The largest deformations have been measured for the Sr isotopes with values of $\beta$=0.40(1) in both $^{98}Sr_{60}$ [2] (just at the onset of deformation) and $^{100}Sr_{62}$ [3] and of $\beta$=0.38(4) in $^{99}Sr_{61}$ [4]. The N=59 isotopes, which are the link between the spherical and deformed regions, exhibit shape coexistence. They have spherical ground states but, at least in the nuclei $^{97}$Sr [5] and $^{98}$Y [6], deformed states have been identified beyond 500 keV excitation energy. In a recent experiment at ISOLDE, the systematics of the very neutron-rich isotopes of Sr has been extended up to $^{101}Sr_{63}$ which is - with 13 neutrons away from the valley of stability - one of the most exotic nuclei in terms of (N-Z)/A investigated so far.

The $^{98}$Sr and $^{100}$Sr nuclei behave like twins (Fig. 1). They have the same deformation, and the transitions between the $6^+ \rightarrow 4^+$ and $4^+ \rightarrow 2^+$ levels of their respective ground state bands have the same energy within 2 keV. This unique behaviour of saturation of deformation already at its onset can be explained in terms of strong quadrupole interaction between protons and neutrons. The microscopic description reveals the importance of interactions between the steeply down-sloping [431]3/2 proton Nilsson orbital (originating from the spherical $g_{9/2}$ shell) and the neutron orbitals from $h_{11/2}$ origin, rather than from the $g_{7/2}$ spin-orbit partner [7]. The specific saturation observed in Sr isotopes seems to result from the cancellation of the contributions from the up-sloping [404]9/2 and down-sloping [532]5/2 neutron orbitals, which appear at the same energy but have opposite quadrupole moments (Fig. 2).

The new data (Fig.3) indicate for both, $^{99}$Sr and $^{101}$Sr, similar deformations since the spacing between their [411]$3/2^+$ and [422]$3/2^+$ neutron single particle levels differs by

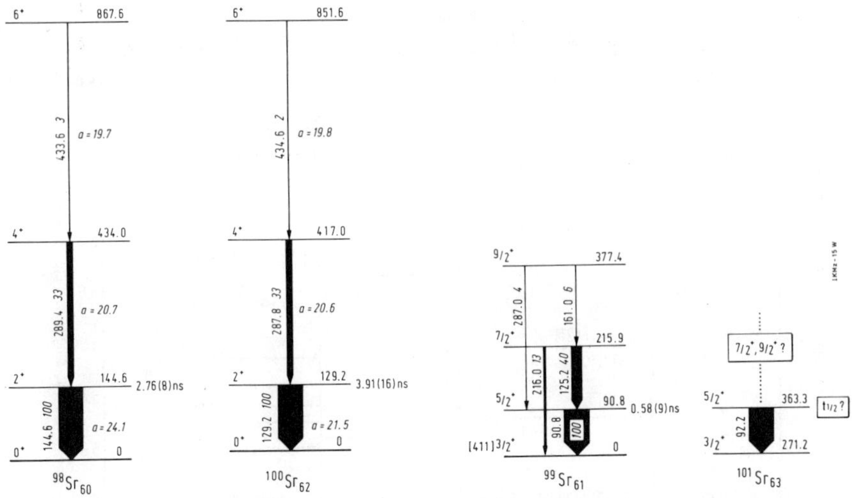


Figure 1: Comparison of the ground state bands in even-even $^{98}$Sr and $^{100}$Sr nuclei [3] and of the $\nu[411]3/2^+$ bands in $^{99}$Sr [8] and $^{101}$Sr.

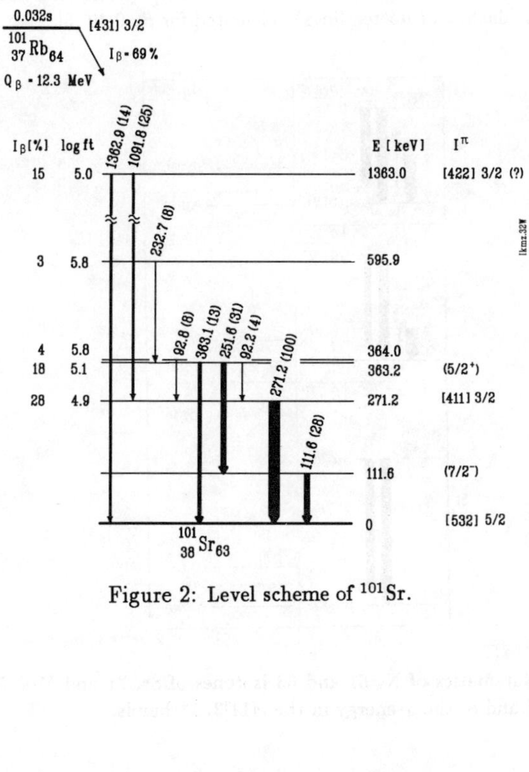


Figure 2: Level scheme of $^{101}$Sr.

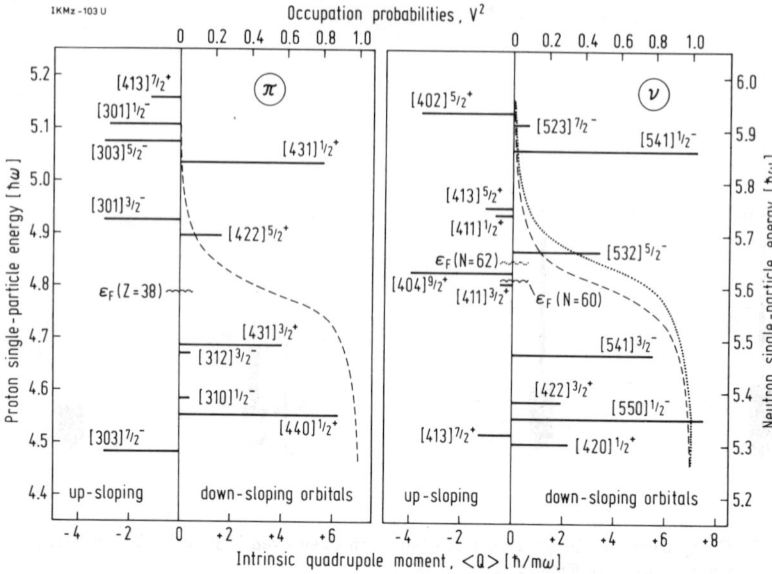

Figure 3: Nilsson orbitals with their quadrupole moments (horizontal bars) and their BCS occupation factors (dashed or dotted lines) calculated for $\beta$=0.40 [3].

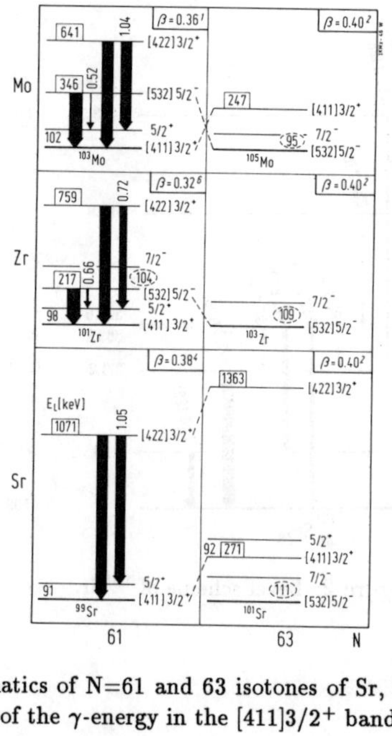

Figure 4: Level systematics of N=61 and 63 isotones of Sr, Zr and Mo. Note the trend of the [422]3/2$^+$ level and of the $\gamma$-energy in the [411]3/2$^+$ bands.

only 20 keV. Moreover, the energies of the 5/2→3/2 transition of their [411]3/2$^+$ bands are very close (Fig. 1). This again points to a probable twin character of these odd-neutron Sr isotopes. In the same experiment, a detailed level scheme of $^{101}$Zr has also been constructed. Level systematics of the N=61 isotones $^{99}_{38}$Sr [8], $^{101}_{40}$Zr and $^{103}_{42}$Mo [9] establish the trend of Nilsson single-particle levels versus deformation (Fig. 4). The moments of inertia in their $\nu$[411]3/2$^+$ bands show a linear variation with the deformation. Since in addition, the twin nuclei $^{98}$Sr and $^{100}$Sr have the same deformation and the same transition energies in their ground state bands, it seems that the $A^{5/3}$ mass-dependence of the moment of inertia in all these isotopes near $^{100}$Sr is cancelled. Thus, the recently observed degeneracies of transition energies within super-deformed bands ($\beta$ ≃0.6) belonging to neighbouring nuclei in the A=150 and A=190 regions [10,11] seem to show up in medium-heavy isotopes already at low spin and low excitation energy and 'only' $\beta$=0.4.

*This work was supported by the German BMFT under contract 06 MZ 106

## References


[1] G. Lhersonneau et al.; Z. Phys. **A323**, 59 (1986)

[2] H. Ohm et al.;Z. Phys. **A327**, 483 (1987)

[3] G. Lhersonneau et al.; Z. Phys. **A337**, 143 (1990) and Inst. Phys. Conf. Ser. 105, 433 (1990)

[4] G. Lhersonneau et al.; Z. Phys. **A332**, 243 (1989)

[5] G. Lhersonneau et al.; Z. Phys. **A337**, 149 (1990)

[6] S. Brant et al.; Z. Phys. **A334**, 517 (1989)

[7] P. Federmann and S. Pittel; Phys. Rev. **C20**, 820 (1979)

[8] B. Pfeiffer et al.; Z. Phys. **A317**, 123 (1984)

[9] K. Shizuma et al.; Z. Phys. **A315**, 65 (1984)

[10] T. Byrski et al.; Phys. Rev. Lett. **64**, 1650 (1990)

[11] F.S. Stephens et al.; Phys. Rev. Lett. **64**, 2623 (1990)

# EXOTIC NUCLEI – THE NUCLEAR STRUCTURE ORIGIN OF ISOTOPIC ANOMALIES IN THE ALLENDE METEORITE?

K.-L. KRATZ[1], R. ANNE[2], D. BAZIN[3], C. BORCEA[4], V. BORREL[5], C. DETRAZ[6], S. DOGNY[5], H. GABELMANN[1], D. GUILLEMAUD-MUELLER[5], W. HILLEBRANDT[7] M. LEWITOWICZ[2], S.M. LUKYANOV[8], P. MÖLLER[9], A.C. MUELLER[5], YU.E. PENIONZHKEVICH[8], B. PFEIFFER[1], F. POUGHEON[5], M.G. SAINT-LAURENT[2], V.S. SALAMATIN[8], F. SCHÄFER[1], H. SOHN[1], O. SORLIN[5], F.-K. THIELEMANN[10], A. WÖHR[1]

[1] Kernchemie Mainz, Germany; [2] GANIL, Caen, France; [3] CEN Bordeaux, France; [4] IFA Bucharest, Romania; [5] IPN Orsay, France; [6] LPC Caen, France; [7] MPI Garching, Germany; [8] JINR Dubna, USSR; [9] Lund University, Sweden; [10] Astrophysics Harvard, USA

## Abstract

*Beta-decay half-lives and β-delayed neutron-emission probabilities of the exotic nuclei $^{44}S$ and $^{45-47}Cl$ have been measured at the achromatic spectrometer LISE installed at GANIL. These isotopes are of astrophysical interest for explaining the correlated Ca-Ti-Cr isotopic anomalies in inclusions of the Allende meteorite.*

Progress in **nuclear astrophysics** is intimately related to continued experimental and theoretical efforts in **nuclear-structure physics**. For example, astrophysical models encountered severe difficulties when they tried to reproduce the solar $^{46,48}$Ca abundances [1] or the correlated Ca-Ti-Cr isotopic anomalies observed in inclusions of the Allende meteorite [2-6]. Such FUN-anomalies (interpreted as superposition of isotopic fractionation and effects of unknown nuclear origin) harbour an incredible wealth of utmost importance for three very different fields of research: (1) the understanding of the conditions in the early solar system, experienced by anomalous matter until its final incorporation into meteorites, (2) the understanding of the mechanisms taking place at the stellar origin of such anomalous matter, and (3) the understanding of the nuclear structure of exotic nuclei as progenitors of the FUN isotopes.

Early theoretical attemps to explain the anomalies in Allende inclusions included two possible approaches: neutron-rich nuclear statistical equilibrium (NSE) scenarios [7] and neutron-capture processes [8], which in both cases were mainly parameter studies not directly linked to astrophysical models. These approaches were later refined by Hartmann et al. [9] and Ziegert et. al. [10], respectively Hillebrandt et al. [11].

Hartmann et al. [9] in their NSE model were able to produce Ca-Ti anomalies by utilizing a mixed-zone approach with varying neutron excess, as it might be expected from the inner core of type II supernova (SN) ejecta. They were able to fit the solar $^{48}$Ca/$^{46}$Ca ratio, but the ratio of $[^{48}Ca]/[^{50}Ti] = 3.6$ observed in the Allende inclusion EK-1-4-1 (square brackets denote abundances relative to their solar system values) could not be reproduced; they only obtained a value of 0.8. Since in their model $^{48}Ca$ is a primary nucleus, strong restrictions on the amount of ejected matter had to be imposed in order to avoid huge overproduction of other neutron-rich isotopes (e.g. $^{66}Zn$). Moreover, $\beta$-decay properties, in particular $\beta$-delayed neutron ($\beta$dn) branchings, of unstable primary nuclei were not taken into account which have been shown to play an important role in the vicinity of $^{48}Ca$ [10].

Neutron-capture scenarios [8,10,11] could take place in the He-zone of massive stars during a SN II explosion. These zones would not attain r-process conditions by ($\alpha$,n)-reactions, as expected initially, but the neutron densities could be large enough to cause nuclear transmutations and non-solar neutron-rich abundance patterns. Sandler et al. [8] suggested that the heavy Ca, Ti and Cr isotopes could have been produced in a high-neutron-density environment of $10^{-7}$mol$\cdot$cm$^{-3}$ with a neutron-exposure time of $10^3$s, in which both neutron captures and $\beta$-decays are effective (n$\beta$-process). They calculated a $^{48}Ca/^{46}Ca$ abundance ratio which was only a factor of 2.5 smaller than the solar value of 53 [1], but the predicted anomalies for $^{46}Ca$ and $^{47}Ti$ were too large by factors of 13 and 5, respectively, compared to the observed ones in the EK-1-4-1 inclusion of Allende [2,3]. In order to reduce the abundances of these isotopes, they had to postulate non-statistical neutron-capture cross sections for the reactions $^{46}K$ and $^{49}Ca(n,\gamma)$, caused by hypothetical low-lying s-wave resonances. In the latter case, indeed, the existence of such a resonance was verified in the "inverse" process, i.e. $\beta$dn-decay from $^{50}K$ [12]. However, subsequent measurements of the partial decay widths of this state exluded the explanation of the meteoritic $^{49,50}Ti$ abundances suggested in [8]. Moreover, the later measured $[^{53}Cr]/[^{52}Cr]$ and $[^{48}Ca]/[^{54}Cr]$ ratios were predicted too high by factors of 3.5 and 7.5, respectively by the n$\beta$-model.
Motivated by those difficulties and by the nuclear-structure related large variations of $\beta$-decay properties of neutron-rich isotopes near the double-magic nucleus $^{48}Ca$, Ziegert et al. [10] and Hillebrandt et al. [11] have investigated possible implications of $\beta$-decay half-lives ($T_{1/2}$) and $\beta$dn-emission probabilities ($P_n$) on the Ca-Ti-Cr abundances. Starting from solar abundances, as did Sandler et al. [8], Ziegert et al. have calculated the build-up of isotopic abundance distributions of short-lived S to Ti nuclei as a function of neutron exposure by using the simplification of treating subsequent neutron capture for each element (Z) separately, without considering competing $\beta$-decay and $(\gamma,n)$-reactions. From these abundance distributions, at freeze-out $\beta$-decay back to the stability line was determined with $\beta$dn-branchings taken into account. They found, that for neutron exposures of about $5 \cdot 10^{-5}$mol $\cdot$ cm$^{-3}$ $\cdot$ s, exposure times of $\leq 1$ s and temperatures of $< 10^9$ K, both the solar abundance ratio $^{48}Ca/^{46}Ca$ as well as the correlated Ca-Ti isotopic anomalies could be explained by the same nucleosynthesis process.
This approach [10] was later improved by Hillebrand et al. [11], who performed complete network calculations which included neutron and charged-particle captures, photodisintegrations as well as $\beta$- and $\beta$dn-decays. In principle, they confirmed the results of Ziegert et al. [10]. With the large odd-even effects in the initial element abundances and with the different build-up times for the $A = 46$ and the $A = 48$ isotopes in each Z chain according

to their distance from $\beta$-stability, specific astrophysical conditions were determined under which $^{46}$Ca was formed with low abundance and $^{48}$Ca with high abundance. Such a situtation was found for neutron exposures of $(6-7)\cdot 10^{-5}\mathrm{mol}\cdot\mathrm{cm}^{-3}\cdot\mathrm{s}$, for which $^{48}$Ca was formed mainly from the even-Z (large initial abundance) $_{18}$Ar progenitors $^{48}$Ar($\beta^-$) $^{48}$K($\beta^-$) $^{48}$Ca and $^{49}$Ar($\beta^-$) $^{49}$K($\beta^-$n)$^{48}$Ca. Of particular importance in this context was the known strong variation in the $P_n$-values of the neighbouring isotopes $^{48}$K (1.1%) and $^{49}$K (86%) which allowed a large enhancement of the $^{48}$Ca abundance and a reduction of $^{49}$Ti without the postulated [8] non-statistical neutron-capture cross sections. Just at the above neutron exposure, the Z − 2 abundance distribution of $_{14}S$ progenitors has – for neutron-exposure times of $\leq 100$ ms – not yet reached A = 46; thus, a low production of $^{46}$Ca originates mainly from the odd-Z (low initial abundance) $_{17}$Cl progenitors $^{46,47}$Cl (see Fig. 1).

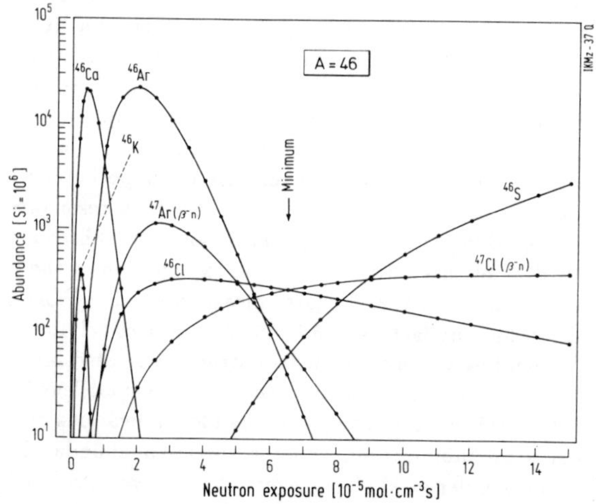


**Figure 1:** Isotopic abundances for A = 46 as a function of neutron exposure [13]. The total abundance of $^{46}$Ca progenitors shows a minimum around $(6-7)\cdot 10^{-5}\mathrm{mol}\cdot\mathrm{cm}^{-3}\cdot\mathrm{s}$, exactly where the overall production of $^{48}$Ca progenitors is largely enhanced. The following theoretical $\beta$dn-branching ratios were used: $P_n(^{46}\mathrm{S}) = 50\%$, $P_n(^{46}\mathrm{Cl}) = 30\%$, $P_n(^{47}\mathrm{Cl}) = 15\%$, $P_n(^{47}\mathrm{Ar}) = 3\%$.

It should be noted, however, that the results of Hillebrandt et al. [11] depended on the – at that time unknown – $T_{1/2}$ and $P_n$-values of $^{44}$S and $^{45-47}$Cl for which model predictions [10,13,14] had to be used. The strong time limitation of the $\beta$dn-process (see Fig. 1 in [11]) was caused by the short $T_{1/2}$ of $^{44}$S of about 310 ms [14]. With the neutron-exposure times becoming comparable to $T_{1/2}(^{44}\mathrm{S})$, neutron capture will continue up to $^{46}$S and nuclei would be fed into the Cl isotopic chain. This would increase the abundances of $^{46,47}$Cl and thus the final $^{46}$Ca production. In the $_{17}$Cl chain itself, successive neutron capture will also proceed up to $^{46,47}$Cl [11,13]. The most important physical quantity of these isotopes was found to be their $\beta$dn-branching ratio. As an example, Fig. 2 demonstrates the influence of $P_n(^{47}Cl)$ on the final $^{48}Ca/^{46}Ca$ ratio. With a fixed theoretical $P_n(^{46}Cl) = 30\%$, the solar ratio of $^{48}Ca/^{46}Ca$ [1] can be reproduced with each $P_n(^{47}Cl)$, whereas the exotic $^{48}Ca/^{46}Ca$ ratio observed in EK-1-4-1 is only reached with a "low" $P_n(^{47}Cl)$. In fact, the effects of the $\beta$-decay properties of $^{44}$S and $^{46,47}$Cl on the resulting $^{48}Ca/^{46}Ca$ ratio are interrelated (see upper part of Fig. 5), and it was strongly requested [11] to **measure** these important nuclear-physics quantities in order to restrict the astrophysical parameter space.

**Figure 2:** Abundance ratio $^{48}Ca/^{46}Ca$ as a function of neutron exposure for various assumptions on $P_n(^{47}Cl)$. For discussion, see text.

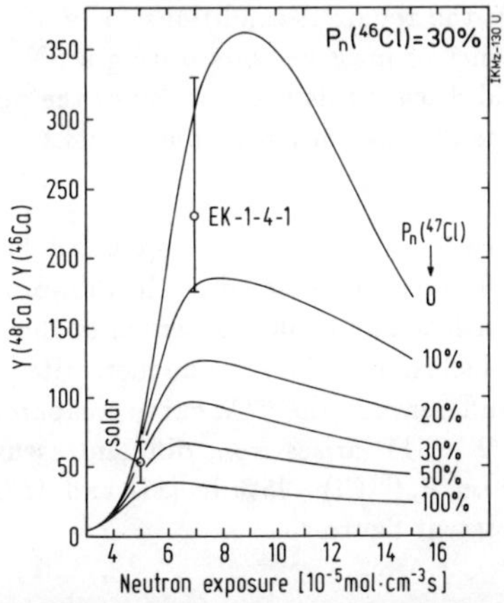


In order to derive meteoritic Ca-Ti abundances, the resulting composition obtained from the network calculations for the $\beta$dn-process was mixed into a reservoir with solar-system abundances, as done by Sandler et al. [8]. From these abundances, isotopic anomalies, as defined in [2,3], were derived. At a neutron exposure of about $(6-7)\cdot 10^{-5}\text{mol}\cdot\text{cm}^{-3}\cdot\text{s}$, best overall agreement with the observed Ca-Ti isotopic anomalies in EK-1-4-1 was obtained (see Fig. 3). Furthermore, the calculated abundance ratio of $[^{48}Ca]/[^{50}Ti] = 3.55$ was in excellent agreement with the measured value of $3.6 \pm 0.2$ [2]. For comparison, the models of Sandler et al. [8] and Hartmann et al. [9] predict ratios of 2.6 and 0.8, respectively. In addition, Hillebrandt et al. predicted abundance ratios for $[^{54}Cr]/[^{52}Cr]$ and $[^{53}Cr]/[^{52}Cr]$, in good agreement with the later measured values [5]. However, the $[^{48}Ca]/[^{54}Cr]$ ratio was strongly overestimated, a deficiency of the $\beta$dn-model which has not been solved so far.

Originally, the results by Ziegert et al. [10] and Hillebrandt et al. [11] gave solutions in a purely mathematical sense for the free parameters neutron density and exposure time. Their results, with the required features for the neutron flux of

$$\int \rho Y_n dt = \rho Y_n \tau \leq 5 \cdot 10^{-5} mol \cdot cm^{-3} \cdot s \tag{1}$$

and the time scale for the neutron irradiation of

$$\tau \leq 100 \text{ ms} \tag{2}$$

have recently been translated into an astrophysical context by Thielemann et al. [15]. The aim was to test whether the requests fromEqs. (1) and (2) can be fulfilled in realistic low-temperature scenarios (e.g. explosive He-burning), based on possible ($\alpha$,n)-neutron sources like $^{13}$C, $^{18}$O or $^{22}$Ne. The authors concluded that with the 1986 data base on exotic nuclei arround $^{48}$Ca and the (still insufficient ?) present knowledge on neutron-capture cross sections and temperature dependent reaction rates for "neutron poisons" (e.g. $^{25}$Mg and $^{56}$Fe), the required conditions for a 100 ms "flash"-scenario might be hard to attain in He-zones of massive stars during a SN explosion. However, this does not mean that one should discard this scenario, since the new experimental results presented below may well change the astrophysical constraints to more moderate conditions.

**Figure 3:** Comparison of the Ca and Ti isotopic anomalies predicted by the $\beta$dn-process [11] with the anomalies observed in the EK-1-4-1 inclusion of the Allende meteorite [2,3]. The difference in the $^{46}$Ca curve compared to Fig. 2 in [11] arises from different assumptions on $P_n(^{37}\text{Cl})$: 35% in [11] and 15% in the present figure.

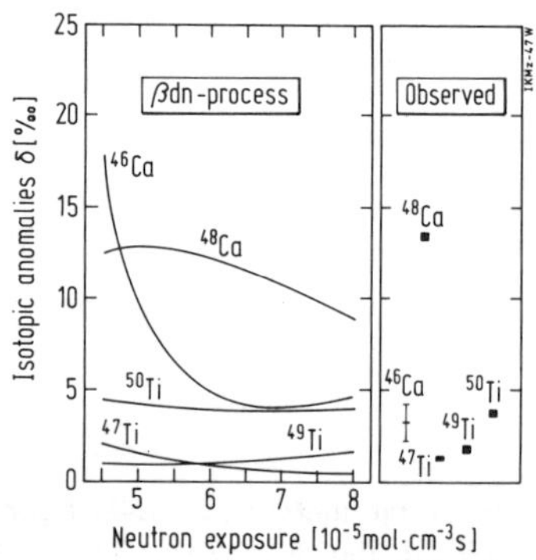


As mentioned above, already in 1986 Hillebrand et al. [11] had requested to measure the missing $\beta$-decay properties of the "key"-isotopes influencing the (low) production of $^{46}$Ca in their $\beta$dn-model, i.e. $^{44}$S and $^{45-47}$Cl. It took several years until these exotic nuclei could be produced by fragmentation of a $^{48}$Ca beam at 60 MeV/u from GANIL onto a $^{64}$Ni target (116 mg/cm$^2$). The separation of the products was provided by a magnetic analysis in A$v$/Z with the doubly achromatic spectrometer LISE [16] (see also Bimbot et al.; these proceedings). In addition, a thick energy degrader (180 g/cm$^2$) of aluminium was used to perform a complementary energy-loss selection in A$^3$/Z$^2$ in order to reduce the number of unwanted nuclei. The projectile-like fragments were redundantly identified through energy-loss, total energy and time-of-flight (TOF) measurements. The identification and implantation telescope was composed of two 300 $\mu$m thick silicon detectors and a 5500 $\mu$m thick Si(Li) detector. The implantation of the nuclei of interest was made in the third detector, which served also for the $\beta$-detection with an efficiency of $\epsilon_\beta = 60 \pm 6$ %. The telescope was surrounded by the Mainz 4$\pi$ neutron long-counter, and a 3" NE213 liquid scintillator was placed in the beam axis. The long-counter device was composed of

two concentric rings of altogether 39 $^3$He-proportional counters embedded in a polyethylene moderator matrix. This detector has a nearly energy independent efficiency of $\epsilon_n = 27\pm2$ % up to $E_n \simeq 2$ MeV and has no low-energy cut-off, as do scintillators. The latter feature is important for a reliable determination of $P_n$-values from $\beta$-neutron coincidences. The NE213 scintillator was added to cover the neutron energy range up to the highest possible energies ($E_n \simeq 10$ MeV for $^{47}$Cl). Each isotope under study was on-line identified by a combined measurement of its time-of-flight through the spectrometer and its energy-loss in the first 300 $\mu$m silicon detector. Two single-channel analysers examined the $\Delta$E and TOF pulses and each time that a nucleus of interest was identified, the primary cyclotron beam was switched off for a time interval related to the expected $T_{1/2}$. During this time, a second data acquisition performed the registration of $\beta$-neutron coincidences at low background. A set of plastic scintillators was used as shieldings on top of and below the neutron long-counter to detect cosmic muons. Thus, in the subsequent off-line analysis it was possible to discriminate between true $\beta$-neutron coincidences and muon induced background. With this set-up, each $\beta$-delayed neutron could be directly correlated to its implanted neutron precursor. The measurements were made for three settings of the magnetic rigidity of the LISE spectrometer in order to optimize the selection of either $^{44}$S, $^{45}$Cl or $^{47}$Cl.
Preliminary results on $T_{1/2}$ and $P_n$-values of $^{44}$S and $^{45,46}$Cl are shown in Fig. 4. The $T_{1/2}$ of $^{47}$Cl could not be deduced from the $\beta$-neutron coincidences because of the very few events which, nevertheless, allowed to limit the $P_n(^{47}\text{Cl})$ to the surprisingly low value of $\leq 3\%$. With the $T_{1/2}$ of $^{44}$S and $^{45}$Cl being considerably shorter than the predictions of [14] used in [11], together with the higher, respectively lower $P_n$-values of $^{46}$Cl and $^{47}$Cl, the conclusions of Hillebrandt et al. concerning the low production of $^{46}$Ca are – in principle – strengthened. As can be seen from Fig. 5, the experimental values may even turn out to be more "favourable" for a $\beta$dn-process than assumed before. Successive neutron capture in the $_{16}$S and $_{17}$Cl chains is halted at the N = 28 "turning-point" nuclei where $\beta$-decay starts to dominate ($T_{1/2}(\beta) < T_{1/2}(n)$). Of particular importance is the result that – due to the short $T_{1/2}(^{45}\text{Cl})$ – feeding of $^{46,47}$Cl is reduced. These isotopes were found to be the main progenitors of $^{46}$Ca in the 1986 approach [11] (see Fig. 1).

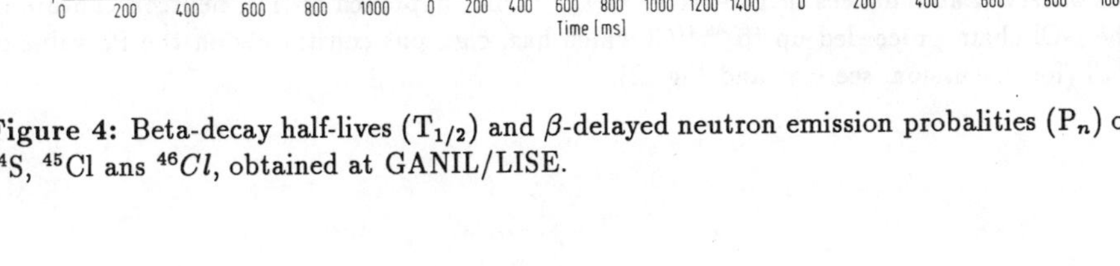


**Figure 4:** Beta-decay half-lives ($T_{1/2}$) and $\beta$-delayed neutron emission probalities ($P_n$) of $^{44}$S, $^{45}$Cl ans $^{46}Cl$, obtained at GANIL/LISE.

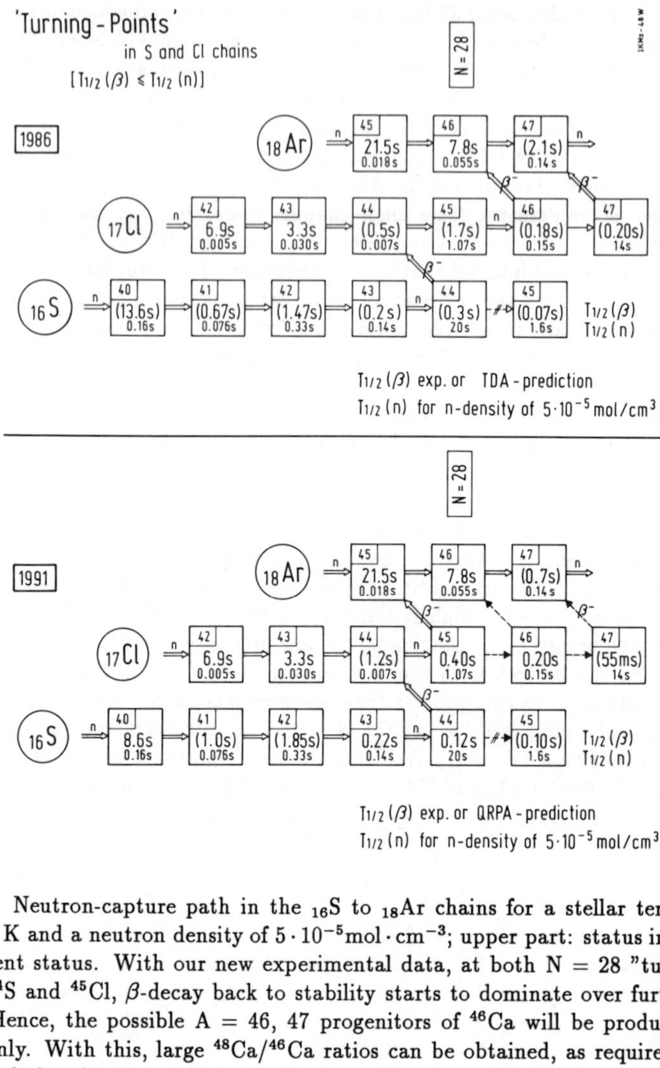


**Figure 5:** Neutron-capture path in the $_{16}$S to $_{18}$Ar chains for a stellar temperature of $T = 8 \cdot 10^8$ K and a neutron density of $5 \cdot 10^{-5} \mathrm{mol} \cdot \mathrm{cm}^{-3}$; upper part: status in 1986, lower part: present status. With our new experimental data, at both N = 28 "turning-point" isotopes, $^{44}$S and $^{45}$Cl, $\beta$-decay back to stability starts to dominate over further neutron capture. Hence, the possible A = 46, 47 progenitors of $^{46}$Ca will be produced in small amounts only. With this, large $^{48}$Ca/$^{46}$Ca ratios can be obtained, as required to explain the observed abundances in EK-1-4-1. In the earlier approach of [11], neutron capture in the $_{17}$Cl chain proceeded up to $^{46,47}$Cl which has, e.g., put constraints on the $P_n$-value of $^{47}$Cl (for discussion, see text and Fig. 2).

Moreover, even if neutron capture in the $_{17}$Cl chain would proceed up to $^{46,47}$Cl, their $P_n$-values would again result in a low $^{46}$Ca abundance (compare upper and lower parts of Fig. 5).
Hence, with the present data, the problem of producing "very little" $^{46}$Ca seems to be solved. This, on the other hand, may moderate the rather strict constraint on the neutron-exposure time requested in [11]. In any case, with the new nuclear-physics parameters determined at GANIL/LISE, the network calculations of [11] for explosive He-burning conditions will have to be repeated – then also including the most recent results on nuclear masses [17] and the improved neutron-capture cross sections from Thielemann (to be published as Appendix in [18].)

Summarizing, we follow the earlier conclusions of [10,11], that structure effects in the $\beta$-decay properties of exotic isotopes may be the nuclear-physics origin of the Ca-Ti-Cr isotopic anomalies in inclusions of the Allende meteorite. However, the question still remains about the possible astrophysical site(s). The presently still existing weaknesses of the NSE model of Hartmann et al. [9] (a small $^{48}$Ca/$^{50}$Ti ratio and a large enhancement of $^{66}$Zn) might be improved by considering freeze-out efects and the new data base on experimental and theoretical $\beta$dn-branching ratios [19]. In explosive He-burning, the neutrons would act on essentially solar-system matter which underwent a previous weak s-process. A strong neutron-flux can produce the Ca-Ti-Cr anomalies as demonstrated in [10,11] and strengthened in this paper. It will have to be shown that the more recently observed enhancements of $^{58}$Fe, $^{64}$Ni and $^{66}$Zn in the EK-1-4-1 inclusion [5,6] can be produced as well in our $\beta$dn-model. This may be subject of further experimental investigations at GANIL/LISE to measure the $\beta$-decay properties of their possible progenitor isotopes.
In addition, our type of n-process scenario would produce heavy r-process nuclei as well, but not in solar proportions. The seed distribution of a week s-process, which produces the s-process nuclei in large amounts only up to A $\simeq$ 90, would lead to a much stronger decrease for the r-process nuclei as a function of A, than found in a standard r-process. This predicted behaviour [15] can be tested by further experiments with meteoritic samples. There is still much which we can learn about stellar models and explosive events from this just emerging new type of astronomy, i.e. **isotopic astronomy** (see e.g. [20]), which astronomical observations will never reveal.

Discussions with C.L. Harper, D. Hartmann and M. El Eid are are gratefully acknowledged. This research was supported by grants from DFG (Kr 806/1), BMFT (06 MZ 106), NSF (AST-8913799) and NASA (NGR-22-007-272).

**References**

[1] A.G.W. Cameron; in Essays in Nuclear Astrophysics (Cambridge Univ. Press), eds. C.A. Barnes et al.; 23 (1982)

[2] T. Lee et al.; Ap. J. **220**, L21 (1978); **228**, L93 (1979)

[3] F.R. Niederer at al.; Geochim. Cosmochim. Acta **45**, 1017 (1981)

[4] J.-L. Birck, C.J. Allegre; Geophys. Res. Lett. **11**, 943 (1984)

[5] D.A. Papanastassiou; Ap. J. Lett. **308**, L27 (1986)

[6] D.A. Papanastassiou, C.A. Brigham; Ap. J. **338**, L37 (1989)

[7] A.G.W. Cameron; Ap.J. **230**, L53 (1979)

[8] D.G. Sandler et al.; Ap. J. **259**, 908 (1982)

[9] D. Hartmann et al.; Ap. J. **297**, 837 (1985)

[10] W. Ziegert et al.; Phys. Rev. Lett. **55**, 1935 (1985)

[11] W. Hillebrandt et al.; in Weak and Electromagnetic Interactions in Nuclei (Springer Verlag), ed. H.V. Klapdor; 987 (1986)

[12] K.-L. Kratz et al.; Astr. Ap. **125**, 381 (1983)

[13] W. Ziegert; Doctoral Dissertation, Univerität Mainz (1985)

[14] H.V. Klapdor et al.; At. Data Nucl. Data Tables **31**, 81 (1984)

[15] F.-K. Thieleman et al.; Proc. Nuclei in the Cosmos, eds. H. Oberhummer, W. Hillebrandt; MPA/P4, 286 (1990)

[16] R. Anne et al.; NIM **A257**, 215 (1987)

[17] G. Audi; The 1991 Nuclear Mass Evaluation, priv. comm.

[18] J.J. Cowan et al.; Phys. Rep. (1991), in press

[19] K.-L. Kratz, P. Möller; unpublished

[20] C.L. Harper et al.; Proc. Nuclei in the Cosmos, eds. H. Oberhummer, W. Hillebrandt; MPA/P4, 138 (1990)

# SPIN-FLIP TRANSITION AT N = 20, THE $^{33}$Al → $^{33}$Si BETA DECAY

M. BOUNAJMA[1], P. BAUMANN[1], A. HUCK[1], G. KLOTZ[1], A. KNIPPER[1], G. MARGUIER[2], C. RICHARD-SERRE[3], G. WALTER[1]
and the ISOLDE Collaboration

[1] Centre de Recherches Nucléaires et Université Louis Pasteur
F - 67037 STRASBOURG, France
[2] Institut de Physique Nucléaire et Université Claude Bernard
F - 69622 VILLEURBANNE, France
[3] IN2P3, F - 75640 PARIS, France
et CERN, CH - 1211 GENEVA 23, Switzerland

**Abstract**

The first study of the $^{33}$Al (N = 20 ) beta decay was made at the CERN on-line mass separator ISOLDE 2. The half-life has been determined ($T_{1/2}$ = 43.6 ± 2 ms) and the distribution of the Gamow-Teller strength has been investigated by beta, gamma and delayed-neutron spectroscopy.

The beta decay is found in agreement with the sd-shell model predictions with a main branch corresponding to a spin-flip (5/2 → 3/2) transition, populating the $^{33}$Si ground state.

For $^{32}$Mg (Z = 12, N = 20), where previous experiments[1)] have revealed a strong deformation, a $(sd)^{n-2}(fp)^2$ component dominates the ground state. The structure of $^{34}$Si (Z = 14, N = 20), recently investigated[2)] by the study of the $^{34}$Al beta decay, corresponds to a transition between the strong deformation (Z = 12) and the spherical region (Z = 16 - 20), the 2p-2h intruder being the first excited state, instead of the ground state. For Z = 12, N = 19 our recent measurements at ISOLDE have confirmed the high density of states at low excitation energy with the strong contribution of intruder states.

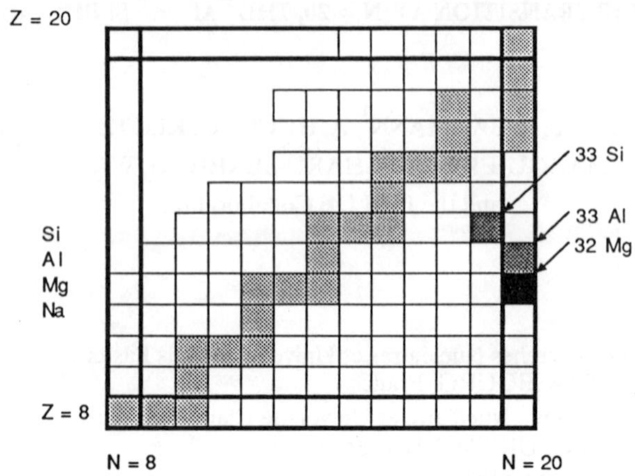


Fig. 1 Part of the chart of nuclides between the lines of shell closure ( 8, 20), Nuclei studied near the strongly deformed $^{32}$Mg, are indicated.

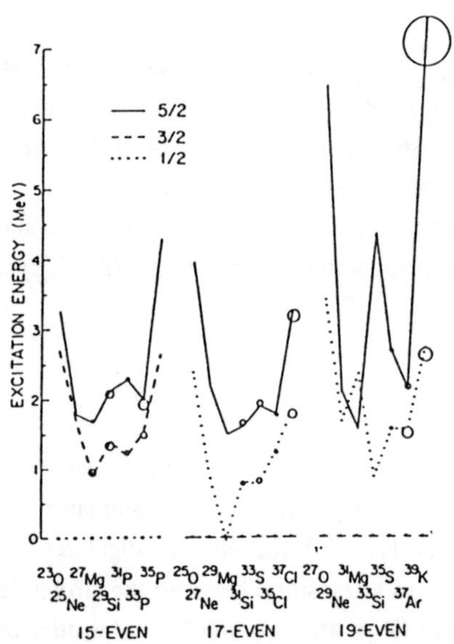


Fig. 2 (Taken from reference 6). Relative energies of the lowest $5/2^+$, $1/2^+$ and $3/2^+$ calculated for odd-even sd-shell nuclei.

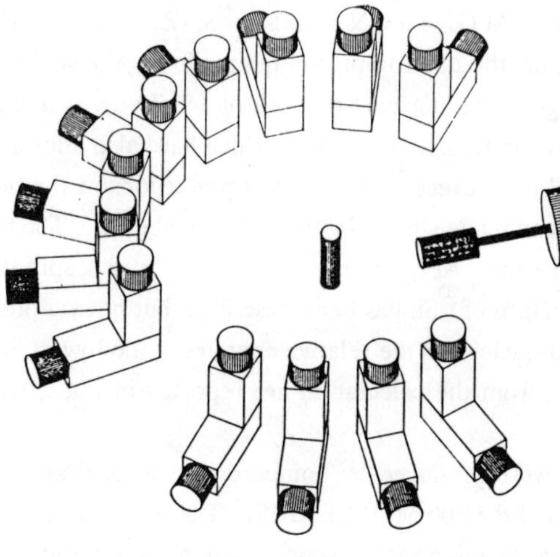

Fig. 3 Multidetector array of 12 neutron counters, surrounding the $4\pi\beta$ detector (start) and associated with one HP Ge counter.

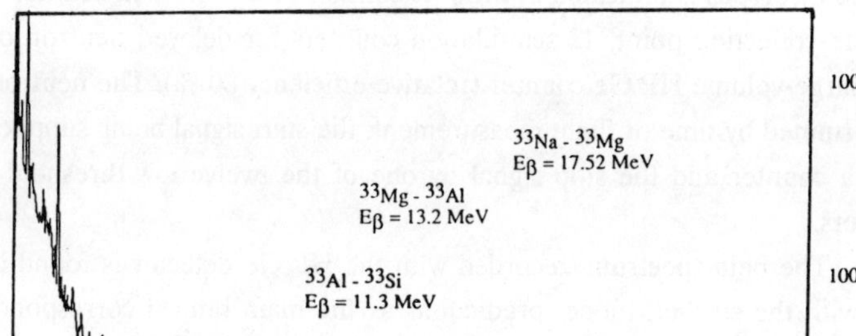


Fig. 4 Beta spectrum from the decay of A = 33 isotopes, measured with the HP Ge detector.

The beta decay of $^{33}$Al (Z = 13, N = 20) to $^{33}$Si (Z = 14, N = 19) is then the logical next step to study the neutron-proton interaction at the sd-fp shell limit (Fig.1). Present knowledge of the nuclear structure of $^{33}$Si is very scarce and limited to the results of heavy ion transfer reactions[3] which have also allowed the first determination of the $^{33}$Al mass excess[4]. Theoretical predictions of the beta decay properties of $^{33}$Al have been made since 1983 by Wildenthal et al[5]. The main beta transition is expected to be the $^{33}$Al (g.s.) → $^{33}$Si (g.s.), $d_{5/2}$ → $d_{3/2}$ spin-flip transition. The low energy structure of $^{33}$Si, has been described, but not yet observed, in the systematic study of sd-nuclei and the relative energies of the lowest $5/2^+$, $1/2^+$ and $3/2^+$ states resulting from the calculation are reported in Fig.2, taken from reference 6.

The $^{33}$Al nuclei were produced by bombarding a thick uranium carbide target with the 2$\mu$A beam of the 600 MeV CERN SC. The Al atoms are ionized by means of the tungsten surface-ionization source and mass separated with the ISOLDE 2 on-line facility. The yield achieved for $^{33}$Al was in the order of 30 atoms/s.

The experimental set-up for observing the decay of short-lived and weakly produced isotopes is indicated in Fig.3. It consists of a ≃ 4$\pi$ beta counter surrounding the collection point, 12 scintillation counters for delayed neutron detection and a large-volume HP Ge counter (relative efficiency 80 %). The neutron energy is determined by time of flight measurement, the start signal being supplied by the 4$\pi$ beta counter and the stop signal by one of the twelve low-threshold neutron detectors.

The beta spectrum, recorded with the HP Ge detector is found in agreement with the sd-shell model predictions as the main branch corresponds to the population of $^{33}$Si ground state (Fig.4). The end point is found close to the value corresponding to the $^{33}$Al - $^{33}$Si mass difference and quite distinct from the $^{33}$Na or $^{33}$Mg values, allowing to assign the recorded spectrum to the $^{33}$Al decay.

The multispectrum analysis of the beta activity, recorded with the Ge detector, indicates for energies higher than 4 MeV, a decay rate of 49 ± 5 ms.

The analysis of the beta activity with the 4$\pi$ beta counter yields an almost pure decay well-fitted by a $T_{1/2}$ = 43.6 ± 2.0 ms exponential (Fig.5). A strong contribution of $^{33}$Na (8.2 ± 0.4 ms) or $^{33}$Mg (90 ± 20 ms)[6] is not observed in our experiment, so that the 43.6 ms half-life can be attributed to the $^{33}$Al decay.

The neutron spectrum, recorded with the multidetector device used as a unique detector (by mixing the signals delivered by each counter) with a flight path of 46 cm, is given in Fig.6. In addition to the strong $\gamma$ peak, corresponding to $\beta$-$\gamma$ coincidences, a weak neutron contribution can be evaluated (Table I).

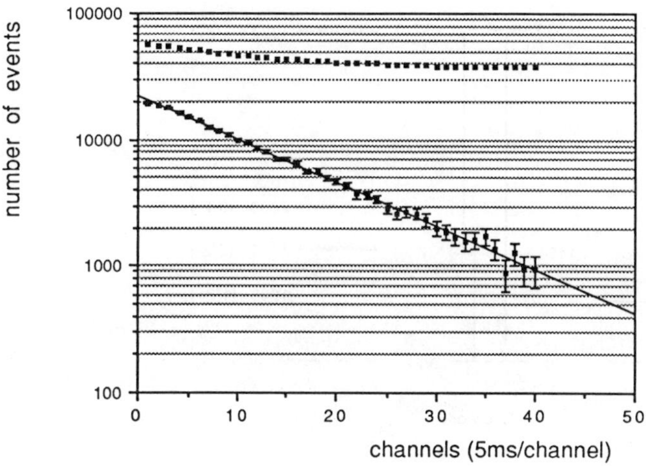


Fig. 5 Multiscaling analysis (4πβ counter) of the decay of A = 33 isotopes. ( 44500 cycles of 200 ms )

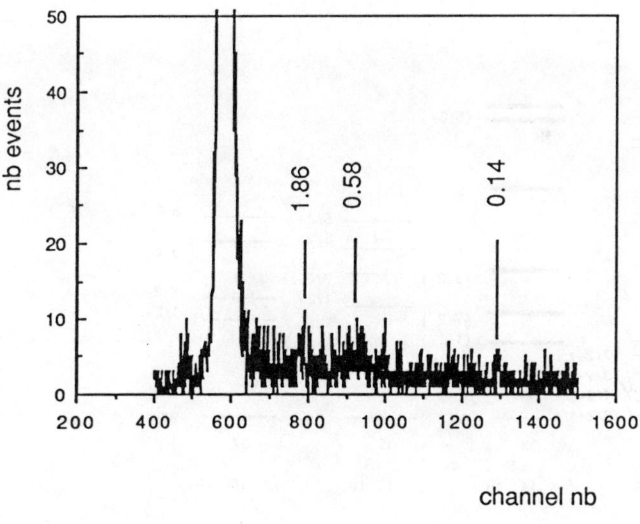


Fig. 6 Neutron time of flight spectrum (flight path = 46 cm, 0.121 ns/channel)

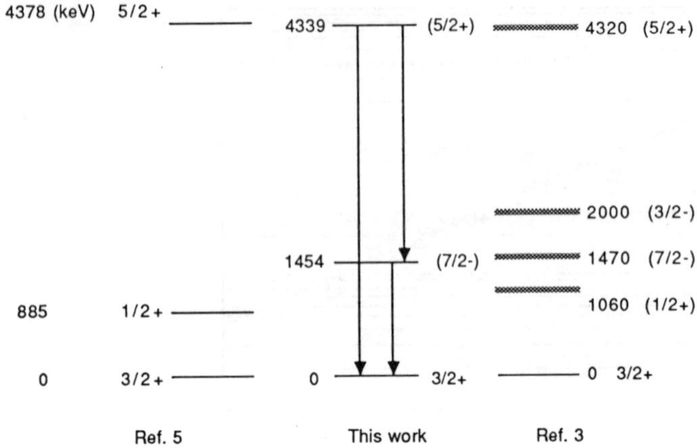


Fig. 7 Low lying levels in $^{33}$Si, resulting from theoretical predictions (ref. 5), $^{33}$Al beta decay (this work) and heavy ion transfer reactions (ref. 3).

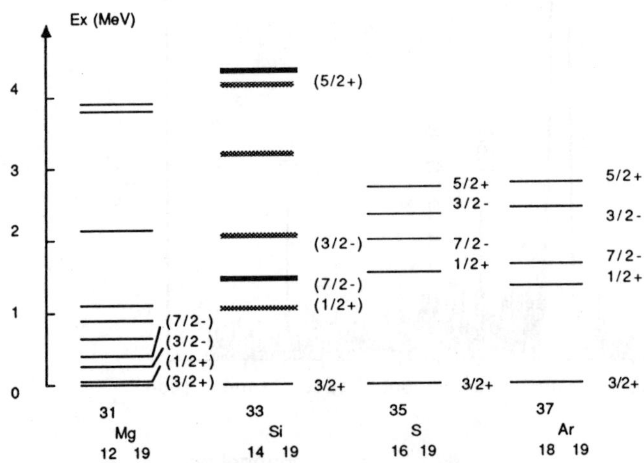


Fig. 8 N = 19 isotones. Indicated $^{33}$Si levels or those reported in heavy ion transfer studies (ref.3) or in this work.

Table I

| $E_n$ (MeV) | $E_X$ $^{33}$Si(MeV) | % $\beta$ | log ft |
|---|---|---|---|
| 0.14 | 4.63 | 0.5 | 5.5 |
| 0.43 | 4.92 | 0.6 | 5.3 |
| 0.58 | 5.08 | 1.1 | 5.0 |
| 0.85 | 5.36 | 0.9 | 5.0 |
| 1.14 | 5.66 | 0.7 | 5.1 |
| 1.86 | 6.40 | 2.0 | 4.4 |

As no gamma line at 1.94 MeV, corresponding to the population of the $^{32}$Si, $2^+$ state, was observed in this experiment, the delayed neutron branches are assigned to transitions to the $^{32}$Si ground state.

The decay rate of the delayed neutron activity was measured by multiscaling of the corresponding events in the time of flight spectra. The theoretical fit to the experimental data was performed, assuming a single exponential decay and a constant background level. A satisfactory agreement is found with $T_{1/2} = 54^{+28}_{-15}$ ms, value less accurate but in agreement with the half-life measured by beta detection.

The gamma spectra, recorded in coincidence with the $4\pi$ beta counter, were taken either as multispectra in two successive 100 ms time bins directly after the collection time or during long collection times (all isotopes in radioactive equilibrium) in order to improve the statistics. In both cases, the analysis was made difficult by the presence of contaminants (mainly $^{132}$Sb) also produced in the target and selected by the separator at the same position of the focal plane, due to their-multiple charge (4+ in the case of $^{132}$Sb). From the list of unassigned gamma rays, only two groups of lines have been tentatively related to the radiative decay of $^{33}$Si levels populated in the $^{33}$Al decay :

- a level at 4339 keV in $^{33}$Si (with radiative decay through 4339.0 and 2884.8 + 1454.0 keV transitions)
- a level at 5435 keV (with lines observed at 5435, 2515 and 2920 keV) .

In the latter case ($E_X$ = 5435 keV), we would observe the radiative decay of a level unbound by ≃ 950 keV. This situation is expected for $7/2^+$ levels, populated by the GT beta decay of $3/2^+$ $^{33}$Al and for which the delayed neutron emission to $^{33}$Si, $0^+$, ground state is severely hampered by the high value ($l \geq 4$) of the angular transfer momentum involved in this decay. In the former case ($E_X$ = 4339 keV), an interesting comparison (Fig.7) can be made with the theoretical

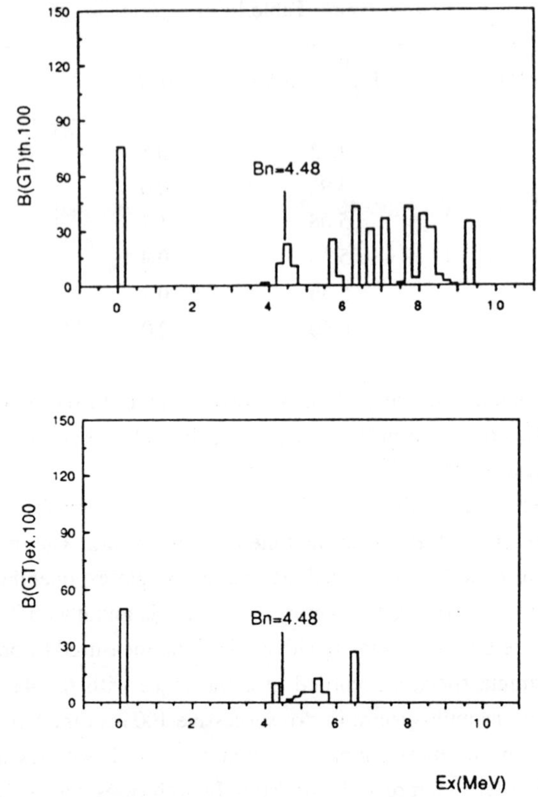


Fig. 9 B(GT) strength distribution for the beta decay of $^{33}$Al.
Theoretical values are taken from ref. 5.

predictions[5] and the results of heavy ion transfer reactions[3]. We note an excellent agreement between the energy of the gamma transitions, the excitation energy of the levels (observed at 1.47 ±0.3 MeV and 4.32 ±0.3 MeV) and the calculated $5/2^+$ level at 4378 keV.

Arguments based on reaction mechanism were given in reference 3 to propose a positive parity for the 4.32 and 1.06 MeV levels, populated by the ($^{11}B$, $^{14}N$) reaction on a $^{36}S$ target whereas the 1.47 MeV level, observed only in the ($^{13}C$, $^{14}O$) two-step reaction on $^{34}S$ has been tentatively identified as a $7/2^-$ level. The relevant data in other N = 19 isotopes are reported in Fig.8 and show up the steep variation of the negative parity 1p-1h neutron states near the Z = 12 value.

The difficulty to discriminate between the contaminant activity, resulting from multicharged ions, and the A = 33 activity, allows to give only an upper limit for the total beta intensity populating bound states in $^{33}Si$. A mean value of two gamma transitions for each beta branch has been assumed and the total beta intensity related to unassigned $\gamma$ lines amounts then to 15 % of the beta decay. In these conditions, taking into account the $P_n$ value determined in our experiment ($P_n$ = 6 %), the spin-flip transition to the ground state amounts to, at least, 79 % of the beta decay. The experimental $\beta$(GT) distribution, obtained in these conditions, is compared with the theoretical one[5] in the Figure 9. The shell model calculations successfully explain the experimental result.

In conclusion, in spite of the difficulty to study a beta decay far from stability with only a weak contribution to excited states, it has been possible to confirm the validity of the sd-shell model for nuclei very close to the region of inversion. The main spin-flip transition from a $\nu d_{5/2}$ to a $\pi d_{3/2}$ state dominates the beta decay. Additionnal evidence for the presence at low energy of 1p-1h intruder states when the neutron excess is important, has also been obtained.

**References**

1) D. Guillemaud-Mueller et al., Nucl. Phys. A 426 (1984) 37
2) P. Baumann et al., Phys. Lett. B 228 (1989) 458
3) L.K. Fifield et al., Nucl. Phys. A 440 (1985) 531
L.K. Fifield et al., Nucl. Phys. A 453 (1986) 497
4) P.J. Woods et al., Phys. Lett. B 182 (1986) 297
5) B.H. Wildenthal et al., Phys. Rev. C 28 (1983) 1343
6) B.H. Wildenthal, Prog. Part. Nucl. Phys. 11 (1983) 5

# ISOTOPIC A- AND ISOBARIC Z- DISTRIBUTIONS IN CHARGED PARTICLE INDUCED FISSION

P.P. JAUHO, J. ÄYSTÖ, A. JOKINEN, J.M. PARMONEN, H. PENTTILÄ
Department of Physics, University of Jyväskylä, SF-40351 Jyväskylä, Finland

K. ESKOLA, M.E. LEINO
Department of Physics, University of Helsinki, SF-00170 Helsinki, Finland

A. ASTIER, R. BÉRAUD, A. BOULDJEDRI, R. DUFFAIT, A. EMSALLEM, M. MEYER, N. REDON
Institut de Physique Nucléaire, Université Claude Bernard Lyon 1,
69622 Villeurbanne Cedex, France

D. BARNÉOUD, J. BLACHOT, J. GENEVEY, A. GIZON, J. INCHAOUH
Institut des Sciences Nucléaires, 38026 Grenoble Cedex, France

P. DECROCK, P. DENDOOVEN, M. HUYSE, G. REUSEN, P. VAN DUPPEN, J. WAUTERS
Instituut voor Kern- en Stralingsfysika, Katholieke Universiteit Leuven, B-3001 Leuven, Belgium

## 1 Abstract

Independent and cumulative yields in nearly symmetric fission have been measured for the first time by bombarding $^{238}$U with 20 MeV protons, 18, 25, and 41 MeV deuterons and 39 MeV $\alpha$-particles. Isobaric charge dispersion has been determined in the A= 110, A= 112, and A= 114 mass chains and, from asymmetric fission, in the A= 80 mass chain. The IGISOL ion guide facility in Jyväskylä, the LIGISOL in Louvain-la-Neuve and another IGISOL in Grenoble were used to separate the isobaric chains from fission products.

## 2 Motivation

The measurement of independent fission yields Y(Z, A) of the nuclides $^{A}$Z allows the determination of the charge dispersion. Mass yield can be studied in a simpler way by determining the cumulative yields in isobaric chains. Extensive studies of thermal neutron induced fission and spontaneous fission yields have been made using for example radiochemical techniques[1] and on-line recoil spectrometers[2]. The study of the width of the charge distribution and of the odd-even effects observed in the yield curves has recently received much attention due to the improved experimental methods available at present[3]. In particular, the effects of excitation energy and fissility parameter $Z^2/A$ of the compound system on the yields are of interest.

Analysis of the yields of fission products provides information on how shell structure and pairing correlations[4] influence the fission process. Systematic knowledge of the yields of very neutron-rich isotopes has, in addition, great practical value in the study of exotic nuclei. In particular, it is of interest to find out whether a given isotope has the highest effective yield in spontaneous fission or fission induced by light particles or heavy ions.

From thermal-neutron-induced fission and spontaneous fission one cannot reliably determine charge dispersion parameters for symmetric products because of their low production rates. When more energy is brought into the system, for instance by charged particles, the symmetric component in the mass distribution is enhanced. In heavy ion induced reactions the mass and charge distributions may be further modified by deep inelastic and transfer reactions competing with complete fusion followed by fission[5].

In this work, charge dispersion was studied for the first time in symmetric or nearly symmetric fission. Independent yields have previously not been determined for symmetric low energy fission, although mass yields around symmetry have been published for thermal neutron induced fission of some nuclides such as $^{235}$U. This lack of experimental data is a result not only of small fission yields but also of the short half-lives of very neutron rich fission products. Even fast on-line radiochemical separation methods are too slow when the halflife is in the region of seconds. On the other hand, use of on-line recoil spectrometers such as LOHENGRIN at ILL requires an extensive experimental effort due to the low efficiency and the need to integrate over charge state and energy distributions of the products. The work described here was performed with on-line mass separators based on the ion guide (IGISOL) principle[6] which have a delay time of some milliseconds and efficiencies on the order of 0.1%.

Our measurements were performed in connection with studies on $\beta$-decay properties and nuclear structure of very neutron-rich nuclei in the mass A= 110-120 region. Among the newest isotopes studied are $^{109}$Tc, $^{112}$Tc, $^{114}$Ru, $^{117}$Rh, $^{119}$Pd, $^{120}$Pd [7,8,9,10,11,12, respectively]. Our new information concerning the decay schemes of these and other neutron-rich isotopes has made it possible to determine the charge distribution for some mass chains as well as cumulative isotopic yields for certain elements.

# 3 Experimental

The IGISOL (Ion Guide Isotope Separator On-Line) facility at the Department of Physics, University of Jyväskylä, was used to separate the mass chain from other products from fission of $^{238}$U induced by 20 MeV protons. The proton beam current from the MC-20 cyclotron was typically 500 nA. The principle of IGISOL is thermalizing the products as singly charged ions in helium gas. The ions are then guided by gas flow and an electric field directly to an ISOLDE II type separator. The mass resolving power of the separation is around 400 under normal operating conditions. This separation method is well suited for the study of the elements Zr, Nb, Mo, Tc, Ru, and Rh which are difficult to separate using mass separators with a conventional on-line ion source. Two other IGISOL facilities were used, one[13] for the study of deuteron induced fission of $^{238}$U, operating at the HI accelerator CYCLONE and the other one on line with the HI SARA accelerator, for the study of $\alpha$-induced fission of $^{238}$U.

# 4 Results

In the following, cumulative isotopic yields in saturation as well as independent isobaric yields will be given. The yields of isomeric states were added to the ground state yields. The total thickness of uranium target material was about 60 mg/cm$^2$ so the results should be considered averaged over the spread in projectile energy.

The cumulative yields of Tc, Ru, Rh, Pd and Ag isotopes in the 20 MeV proton induced fission of $^{238}$U are shown in Fig. 1. A typical uncertainty of $\pm 50$ % is marked on some data points. One source of error is in the determination of the ground state $\beta$-branch, zero branching was assumed when it was not known. This gives lower limit for yield. The corresponding yields in the 39 MeV $\alpha$-particle induced fission are shown in Fig. 2. No results are given for isotopes where the absolute $\gamma$-intensity is not known.

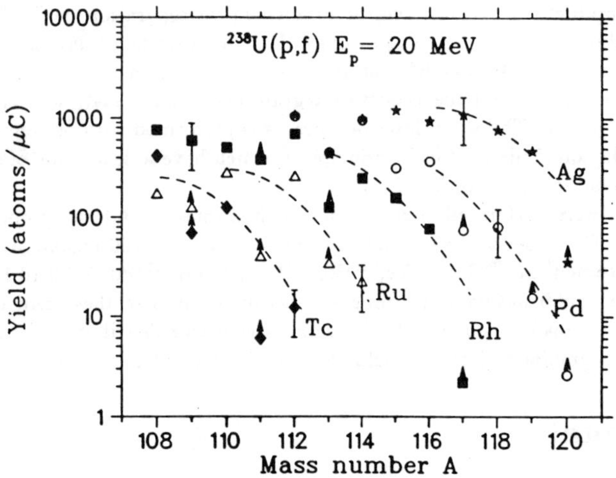


FIG. 1 Cumulative yields in saturation of Tc, Ru, Rh, Pd, and Ag isotopes. The dashed lines are drawn to guide the eye. Lower limits are given when the ground state $\beta$-decay branch is unknown in which case zero branching was assumed.

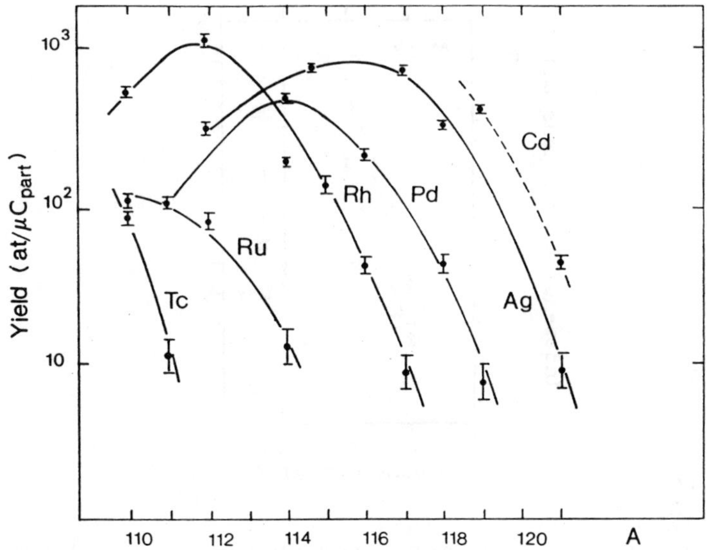


FIG. 2 Cumulative yields of Tc, Ru, Rh, Pd, and Ag isotopes in the reaction $^{238}U(\alpha,f)$, $E_\alpha = 39$ MeV.

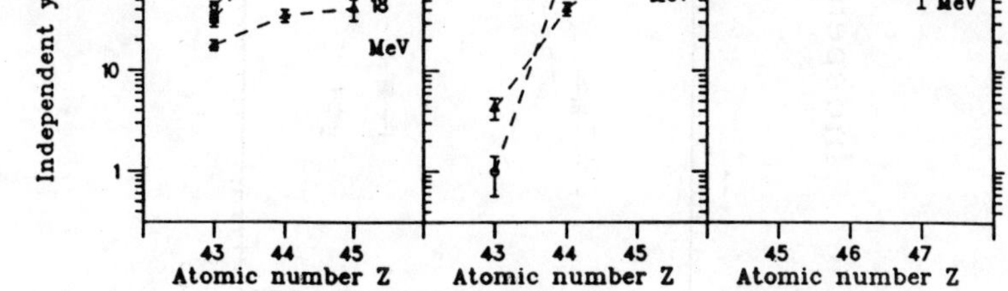


FIG. 3 Independent yields for A= 110, 112, and 114 from the d+$^{238}$U reaction determined at three bombarding energies of 18, 25, and 41 MeV as indicated in the figure. At A= 112, no measurement was performed with $E_d = 25$ MeV. The $^{114}$Pd data are not shown because of too large uncertainty in the $\gamma$-branch for extracting the independent yield from saturated yields.

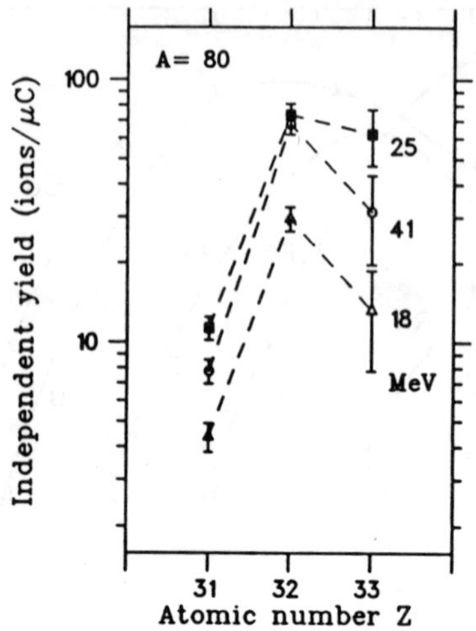


FIG. 4 Independent yields for A= 80 from the d+$^{238}$U reaction determined at three bombarding energies of 18, 25, and 41 MeV as indicated in the figure.

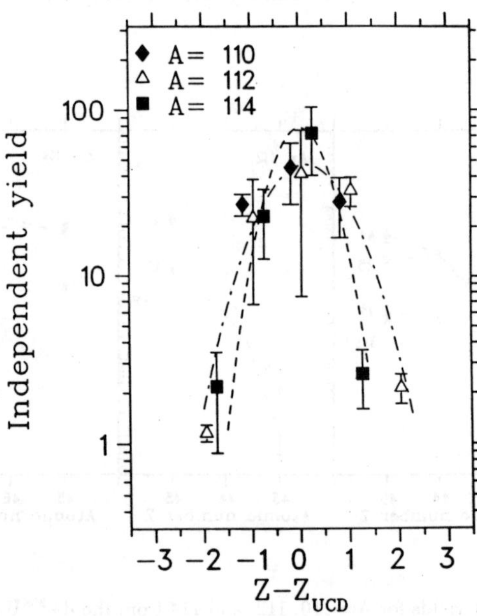


FIG. 5 Relative independent yields of isobars with A= 110, 112, and 114 from the 20 MeV p+$^{238}$U reaction. The data for each isobaric chain was shifted using the respective $Z_{UCD}$ values.

# References


[1] T.M. Semkow, A.C. Wahl, and L. Robinson, Phys. Rev. C 30, 1966 (1984)

[2] U. Quade, K. Rudolph, S. Skorka, P. Armbruster, H.-G. Clerc, W. Lang, M. Mutterer, C. Schmitt, J.P. Theobald, F. Gönnenwein, J. Pannicke, H. Schrader, G. Siegert, and D. Engelhardt, Nucl. Phys. A487, 1 (1988)

[3] J.P. Bocquet and R. Brissot, ILL-report

[4] R. Vandenbosch and J.R. Huizenga, Nuclear Fission, Academic Press, New York (1973)

[5] W. Reisdorf, M. de Saint-Simon, L. Remsberg, L. Lessard, C. Thibault, E. Roeckl, and R. Klapisch, Phys. Rev C 14, 2189 (1976)

[6] P. Taskinen, H. Penttilä, J. Äystö, P. Dendooven, P. Jauho, A. Jokinen, and M. Yoshii, Nucl. Inst. and Meth. A281, 539 (1989)

[7] H. Penttilä, H. Salminen, J. Äystö, K. Eskola, Z. Janas, P.P. Jauho, A. Jokinen, M.E. Leino, J.M. Parmonen, University of Jyväskylä, Department of Physics, Annual Report 1990

[8] J. Äystö, P.P. Jauho, Z. Janas, A. Jokinen, J.M. Parmonen, H. Penttilä, P. Taskinen, R. Béraud, R. Duffait, A. Emsallem, J. Meyer, M. Meyer, N. Redon, M.E. Leino, K. Eskola, and P. Dendooven, Nucl. Phys. A515, 365 (1990)

[9] A. Jokinen et al., to be published

[10] H. Penttilä, P.P. Jauho, J. Äystö, P. Decrock, P. Dendooven, M. Huyse, G. Reusen, P. Van Duppen, and J. Wauters, University of Jyväskylä, Department of Physics, Annual Report 1990

[11] H. Penttilä, J. Äystö, K. Eskola, Z. Janas, P.P. Jauho, A. Jokinen, M.E. Leino, J.M. Parmonen, and P. Taskinen, submitted to Z. Phys. A

[12] Z. Janas, J. Äystö, K. Eskola, P. Jauho, A. Jokinen, J. Kownacki, M. Leino, J.M. Parmonen, H. Penttilä, J. Szerypo, and J. Żylicz, University of Jyväskylä, Department of Physics, Annual Report 1990

[13] K. Deneffe, B. Brijs, E. Coenen, J. Gentens, M. Huyse, P. Van Duppen, and D. Wouters, Nucl. Instr. Meth. B26, 399 (1987)

## References

# IV. Various Aspects of Heavy-Ion Reactions

# LIMITING EXCITATION ENERGY IN HEAVY ION FUSION REACTIONS

W. BOHNE, H. MORGENSTERN, S. PROSCHITZKI

Hahn-Meitner-Institut, Glienicker Str. 100, D-1000 Berlin 39, F.R.G.

Abstract

Values of the critical excitation energy, above which a hot nucleus cannot decay any more by particle evaporation, have been determined for medium mass compound nuclei ($65 < A < 110$) with the aid of a new integral method, by measuring the width of the ER-mass distribution as function of the residue velocity. The mass dependence of this critical excitation energy was found to be stronger than expected by various models. A predicted dependence on the Z/A ratio of the compound nucleus seems to exist. First coincidence measurements yield the same value of the critical excitation energy for the fission decay channel.

## I. Introduction

Fusion reactions with heavy ions are a good tool to create nuclei at high excitation energies. With increasing beam energy it is possible to heat up the compound nucleus more and more. Then, of course, the question rises which is the maximum amount of energy a thermal equilibrated nucleus is able to absorb. As a first answer one could say that the absolute upper limit for the excitation energy has to be in the order of the total binding energy, i.e. about 8 MeV/u, because beyond this value the nucleus can be broken down into its individual nucleons. If this would be true one would expect that hot compound nuclei with excitation energies up to this value should be detectable by means of their normal decay products, i.e. evaporation residues and/or fission fragments. In reality this upper limit for the conventional decay (evaporation of nucleons and light particles up to alpha particles and binary fission) is expected to be substantially lower, following the predictions of several theoretical models [see e.g. refs. 1-8]. All these models have in common that, above a certain excitation energy - the critical excitation energy - it is more likely for the nucleus to emit sequentially or even simultaneously more than two heavier fragments. All models yield for the onset of this new de-excitation process - called multifragmentation, cracking, or emission of intermediate-mass fragments (IMF), etc. - values of about 70% of the total binding energy. Some experiments, measuring the decreasing cross section of fusion like events as function of the projectile energy [see e.g. refs. 9, 10] or the saturation of the multiplicity of evaporated neutrons [11] with increasing energy, seem to corroborate this predicted behaviour.

All these experiments are rather costly because cross sections have to be determined over a wide range of incident energies. We are not going to discuss the pros and cons of the different techniques used up to now, but restrict ourselves to present the results of a new, simple and integral method to derive values for the critical excitation energy of medium mass compound nuclei.

This method is described in section II. A short presentation of the used experimental setup is given in section III. The results are shown in section IV, followed by a short summarizing final section.

## II. Method

To determine the total excitation energy deposited in a compound nucleus the energies and momenta of all particles - including γ-rays - in the final state should be measured. Such a complete experiment is not yet performed and is up to now not feasible with the existing experimental equipments, at least not with sufficiently high accuracy. Therefore we are going to extract at least some averaged values of this quantity from the mass width of precisely measured inclusive evaporation residue mass distributions. In addition we measured coincidences of evaporation residues (and fission fragments) with evaporated light particles as well as with projectile like fragments.

The idea is simple. From the systematic study of fusion reactions [see e.g. ref. 12] it is well known that at sufficiently high beam energy besides the normal complete fusion many different incomplete fusion reactions with different momentum transfers contribute to the total fusion-like cross section. This means, in one experiment different compound nuclei with different excitation energies can be formed. The more complete the process the higher the excitation energy of the corresponding compound nucleus should be. As a consequence, the number of statistically evaporated light particles and therefore the width of the ER's (in velocity and also in mass) should also increase. It is also well known [12] that the velocity of the compound nucleus or, due to the statistical nature of the evaporation, the mean velocity of the residues is a measure of the linear momentum transfer respectively of the excitation energy of the corresponding compound nucleus. This means, that $\sigma_A$, the width of the mass distribution, which is formed by a superposition of all the contributing residue distributions, should vary with $v_R$, the velocity of the residues. In the case of "normal kinematics" (projectile lighter than the target) the mass width $\sigma_A$ is expected to increase steadily with increasing $v_R$ until $v_R$ reaches $v_{CN}$, the well defined value for complete fusion if complete fusion can occur or if the corresponding compound nuclei can de-excite mainly by light-particle evaporation. Above $v_{CN}$ the value for $\sigma_A$ should stay constant because the ER cross section at higher velocities exists only due to the width of the velocity distribution originating from the statistical evaporation. To check whether the excitation energy is really limited according to the predicted values one has to use sufficiently high beam energies to produce compound nuclei in complete fusion reactions with excitation energies above the 70% value of the total binding energy. On the other hand one has to be careful that the fusion partners will not be too fast, because according to our systematics [13] the fraction of complete fusion decreases linearly with increasing beam (and target) velocity and disappears when the available c.m. velocity of one partner (the lighter one) overcomes a value of about 19% of the velocity of light. Both conditions we can fulfill with the $^{32}S$ beam of the VICKSI accelerator at the HMI Berlin.

If a maximum excitation energy for light particle evaporation exists and if the excitation energies of the compound nuclei formed with complete or almost complete fusion reactions are higher than this maximum value, then we expect that the mass width starts to stay constant (saturates) at a residue velocity below $v_{CN}$. This value we will call the critical velocity $v_{crit}$. The excitation energy of the corresponding heaviest compound nucleus $A^{crit}$ of the studied system, which contributes to the ER mass distribution, will be labelled critical excitation energy $E^*_{crit}$ or critical excitation energy per nucleon $\varepsilon^*_{crit} = E^*_{crit}/A^{crit}$, respectively. How can we determine this excitation energy? In general the transformation of the velocity scale into an excitation-energy scale is not possible (except for $v_{CN}$, the well defined value for complete fusion) because in all cases of incomplete fusion we don't have a precise knowledge of the compound-nucleus mass $m_R = m_P - \Delta m_P + m_T - \Delta m_T$, where $\Delta m_P$ and $\Delta m_T$ are the non-fusing fractions of the projectile with mass $m_P$ and the target with mass $m_T$, respectively. But using the simple assumptions of a participant-spectator picture (psp) where the non-fusing parts $\Delta m_p$ and $\Delta m_T$ are not excited and keep their initial velocities $v_P$ and $v_T$, respectively, the energy and momentum conservation gives a connection between the excitation energy per nucleon and the residue velocity $v_R$:

$$\varepsilon^* = 1/2\,(v_P - v_R)\,v_R + Q/m_R \qquad (1)$$

where Q is the groundstate Q value.

Unfortunately, nature is not that simple. Due to the interaction of projectile and target the non-fusing spectators do not keep their initial velocities and can be excited themselves.
Consequently the total excitation energy E* of the system is:

$$E^* = E^*_{ICN} + E^*_{PLF} + E^*_{TLF} \qquad (2)$$

where $E^*_{ICN}$ is the excitation energy of the reduced compound nucleus formed with the reduced mass $A^{crit}$ in the incomplete fusion reaction. $E^*_{PLF}$ and $E^*_{TLF}$ are the excitation energies of the non-fusing projectile and target spectators (projectile-like fragments and target-like fragments), respectively. Thus equation **1** for the total excitation energy has to be extended:

$$\begin{aligned} \varepsilon^* &= 1/2\,(\cos\theta_R \cdot v_P - v_R)\,v_R + Q/m_R \\ &+ 1/2\;\Delta m_P/m_R \cdot (\cos\theta_P \cdot v_P - v'_P)v'_P \\ &+ 1/2\;\Delta m_T/m_R \cdot (\cos\theta_T \cdot v_P - v'_T)v'_T \end{aligned} \qquad (3)$$

where $v'_P$ and $v'_T$ and $\theta_P$ and $\theta_T$ are the final velocities and the angles relative to the beam axis of the non-fusing parts $\Delta m_P$ and $\Delta m_T$, respectively. $\theta_R$ is the detection angle of the residues relative to the beam axis. Since the masses of the spectators are much smaller than $m_R$ and since $v'_P$ should not be too different from the beam velocity and since $v'_T$ should be small (in the case of normal kinematics), the last two terms are relatively small corrections. But never the less, we measured, in coincidence with the residues, the properties of the PLF's to include at least the corrections due to the projectile spectator. Unfortunately, the target spectators elude detection because of their small velocity in the laboratory system. But, according to our systematic study of incomplete fusion [13], the contribution of the break up of the slower target to the incomplete fusion cross section is much smaller than that of the projectile break up. Therefore, we neglected the last term of eq. 3 and also $E^*_{TLF}$ in eq. 2. Averaged Q values for the different momentum transfers have been included in the calculation of $\varepsilon^*$.

This method gives only the energy transferred from initial relative motion into other degrees of freedom. Since not all of this energy is necessarily transformed into thermal energy (rotational energy for instance is estimated to about 20 MeV), the extracted excitation energy per nucleon is an upper limit when compared with the temperature of the compound nucleus.

## III. Experiment

The experiments were performed with $^{32}S$ beams of 710, 840 and 950 MeV from the heavy-ion accelerator VICKSI. We used self supporting targets made from highly enriched $^{45}Sc$, $^{51}V$, $^{58}Fe$, $^{58}Ni$ and $^{80}Se$ with thicknesses between 0.3 and 1 mg/cm$^2$. The residues have been measured with the time-of-flight method, using a new compact type of a gridless channel-plate start detector with a central hole [14]. The flight paths between these detectors and the surface-barrier stop detectors were 1175, 1340 and 295 mm for the three TOF systems we used simultaneously at angles θ between 3° and about 50° relative to the beam axis. The total time resolution amounted to about 50 ps. Fig. 1 shows a typical mass-velocity spectrum.

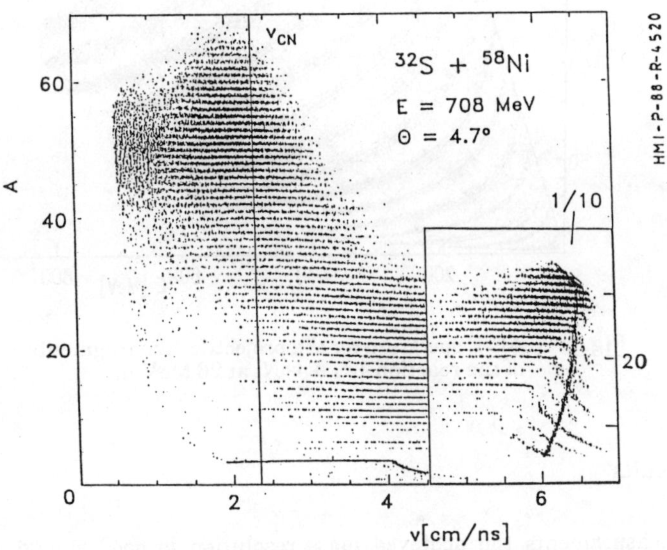


Fig. 1: Mass vs. velocity spectrum of the fusion reaction $^{32}S + ^{58}Ni$ at 22 MeV/u. The vertical line $v_{CN}$ gives the velocity for complete fusion.

At the opposite side of the scattering chamber relative to the beam direction (in the following denoted as negative angles) we installed some Si detectors (450 mm$^2$) at fixed angles between -70° and -150° (150 mm from the target) in order to detect the evaporated light particles. The particle identification was achieved also by time of flight, using the pulsed beam structure of VICKSI ($\Delta t \approx 0.5$ ns). At forward angles between -5° and -40° we used some ΔE-E Si telescopes to identify in addition also the PLF's. The ΔE detectors were about 100 μm thick and allowed to resolve all elements up to the projectile. For the lighter elements mass identification was possible. A typical example is depicted in Fig. 2. All measured energies and times were corrected for ionization defect [15] and plasma delay [16], respectiveley.

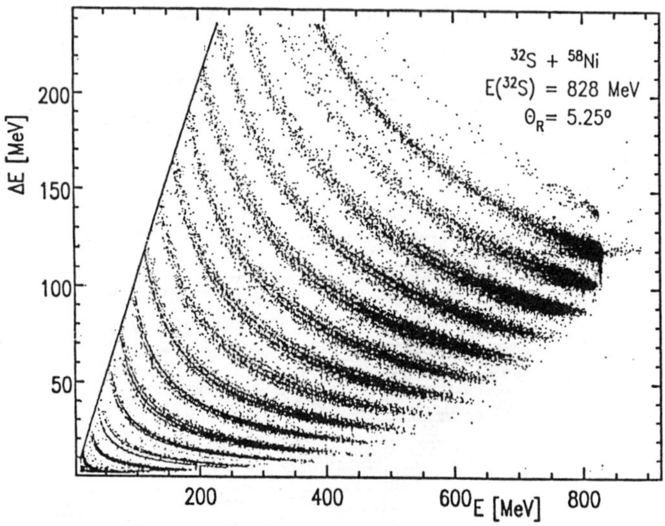


Fig. 2: ΔE-E spectrum for the projectile-like fragments of the reaction $^{32}S + ^{58}Ni$ at 26 MeV/u.

## IV. Results

In all measurements the achieved mass resolution is good enough to obtain velocity spectra of clearly separated masses. This is absolutely necessary in order to extract reliable values for the mass widths and to be able to subtract the contributions of quasi-elastically scattered target-like recoils at small velocities. For small velocity intervals we determined the mass widths $\sigma_A$ with the aid of normal distributions. To avoid or at least to minimize the contribution of fusion-fission events we took into account in the fit procedure only the cross section for masses heavier than the mean mass of the ER's. These values for $\sigma_A$ are plotted as a function of $v_R$, the centroids of the velocity intervals. Typical examples at two different beam energies and a $^{58}$Fe target are shown in fig. 3. The vertical lines give the ER velocity for complete fusion. The small arrows indicate the critical velocities $v_{crit}$, where $\sigma_A$ starts to saturate. (The exact value of $v_{crit}$ was determined from the velocity interval in which $\sigma_A$ the first time falls below the mean value of $\sigma_A$ in the saturation area by more than three standard deviations.)

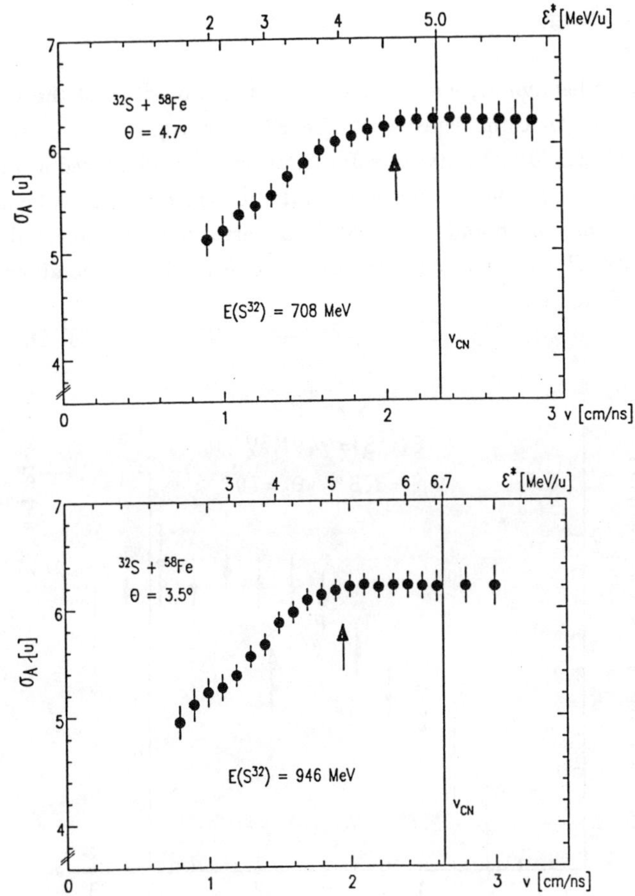


Fig. 3: Mass width $\sigma_A$ of the evaporation residues from the reactions $^{32}$S + $^{58}$Fe at 22 and 30 MeV/u as a function of the residue velocity (lower abzissa). The upper abzissa refer to the excitation energy per nucleon of the corresponding compound nuclei. The vertical lines $v_{CN}$ indicate the velocities for complete fusion.

We see that the width functions show indeed the expected behaviour as discussed in section II. At the low incident energy the saturation starts very close to $v_{CN}$ the velocity for complete fusion. But, at 949 MeV $\sigma_A$ reaches its saturation value at a velocity much below $v_{CN}$. This means, at this energy compound nuclei formed by fusion reactions above a certain momentum transfer up to complete fusion do not contribute to the ER cross section, even though, according to our systematics [13], the fusion-like cross section should have a portion of complete fusion of about 25 percent at this energy.

To confirm this behaviour, we also measured the properties of the evaporated charged particles in coincidence with the ER's. In fig. 4 the relative proton multiplicity $(N_p(\pm 70°) / N_{ER}(v_{ER}) \times \text{Jacobian}(v_{ER}))$ is displayed as function of the residue velocity $v_R$. Besides the worse statistics, the multiplicity behaves very similar to the mass width and yields within the error bars the same value for the critical velocity. The arrow in fig. 4 indicates the result for $v_{crit}$ obtained from the mass widths of the ER's.

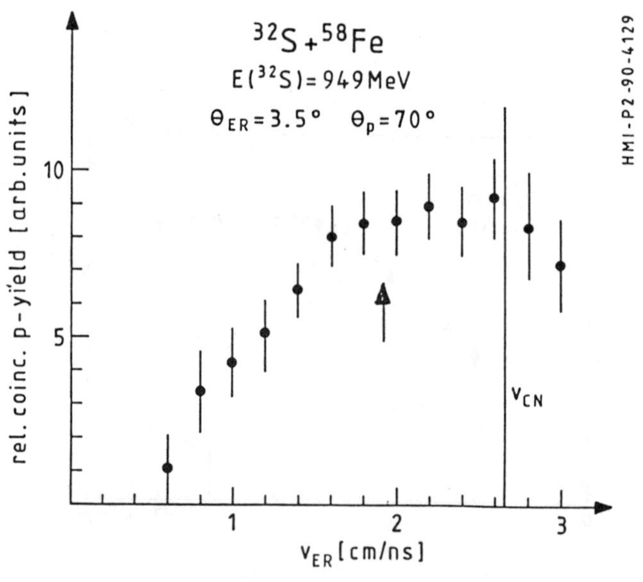


Fig. 4: Relative proton multiplicity in coincidence with ER's of the reaction $^{32}S + ^{58}Fe$ at 30 MeV/u as function of the residue velocity. The vertical line gives the velocity for complete fusion. The arrow shows the value for the critical velocity obtained from the mass widths of the ER's.

Another possibility to obtain the same information is to measure the energy spectra or only the mean energy of evaporated particles, again as function of the residue velocity. The higher the velocity of the ER's the more complete is the fusion reaction (in our case of normal kinematics), i.e. the excitation energy of the corresponding nuclei grows. Therefore the mean energy of an evaporated particle $<E_{evap}> \approx 2T=2\ [aE^*/A]^{1/2}$ also increases. In the case of charged particles we have to add the Coulomb energy which also grows towards complete fusion, because the compound nuclei become heavier. For the measurement in the laboratory system we have to include the velocity of the emitter, i.e. the velocity of the ER's. All three components are growing with increasing momentum transfer. Therefore, when measuring in forward direction, we are also expecting an increasing function for the mean energy of evaporated particles with increasing ER velocity. In fig. 5 we present an example of an alpha-energy spectrum in coincidence with ER's of a narrow velocity interval to show that a clear distinction between evaporated alphas (labelled $\alpha_{EVAP}$ in fig. 5) and pre-equilibrium alphas ($\alpha_{PLF}$) is possible. In fig. 6 some results for the centroids of $\alpha_{EVAP}$ are shown, again for the reaction $^{32}S$ (949 MeV) + $^{58}Fe$. Indeed the functions show the expected behaviour with a saturation at a velocity much below $v_{CN}$, the value for complete fusion.

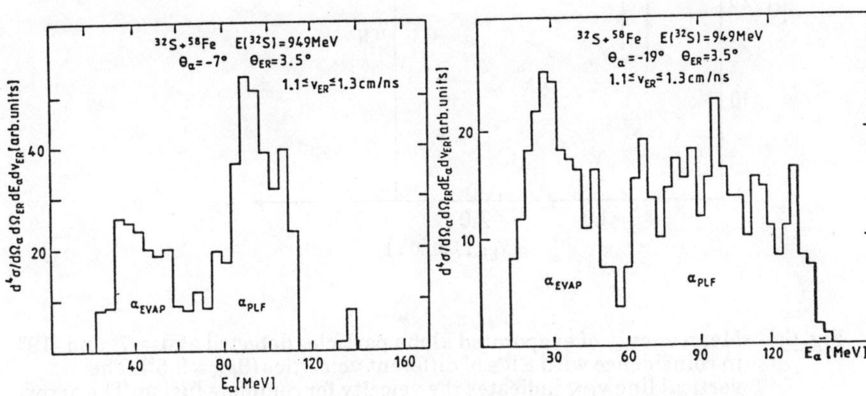


Fig. 5: Energy spectra of alphas from the reaction $^{32}S+^{58}Fe$ at 30 MeV/u measured at -7° and -19° in coincidence with evaporation residues of the velocity $v_{ER}$ = (1.2 ± 0.1) cm/ns. The residues were detected at $\theta_{ER}$=3.5° relative to the beam axis. The peaks at lower energies are due to evaporation alphas ($\alpha_{EVAP}$). The high energetic pre-equilibrium alphas are labelled $\alpha_{PLF}$.

Again, within the experimental uncertainties, the values of $v_{crit}$ extracted from this method are compatible with those obtained from the mass-width measurements, indicated by an arrow in fig. 6. The steep rise of <$E_{evap}$> can be reproduced with calculations using a level density parameter of A/10. Assuming the same value also for complete fusion, the mean energy of the alpha particles should amount to about 84 MeV at $v_{CN}$ and $\theta_\alpha$=-7°.

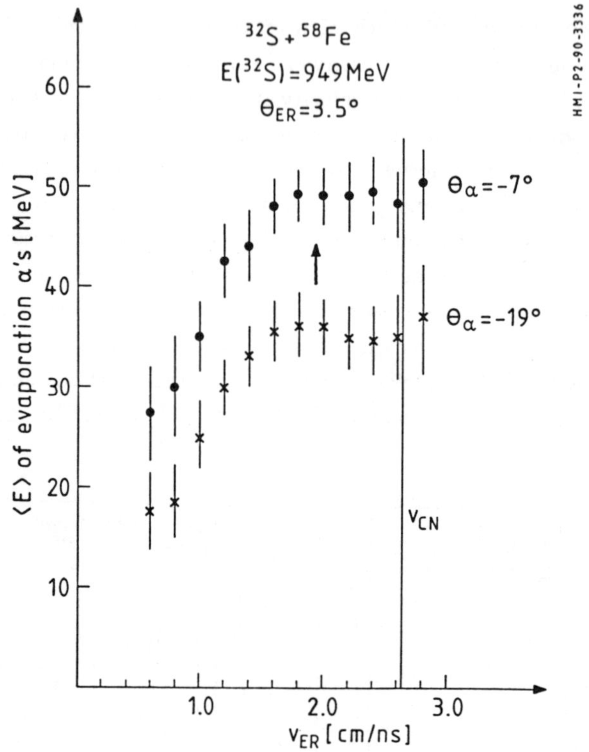


Fig. 6: Mean energy of evaporated alpha particles detected at θ=-7° and -19° in coincidence with ER's of different velocities ($\theta_{ER}$=3.5°). The vertical line $v_{CN}$ indicates the velocity for complete fusion. The arrow shows the value for the critical velocity obtained from the mass widths of the ER's.

Now we are going to transform the residue-velocity scale (cf. fig. 3) into an excitation-energy scale of the compound nuclei. To be as exact as possible, we include the second line of eq. 3. Therefore we have measured the coincidences between the ER's and all projectile-like fragments with $1 \leq Z \leq 12$. (The cross section for coincident PLF's with $Z>12$ is negligible.) For all elements we found a very steeply forward peaked angular distribution with a mean deflection angle of $<\theta_P>=(10\pm2)°$ for the centroid of $d^2\sigma/d\theta_P dZ$. On the average these fragments move only with $v'_P=(93\pm1)\%$ of the beam velocity. Assuming corresponding values for $\theta_T$ and $v'_T$, the correction resulting from the target-break up (third line of eq. 3) can be estimated to be less than 10% of the already small value due to projectile-break up. Therefore we neglected the third line.

To get an idea about a possible excitation energy of the spectators we analysed the velocity spectra of all ER's measured in coincidence with PLF's of one fixed mass (actually we summed over all isotopes of one PLF element in order to improve the statistics). A typical ER-velocity spectrum obtained in coincidence with all PLF's with $Z=8$ is shown in fig. 7.

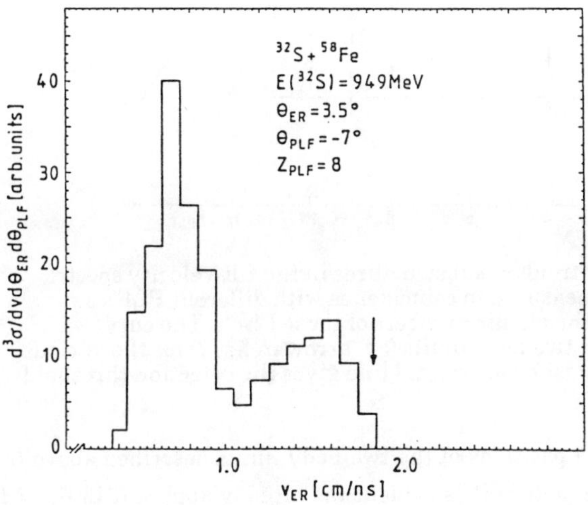


Fig. 7:

Velocity spectrum of ER's measured at 3.5° in coincidence with oxygen detected at -7°. The arrow indicates the position of the velocity centroid which is calculated in two-body kinematics with oxygen as primary spectator.

Assuming that the measured oxygen is the emitted spectator and assuming a two-body kinematics, one would expect the residues having a mean velocity as indicated by the arrow in fig. 7. But there the cross section is negligible. To explain the shift towards smaller velocities, one has to assume less momentum transfer. This means, the spectator must have been heavier. Possible explanations are: The detected PLF is a secondary product formed by sequential decay of an excited heavier spectator or more than one spectator have been formed. The positions of the distinct substructures can be explained within the two-body kinematics if one increases the mass of the detected PLF by 4, 8 or 12 mass units (1α-, 2α- or 3α-particles). This is also true for all the other PLF isotopes. This is demonstrated in fig. 8, where we plotted the velocity centroids of these peaks in the ER-velocity spectra versus the atomic number of the PLF's measured in coincidence.

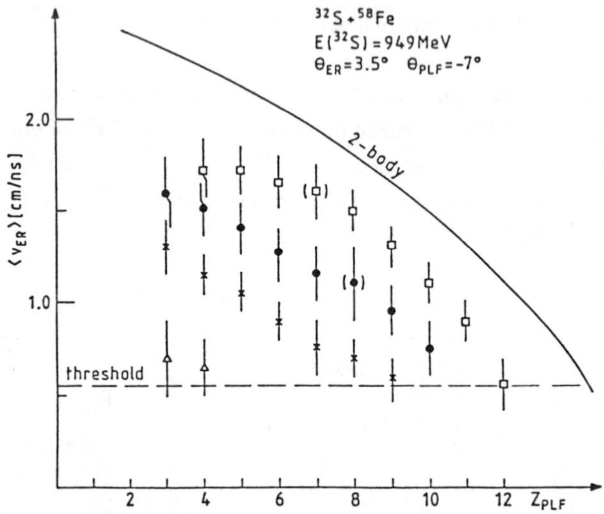


Fig. 8:: Velocity centroids of substructures in the ER-velocity spectra (see fig. 7) measured in coincidence with different PLF's as function of the atomic numbers of these PLF's. The curve connects the two-body limits (c.f. arrow in fig. 7) for the various PLF's. The dashed horizontal line gives the detection threshold

The full curve connects the positions of the two-body limits described above (e.g. arrow in fig. 7). From this picture it is evident (as already apparent in fig. 7 for Z=8) that there is no or at least no considerable coincident ER-cross section with the spectator being identical with the detected PLF. In most cases we are missing the mass of one to three alpha particles. This means, the participants always carry away some excitation energy. This excitation energy can be estimated to about 20 MeV on the average. The distribution should not be too broad (in the order of 10 MeV) because there are no cold spectators detectable on the one hand and on the other hand there still exist complex PLF's measured in coincidence with ER's. This small amount of energy was taken into account (second term of eq. 2) when calculating the critical excitation energy or, in general, when transforming the $v_R$ scale into an excitation-energy scale (see e.g. the upper abzissa in fig. 3).

This procedure we performed for all studied systems. The extracted critical excitation energies are summarized in table I together with other interesting parameters like the maximum possible excitation energy for complete fusion $\varepsilon^*_{CN}$, the critical velocity $v_{crit}$ and the mass of the corresponding heaviest compound nucleus $A^{crit}$ which contributes to the ER cross section.

Table I: Critical excitation energy per nucleon $\varepsilon^*_{crit}$ measured at the cricial velocity $v_{crit}$ with the calculated mass of the corresponding compound nucleus $A^{crit}$. Maximum excitation energy $\varepsilon^*_{CN}$ for complete fusion expected at $v_{CN}$ with the compound nucleus mass $A_{CN}$.

| $E(S^{32})$ [MeV] | Target | $\varepsilon^*_{CN}$ [MeV/u] | $v_{CN}$ [cm/ns] | $A_{CN}$ [u] | $v_{crit}$ [cm/ns] | $\varepsilon^*_{crit}$ [MeV/u] | $A^{crit}$ [u] |
|---|---|---|---|---|---|---|---|
| 708 | $^{58}$Ni | 4.8 | 2.3 | 90 | 2.05±0.05 | 4.4±0.3 | $84\pm^{1}_{3}$ |
| 708 | $^{58}$Fe | 5.0 | 2.3 | 90 | 2.05±0.05 | 4.5±0.3 | $84\pm^{1}_{3}$ |
| 828 | $^{80}$Se | 5.1 | 1.9 | 112 | 1.7 ±0.1 | 4.3±0.3 | 106±3 |
| 828 | $^{58}$Ni | 5.7 | 2.5 | 90 | 2.05±0.05 | 4.8±0.3 | $81\pm^{1}_{3}$ |
| 839 | $^{58}$Fe | 5.9 | 2.5 | 90 | 2.0 ±0.05 | 4.9±0.3 | $80\pm^{1}_{3}$ |
| 828 | $^{45}$Sc | 6.3 | 2.9 | 85 | 2.5 ±0.1 | 5.9±0.3 | $69\pm^{2}_{3}$ |
| 955 | $^{58}$Ni | 6.6 | 2.7 | 90 | 1.94±0.05 | 5.1±0.3 | $77\pm^{1}_{3}$ |
| 946 | $^{58}$Fe | 6.7 | 2.6 | 90 | 1.95±0.05 | 5.2±0.3 | $77\pm^{1}_{3}$ |

In order to demonstrate the mass dependence of the critical excitation energy, we plotted in fig. 9 the results as a function of the compound nucleus mass together with predicted values according to the work of Leray [1], shown as full curve. The dashed curves indicate the possible upper and lower limits of this calculation. The experimental values are not very different from those within the predicted area, but it is evident that they drop much faster than the calculated curves. An extrapolation to larger masses would yield a value of zero for the critical excitation energy for evaporation at a compound-nucleus mass of about 200. That would be compatible with the vanishing cross section of evaporation residues for these heavy compound nuclei, even at low excitation energies.

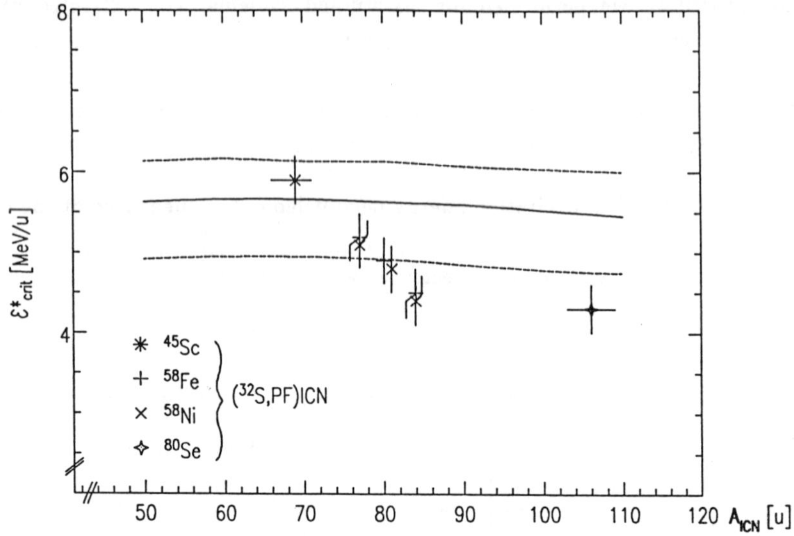


Fig. 9: Critical excitation energies of compound nuclei formed with the reactions $^{32}S+^{45}Sc$, $^{58}Fe$, $^{58}Ni$, $^{80}Se$ as function of their mass number. The full curve shows the prediction of Leray [1] with possible upper and lower limits (dashed curves).

Of course, the picture of Leray [1] is very simple. The mass of the compound nucleus alone is perhaps not sufficient to describe the behaviour of the critical excitation energy. In the model of Suraud [6], for example, the ratio of Z/A of the compound nucleus is also included. Besides the mass dependence (decreasing $\varepsilon^*_{crit}$ with increasing mass) this model shows a non-neglectable decreasing of $\varepsilon^*_{crit}$ with growing Z/A ratio. In order to investigate this Z/A dependence we studied the two isobaric systems $^{32}S+^{58}Ni$ and $^{32}S+^{58}Fe$. The difference between the two Z/A ratios is not very large. Using this difference and extrapolating the calculations of Suraud [6] to the mass region studied here, one can estimate that the critical excitation energy in the case of the S+Fe reactions should be about 0.1 MeV/u higher than in the reactions with the Ni target. That is exactly the difference, in value as well as in sign, we measured for all three energies. Unfortunately the experimental uncertainty is still larger than this difference.

Up to now we only discussed the critical excitation energy concerning light-particle evaporation. But at these high excitation energies the fission-decay channel is by far not negligible, even in the case of such light compound nuclei with A≈65. At the highest energy (949 MeV) and, for example for the Fe target the fission cross section is roughly of the same strength as the one for ER's. Therefore, we will try to check whether there exists as well a critical excitation energy for the fission-decay channel. To do so, we are going to apply the method of measuring the mean energy of evaporated alpha particles to fission fragments. Because we did not measure the folding-angle correlations of the fission fragments to determine the momentum transfer we have to find another indication for the degree of completeness of the fusion-fission reaction. Again, the idea is very simple. We know from the inverse fusion reaction $^{58}Ni + ^{27}Al$ with a Ni beam of 840 MeV, forming a very similar compound nucleus, that the major part of its excitation energy is carried away by particle evaporation before fission occurs [17]. For heavier systems with mainly neutron evaporation this fact is well known [18].

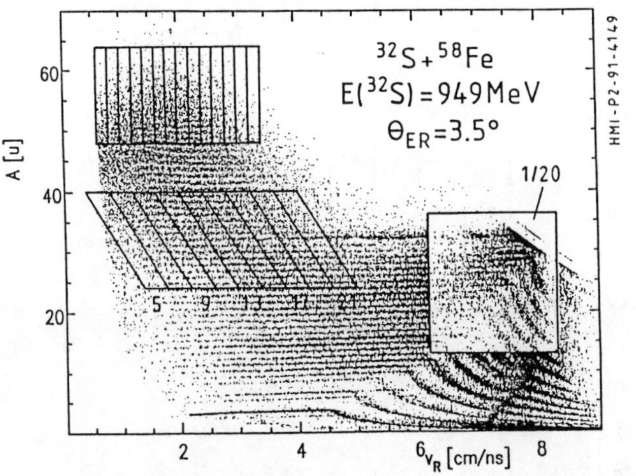


Fig. 10: Mass-velocity spectrum of the fusion reaction $^{32}S + ^{58}Fe$ at 30 MeV/u. The gates with numbers 5 to 21 indicate the coincidence conditions for the alpha particles and FF's. For comparison, the vertical gates above represent the conditions for the coincidences with the ER's.

This means, the averaged alpha-energy spectrum in coincidence with fission fragments should be very similar to the one measured in coincidence with ER's. Furthermore we know that the mass centroids of the ER distributions originating from different momentum transfers have almost the same value. Approximately from these centroids at different velocities the corresponding fission branches start. The measured total fission distribution is a superposition of these almost parallel branches. Therefore, velocity bins parallel to the maximum of the total fission branch (the gradient is roughly given by the Viola systematics extended to asymmetric mass split [19]) are a measure of the momentum transfer. With such conditions on the fission events, depicted as sloping gates with the numbers 5 to 21 in fig. 10, we have analysed again the mean energy of the evaporated alpha particles. The values obtained for the different gates are shown in fig. 11.

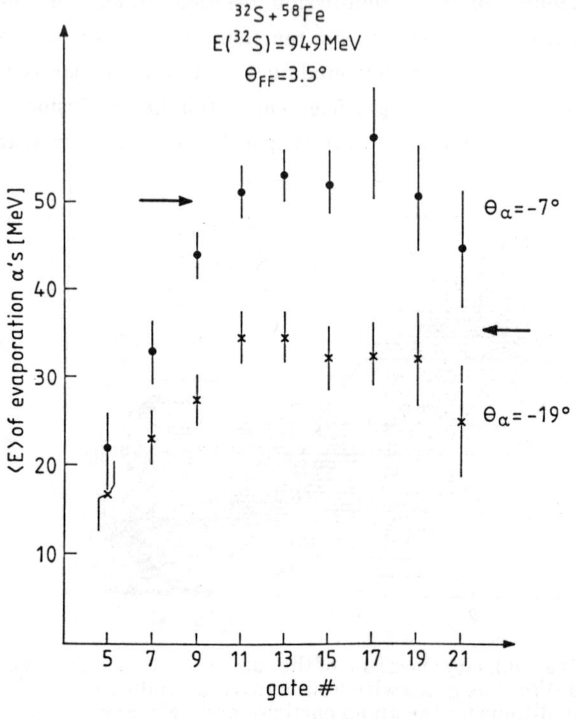


Fig. 11: Mean energies of evaporated alpha particles detected at -7° and -19° in coincidence with FF's inside the different gates (see fig. 10). The horizontal arrows indicate the averaged maximum alpha energies obtained with coincident ER's (see fig. 6).

The distribution looks very much the same as the one of the corresponding results for the coincidence measurements with the ER's. Especially the mean values of the maximum alpha energies (the results from the coincidences with the ER's are indicated by horizontal arrows in fig. 11) are within the error bars the same, i.e. the sources for the alphas with the highest energy are the same. From this we can conclude that the critical excitation energy for the composite system which finally fissions is more or less the same as the limiting excitation energy for evaporation, at least for this system analysed up to now. This means, the compound nuclei formed with higher excitation energies up to complete fusion have to decay differently than by normal evaporation and fission.

## V. Summary

We have shown that the inclusive measurement of the evaporation residue mass widths as a function of the residue velocity is a new and simple method to check whether there exists a limiting excitation energy for normal evaporation of light particles. The existence of such a critical excitation energy is evident for all the reactions we have studied. The behaviour of the multiplicities and energies of the evaporated charged particles corroborates the findings obtained with our new integral method.

The extracted values of the critical excitation energy are compatible with the predictions of model calculations. But the dependence on the mass of the formed compound nuclei is remarkably stronger than expected, at least in the studied mass region $65 < A < 110$. A predicted [6] influence of the Z/A ratio of the compound nucleus on the critical excitation energy seems to exist.

Finally we attempted to show that also the de-excitation via binary fission is only possible up to a critical excitation energy, which seems to have the same value as the one determined for evaporation of light particles, at least for the case analysed up to now. The existence of a (possibly common) maximum excitation energy, above which a de-excitation by normal evaporation of light particles and/or binary fission is no longer possible, requires a new decay mechanism for all the compound nuclei with excitation energies higher than $\varepsilon^*_{crit}$. This is of course not a proof for the existence of multifragmentation, which in all the models cited above is the reason for the appearance of $\varepsilon^*_{crit}$.

Further experiments are needed to complete our knowledge of the dependence of $\varepsilon^*_{crit}$ on the parameters of the compound nuclei (like mass dependence, dependence on the Z/A ratio, etc.), especially for the fission-decay channel, to be able to have a more detailed comparison with the various models.

References


1) S. Leray, J. de Phys. **47**, C4 (1986) 275.
S. Leray et al., Z. Phys. **A320** (1985) 533.

2) S. Levit and P. Bonche, Nucl. Phys. **A437** (1985) 426.

3) P. Bonche et al., Nucl. Phys. **A436** (1985) 265.
P. Bonche et al., J. de Phys. **47**, C4 (1986) 339.

4) J. Bondorf et al., Nucl. Phys. **A444** (1985) 460.

5) D.H.E. Gross et al., Phys. Rev. Lett. **56** (1986) 1544.
D.H.E. Gross and H. Massmann, Nucl. Phys. **A471** (1987) 339c.

6) E. Suraud, Nucl. Phys. **A462** (1987) 109.

7) J. Desbois et al., Z. Phys. **A325** (1986) 245.
J. Desbois et al., Z. Phys. **A328** (1987) 101.

8) H.W. Barz et al., Phys. Lett. **184B** (1987) 125.

9) A. Fahli et al., Phys. Rev. **C34** (1986) 161.

10) M. Conjeaud et al., Phys. Lett. **B159** (1985) 244.

11) U. Jahnke et al., this conference and references therein.

12) H. Morgenstern et al., Phys. Lett. **113B** (1982) 463 and references therein.

13) H. Morgenstern et al., Phys. Rev. Lett. **52** (1984) 1104.

14) T. Nakagawa and W. Bohne, Nucl. Instr. and Meth. **A271** (1988) 523.

15) S.B. Kaufmann et al., Nucl. Instr. and Meth. **115** (1974) 47.

16) W. Bohne et al., Nucl. Instr. and Meth. **A240** (1985) 145.

17) H. Morgenstern et al., Annual Report 1990, HMI (1991) contribution 2.2.7.

18) E. Mordhorst et al., Phys. Rev. **C43** (1991) 716 and references therein.

19) V. Viola et al., Phys. Rev. **C31** (1985) 1550.

# NON-EQUILIBRIUM EMISSION AND THE DECOMPOSITION OF THE CROSS SECTION IN FUSION-LIKE REACTIONS

H.K.W. Leegte, R.H. Siemssen, K. Siwek-Wilczynska, J. Wilczynski, H.W. Wilschut
*Kernfysisch Versneller Instituut, Zernikelaan 25, 9747AA Groningen, The Netherlands*

March 23, 1991

**Abstract**

The reaction $^{14}N+^{232}Th$ at 30 MeV/nucleon has been studied using large area particle and fission detectors. The correlation between momentum transfer and charged particle emission has been measured. It is shown that in fusion-like reactions the incomplete momentum transfer can be attributed to reactions where mainly a single beam-like particle is emitted. The distribution of the momentum transfer is determined by the mass distribution of these particles and not by the multiplicity distribution of emitted nucleons. The experimental techniques are discribed. The consequences of the importance of complex particles emission for a theoretical interpretation of fusion-like reactions are discussed.

# 1 Introduction

One of the basic problems in nucleus-nucleus reactions in the energy domain between 10 and 100 MeV/nucleon is the mechanism by which the relative motion of the colliding nuclei is dissipated into heat. For asymmetric systems where the projectile is much lighter than the target one observes that with increasing beam energy the projectile cannot be stopped completely in the target because the composite system does not obtain the full beam momentum. In fig. 1 (from ref. [1]) this effect is shown for a range of reactions. From this compilation it is also seen that the missing momentum seems to depend only on the beam velocity at impact. In the work presented here it was the aim to find by which particles the missing momentum is carried away. For this we chose to study the reaction $^{14}N + ^{232}Th$ at 30 MeV/nucleon. With a thorium target all non-peripheral reactions will lead to fission, and in this way the momentum transfer can then be measured with the folding angle method.

Conceptually it would be convenient if the missing momentum is carried by non-equilibrium nucleons for reasons we will shortly discuss in the next section. We find, however, that emission of nucleon clusters is very important (d, t, $\alpha$). The data seem to indicate two types of mechanisms for non-equilibirum emission: one associated with beam-like particles presumably from the first stage of the reaction and one associated with a later stage but prior to complete equilibration, i.e. pre-equilibrium emission.

## 1.1 Theoretical concepts

The projectile nucleons entering the target with low velocity are stopped when they cannot escape from the mean-field of the target. At large velocity they must collide with target nucleons until their velocity is reduced enough before they can be trapped. At low beam energy collisions between the nucleons inside the nuclei are forbidden due to Pauli blocking (one-body disspation). In that case projectile nucleons with energies larger than $(\sqrt{\epsilon_F} + \sqrt{E_B})^2$ will escape from the target. This process goes by the name of Fermi-jetting. The Fermi energy ($\epsilon_F$) and the binding energy of a nucleon ($E_B$) are nearly independent of the nuclear mass. In this respect the incomplete momentum transfer should be a universal function of the beam velocity independent of the reaction as fig.1 may indicate. An upper limit for Fermi jetting based on these simple arguments [2] is indicated by the dashed line in fig. 1, it underestimates the momentum transfer but gives the overall trend. An improvement is found using window-wall dissipation, which is an elegant way to describe both Fermi-jetting and one-body dissipation [3]. An *upper limit* for Fermi-jetting using eq. 30 of ref. [3] is shown in fig.1 as the drawn line.

With increasing energy Pauli blocking is not complete, and collisions between the nucleons from projectile and targets are allowed (two-body dissipation). Therefore more momentum transfer can occur. A model that includes two-body dissipation is the exciton model [4]. It has been developped for nucleon induced reactions but was extended for heavy ion reactions [5]. The latter model is based on the Boltzmann master equation (BME), it describes the time evolution of the particle-hole excitations in the composite system allowing emission of nucleons into the continuum during the relaxation process. For this work we will make use of the code of Blann et al. (see ref. [5] for more details).

More recently attempts are made to describe the time evolution of the complete phase space population $f(\vec{r}, \vec{p}, t)$ using the Vlasov-Uhling-Uhlenbeck equation. For illustrative purposes we will make use of the code used by Cassing and co-workers, for details see ref.

[6]. VUU can also be used to study aspects of the more phenomenological models.

In all models sketched here, only nucleon emission can be described. Although a simple model for one-body dissipation appeared to explain the general trend of the data (fig. 1) arguments were given that stopping should be more effective when two-body dissipation is taken into account. Therefore, the nature of the processes leading to incomplete momentum transfer may be more complex than the emission of non-equilibrium nucleons only. For this reason we like to establish a correlation between momentum transfer and the associated particle emission.

## 1.2 Experimental details

For the experiment a $^{14}N^{7+}$ ions from the ECR source was accelerated by the KVI cyclotron to 420 MeV. A natural Th target was used of 550 $\mu g/cm^2$ areal density. In fig. 2 a schematic view is given of the setup. Left and right of the beam two multi-wire avalanche detectors [7] were placed to detect fission fragments. In the forward angle region 30 phoswich detectors were put in close geometry to measure the charged particles in coincidence with the fission fragments. These detectors consist of 1mm fast plastic scintillator (NE102) and about 5 cm of slow scintillator (NE115) [8]. The detectors were calibrated in the focal plane of a spectrograph, and were found to be linear over the full dynamic range where particles could be identified. The calibration constants depend both on Z and A. The $\Delta E, E$ information was obtained in the usual way with a short and long gate on a charge sensitive ADC. Common gate ADC's were used (FERA). If more than one detector fires the differences in the gate timing affect mainly the values for $\Delta E$. A simple software procedure was developped to correct for this effect using the recorded time and energy information.

For the analysis only those events were used where at least one of the fission fragments hits a particular region of the detector. The region is chosen such that it garantees detection of the other fragment in the opposite detector. Experimentally this can be achieved by requiring that the full angular correlation of the fission fragments is observed in one detector when gating on a small part of the other. This method of selecting events facilitates comparison of reaction channels with minimal bias. In order to determine the momentum transfer the folding angle was determined from the two impacts on the MWAD's with $\Theta_{ff} = acos(\vec{F_1} \cdot \vec{F_2})$ where $\vec{F_1}$ and $\vec{F_2}$ are unit vectors in the direction of the two fission fragments. Although this definition is straight forward, a disadvantage is that the folding angle distribution is per definition zero when $\Theta_{ff} = 180°$ (cf. the angular distribution $d\sigma/d\Theta = 2\pi sin\Theta d\sigma/d\Omega = 0$ at 0°). In the literature other definitions have been used, although never explicitly and often quite unclear. However, in the region of interest (fusion-like reactions) the various definitions are expected to give nearly identical values for $\Theta_{ff}$. As an example, we show in fig. 3 folding angle distributions for cases where only one particle (C, Li or H) was detected in the forward wall. Note that for hydrogen particles a the distribution has a double structure. The structure at small $\Theta_{ff}$ is associated with fusion-like reactions, whereas the component at large $\Theta_{ff}$ is due to processes such as $^{14}N + {}^{232}Th \rightarrow {}^{13}N^* + {}^{233}Th^* \rightarrow {}^{12}C + {}^{1}H + F_1 + F_2$. In the latter, the larger fragment has escaped detection due to the opening in the detection wall around 0°. This point illustrates that the effects of the opening for the beam at 0° must be considered. The data shown here are not corrected for this effect unless stated otherwise. Correction methods will be discussed elsewhere [9]. Because of the hole around 0° the forward wall is not effective to select central collisions and additional measurement of an observable associated with the target-like fragment will always be necessary to remove peripheral events [10, 11]. In the present case the fission fragments are used.

## 2 Experimental results

Using fission as a trigger we obtain the fold distribution of hits in the forward wall, as shown in fig. 4. The largest contributions are reactions in which no - or only one - particle is observed. Charged particle evaporation from excited target-like fragments is not important for this distribution, as can be concluded on basis of the forward peaked angular distributions observed for single hit events of protons and $\alpha$-particles. Therefore, the multiplicity distribution shown in fig. 4 is almost entirely due to non-equilibrium particles. In fig. 5 the partition for M=1 events is shown. Here the largest contributions are due the light particles. Note that the sum of the deuteron and triton yield is as large as the proton yield. Although, the ratio $p : d : t = 10 : 6 : 4$ is near what is expected for statistical emission from the compound nucleus (10 : 5 : 4), we will show later that the spectra are non-evaporative. Figs. 4 and 5 clearly indicate qualitativley the main result of the experiment, i.e. that per event about one charged particle is emitted prior to equilibration, and that the emitted fragment has mostly $A > 1$.

In order to obtain the momentum transfer the angular information from the fission detectors was used. Using various approximations [9, 7] we obtain for the momentum transfer relative to the beam momentum

$$\frac{< \Delta p >}{p} = \sqrt{\frac{M_T < E_{Viola} >}{M_P E_P}} \frac{sin\Theta_{ff}}{\sqrt{1 + cos^2\Theta_{ff} - 2cos\Theta_{ff}cos(\Theta_{ff} - 2\Theta_1)}}. \quad (1)$$

Here, $M_T$ is the mass of the composite system, $< E_{Viola} >$ the total kinetic energy expected for fission fragments [12], $M_P$ and $E_P$ the mass and energy of the projectile, and $\Theta_1$ is the emission angle of one of the fragments. The result is shown in fig. 6 as a function of the total detected charge for various multiplicities. The drawn line indicates what would be expected if the particles were emitted with beam-velocity. The method of eq.(1) is poor when the momentum transfer is small (large Z). If no particle is observed the momentum transfer is not 100%, this is due to two effects: The limited coverage at 0° and the emission of non-equilibrium neutrons. For a given value of total detected charge, $Z_{det}$, the momentum transfer increases. The larger multiplicity apparently requires a "harder" impact, a similar observation was made for $\gamma$-ray multiplcity measurements [13].

From the inclusive folding angle distribution (not shown) we find the most probable momentum transfer to be 82%, which is slightly above the trend shown if fig. 1 but agrees very well with the value found for $^{14}N + ^{238}U$ [15] and $^{197}Au$ [16] at 30 MeV/nucleon.

We define the fusion-like cross section by the minimum in the folding angle distribution (corresponding to 50% of the beam momentum). With this definition and a preliminary estimate for the limited coverage of the forward wall we find on the average 1.1 emitted protons (including the ones bound in clusters) and an average observed mass of 2.0 nucleons. The average number of neutrons bound in clusters is thus 0.9. Unobserved are the unbound neutrons. From the systematic of neutron measurements of Hilscher and others [14] one may estimate approximately 2.2 pre-equilibrium neutrons.

# 3 Comparison with theory

In the following we compare the average momentum transfer and the pre-equilibrium nucleon multiplicity to the results of various theories mentioned in the introduction. (There are presently no models that can predict the fluctuations.) The main difficulty is to include the fact that cluster emission is so important. In the present comparison we assume that clusters are formed in some sort of coalesence. The results are presented in table 1, where we have listed values for $< \Delta p/p >$, the average number of protons and the total emitted mass. Note, that the latter experimental value does not include unbound neutrons. The jetting model of ref. [3] underestimates the nucleon multiplicities. Better agreement is obtained if the result for $b = 0$ is taken instead of the angular momentum averaged values. The Blann model underestimates the momentum transfer, this is mainly because the assumed angular distribution of emitted particles is taken too narrow [5]. The entry labeled "Bertsch" is taken from ref. [18] and concerns VUU type calculations, the values are for the reaction $^{14}N + ^{154}Sm$ at 35 MeV/nucleon. The stopping is surprisingly effective (large momentum transfer). This is also the case for the prelimenary value we obtained with the code of Cassing et al. [6]. The nucleon multiplicities predicted by these models are large. We have only considered the emission between 0° and 90°. The angular distributions are very wide also the energy spectra peak at small energies, in how far this is due to the calculational technique is not clear. Finally, as an exercise, we made a calculation without allowing for nucleon-nucleon collisions (mean field only). In this case the momentum transfer is very small and compares nicely to the one-body dissipation limit.

**Table 1, Comparison to theory**

| | $< \Delta p/p >$ (%) | $< M_p >$ | $< Mass >$ | ref. |
|---|---|---|---|---|
| experiment | 82 | 1.1 | 2.0 | this work (prelim.) |
| Möhring et al. | - | 0.4 | 2.1 | [3] |
| idem, $b = 0$ | 83 | 0.8 | 2.5 | |
| Blann | 72 | 0.6 | 3.1 | [5, 15] |
| Bertsch | 85 | - | 11.0 | [18] |
| Cassing | 89 | - | 3.8 | [6] |
| idem, mean field | 48 | - | 8.3 | |
| one-body limit | 48 | - | - | [2] |

In fig. 7 we compare the measured energy spectra (transfered to CM and integrated over angle) to the prediction of the BME model [5]. The proton and alpha particle energy spectra shown here are absolute (i.e. per fission event). The drawn line in the proton spectrum is the result of the calculation. It overpredicts the number of emitted protons. According to the interpretation given above, the excess proton yield should be found in the cluster yield. We have therefore applied the coalesencse method of ref. [17] to the theoretical data. The result is shown for $\alpha$-particles, where we have normalized the calculation to the maximum experimental yield. Clearly the higher energy side is underpredicted, which is consistent with the result in table 1, where it shown that only part of the total emitted charge can be accounted for by the BME model.

The excess on the high energy side of the $\alpha$-particle energy spectrum can also be interpreted as the result of more peripheral reactions. The fusion-like reactions are about 60% of the total reaction cross section, therefore the overlap of the projectile with the target is not always complete. In this case part of the projectile may escape as a cluster. Such reaction channels are of course not included in the simple models. In the VUU case the details concerning formation of small clusters can not be described very well [19]. Also in the Quantum Molecular Dynamics approach the formation of an $\alpha$-particle as the only product of an peripheral reaction would be difficult to describe [20]. It is therefore worthwile to persue studies where the distinction between "head-on" and more "off central" reactions could be made more cleanly.

# 4 Summary

By measuring the fission folding angle distribution in coincidence with the associated particles, we have found a correlation between momentum transfer and the number and type of particles emitted. In fusion-like reactions the non-equilibrium particles are often emitted in the form of a cluster $(d, t, \alpha)$. The momentum transfer distribution of the fusion-like reactions are not due to variations in the multiplicity of unbound nucleons, but rather seems to be related to the distribution in the cluster size, because mostly only one charged particle is emitted in reaction of this type. Models for fusion-like reactions in the present energy domain are based on the "head-on" reaction picture. They predict only nucleon emission, whereas experiment shows the importance of complex particle emission. With the detailed partition of the cross sections in combination with the momentum transfer a better understanding of the dynamics of fusion-like reactions may become possible.

This work performed as part of the research program of the Stichting voor Fundamenteel Onderzoek der Materie (FOM), with financial support from the Nederlandse Oraganisatie voor Wetenschappelijk onderzoek (NWO).

# References

[1] J. Jastrzebski, P.P. Singh, T. Mroz, H.J. Karwowski, S.E. Vigdor, M. Fatyga, Phys. Lett. B136(1984)153.

[2] C. Gregoire, F. Scheuter, Phys. Lett. B146(1984)21.

[3] K. Möhring, W.J. Swiatecki, and M. Zielinska-Pfabé, Nucl. Phys. A440(1985)89, and refs. therein.

[4] G.D. Harp, J.M. Miller, B.J. Berne, Phys. Rev. 165(1968)1116.

[5] M. Blann, Phys. Rev. C31(1985)1245.

[6] W. Cassing, V. Metag, U. Mosel, K. Niita, Phys. Rep. 188(1990)363.

[7] P.C.N Crouzen, thesis 1988, Groningen.

[8] E.E. Koldenhof, KVI annual report 1986, p.132.

[9] H.K.W. Leegte et al., to be published.

[10] M.D. Tsang, Y.D. Kim, N. Carlin, Z. Chen, R. Fox, C.K. Gelbke, W.G. Gong, W.G. Lynch, T. Murakami, T.N. Nayak, R.M. Ronningen, H.M. Xu, F. Zhu. L. Sobotka, D. Stracener D.G. Sarantites, Z. Majka, V. Abenante, H. Griffin, Phys. Lett. B220(1989)492.

[11] J.D. Hinnefeld, E.E. Koldenhof, H.K.W. Leegte, R.H. Siemssen, Z. Sosin, H.W. Wilschut, Z. Zhang, Phys. Lett. B225(1989)308.

[12] V.E. Viola, K. Kwiatkowski, M. Walker, Phys.Rev. C31(1985)1550.

[13] G.J. Balster, H.W. Wilschut, R.H. Siemssen, P.C.N. Crouzen, P.B. Goldhoorn, Z. Sujkowski, Nucl.Phys. A468(1987)131.

[14] D. Hilscher, H. Rossner, A. Gamp, U. Jahnke, B. Cheynis, B. Chambon, D. Drain, C. Pastor, A. Giorni, C. Morand, A. Dauchy, P. Stassi, G. Pettit, Phys. Rev. C28(1987)208, and refs. therein.

[15] M. Fatyga, K. Kwiatkowski, V.E. Viola, C.B. Chitwood, D.J. Fields, C.K. Gelbke, W.G. Lynch, J. Pochodzolla, M.B. Tsang, and M. Blann, Phys. Rev. Lett. 55(1985)1376.

[16] M.B. Tsang, C.B. Chitwood, D.J. Fields, C.K. Gelbke, D.R. Klesch, W.G. Lynch, K. Kwiatkowski, V.E. Viola, Phys. Rev. Lett. 52(1984)1967.

[17] T.C. Awes, G. Poggi, C.K. Gelbke, B.B. Back, B.G. Glagola, H. Breuer, V.E. Viola, Phys. Rev. C24(1981)89.

[18] M.B. Tsang, G.F. Bertsch, W.G. Lynch, M. Tohyama, Phys. Rev. C40(1989)1685.

[19] A. Bonesera, M. Colonna, M. Di Toro, F. Gulminelli, H.H. Wolter, Phys. Lett. B244(1990)169.

[20] G. Peilert, H. Stöcker, W.Greiner, A. Rosenhauer, A. Bohnet, J. Aichelin, Phys. Rev. C39(1989)1402, and J. Aichelin private communication.

## Figures

Fig. 1 Systematic for momentum transfer in fusion-like reactions, taken from [1]. The dashed curve is for the one-body dissipation limit according to [2] and the drawn curve for jetting using the window formalism [3].

Fig. 2 Schematic setup of the position sensitive avalanche detectors (PSD) and phoswhich detectors.

Fig. 3 Folding angle distribution obtained by gating on various particles. On the right, the effect of mass transfer on the folding angle is shown schematically (exaggerated).

Fig. 4 Multiplcity distribution of charged particle observed in the forward wall.

Fig. 5 The partition of reaction channels when only one particle is observed in the forward wall.

Fig. 6 Average momentum transfer as function of total detected charge, $Z_{det}$ and multiplicity $M$.

fig. 7 Experimental energy spectra (histogram) and calculated spectra, see text.

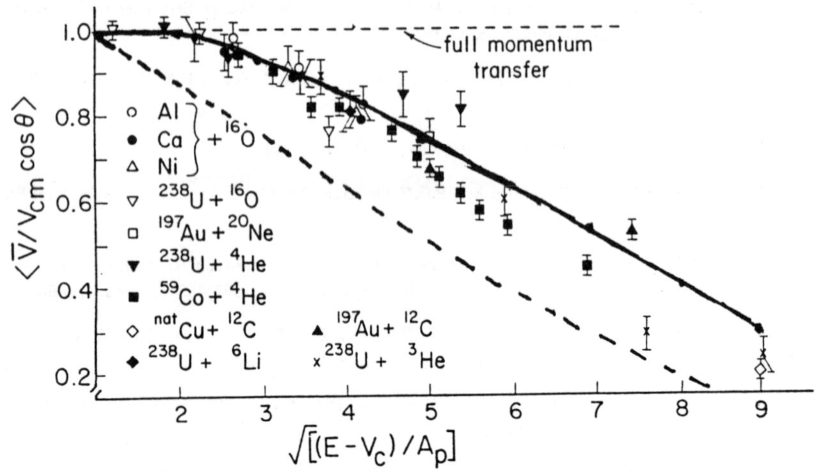

full momentum transfer
Al
Ca
Ni
+ $^{16}$O
$^{238}$U + $^{16}$O
$^{197}$Au + $^{20}$Ne
$^{238}$U + $^{4}$He
$^{59}$Co + $^{4}$He
$^{nat}$Cu + $^{12}$C
$^{238}$U + $^{6}$Li
$^{197}$Au + $^{12}$C
$^{238}$U + $^{3}$He
$\langle \bar{V}/V_{cm} \cos\theta \rangle$
$\sqrt{[(E-V_c)/A_p]}$


Fig. 1

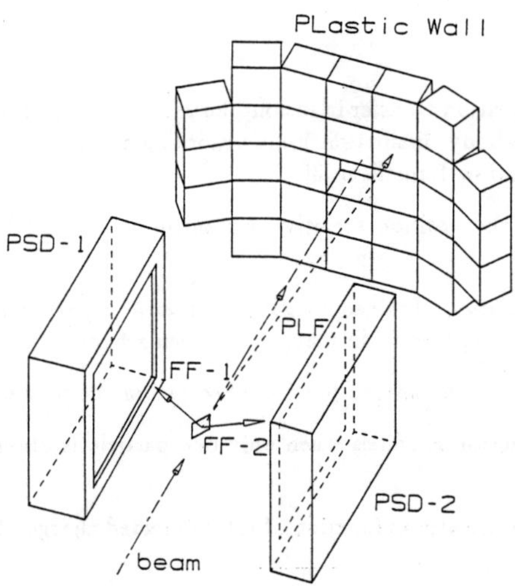

PLastic Wall
PSD-1
PLF
FF-1
FF-2
PSD-2
beam


Fig. 2

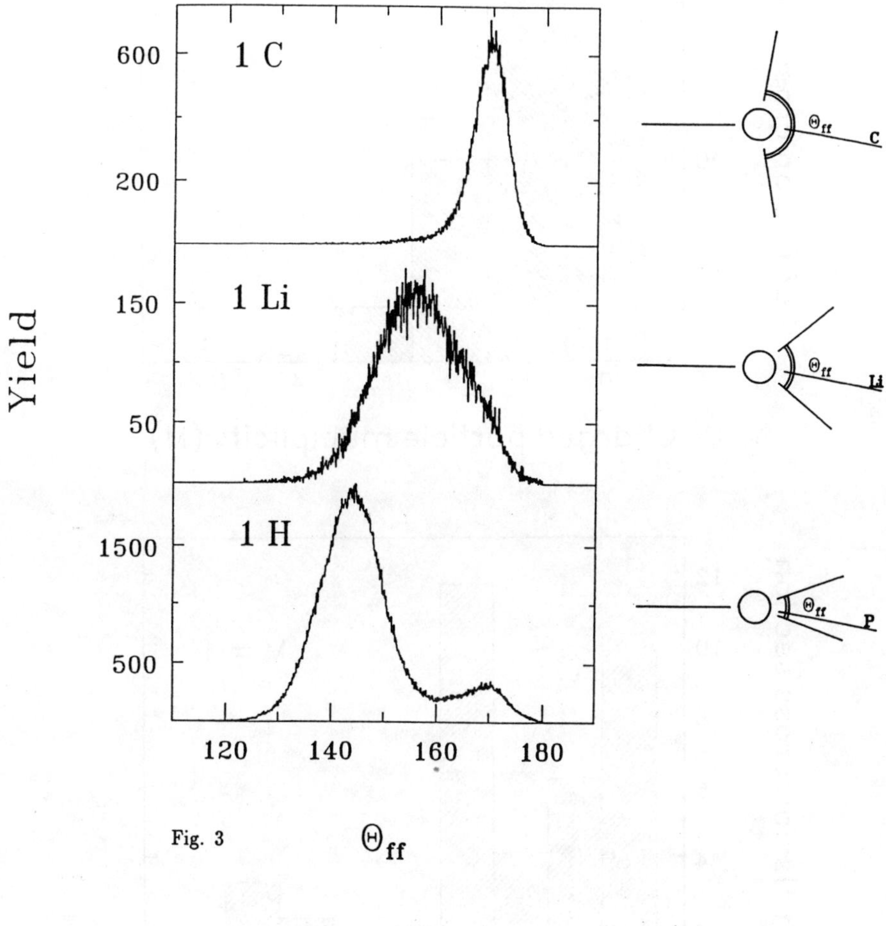
600
1 C
200
1 Li
150
50
1 H
1500
500
Yield
120
140
160
180
$\Theta_{ff}$
C
Li
P

Fig. 3

Measured partition of the fission cross section.

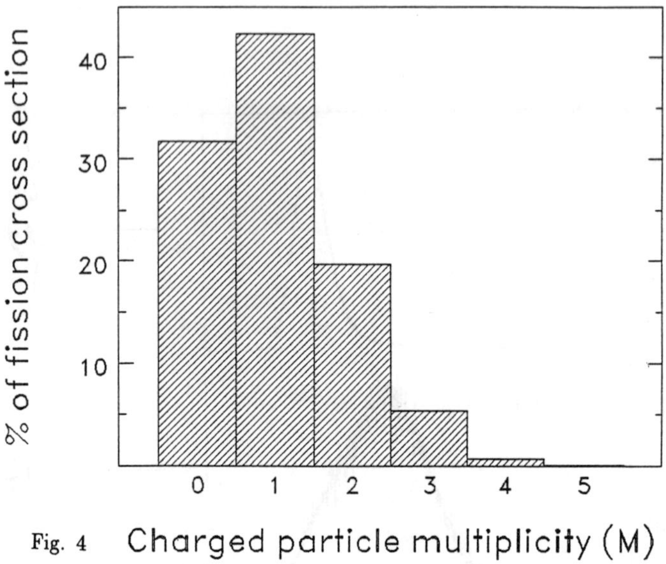


Fig. 4

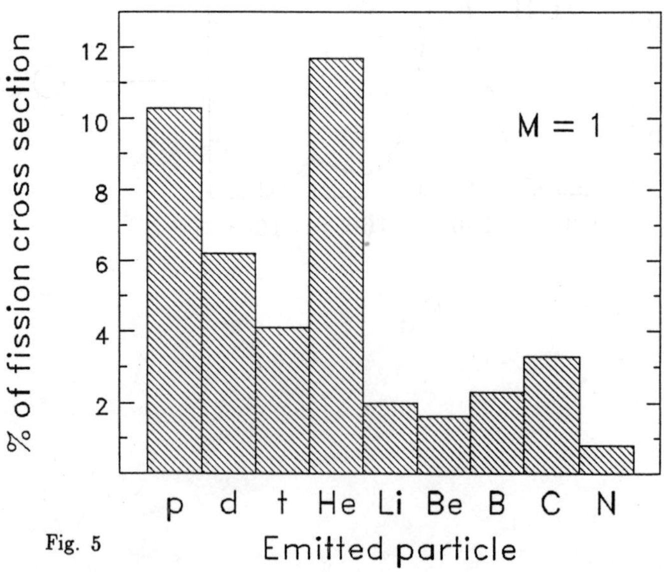


Fig. 5

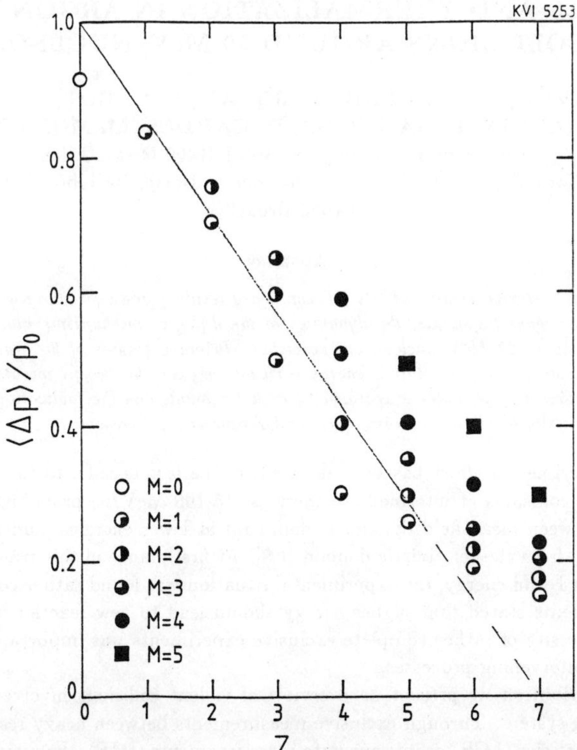
KVI 5253
⟨Δp⟩/p0
Zdef
M=0
M=1
M=2
M=3
M=4
M=5

Fig. 6

fig. 7

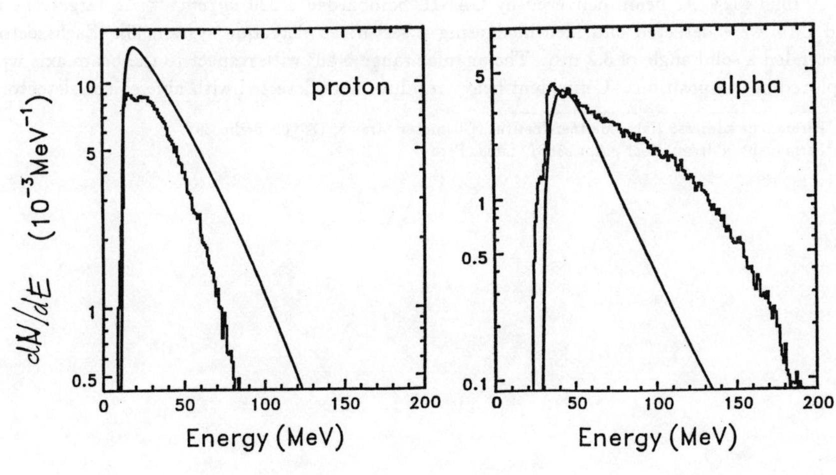
proton
alpha
dN/dE (10^-3 MeV^-1)
Energy (MeV)
Energy (MeV)

# DYNAMICS AND THERMALIZATION IN ARGON INDUCED COLLISIONS AROUND 30 MeV/NUCLEON

M. F. RIVET, B. BORDERIE, D. JOUAN, C. CABOT, H. FUCHS[1] ,
H. GAUVIN, F. HANAPPE*, D. GARDES, M. MONTOYA[2]
Institut de Physique Nucléaire, F91406 Orsay Cedex
*Fonds National de la Recherche Scientifique et Université Libre de Bruxelles,
B-1050 Bruxelles

**Abstract**

*Through exclusive measurements between heavy residues and light charged particles or intermediate mass fragments, the dynamics of the different mechanisms involved in the* $^{40}Ar +^{nat} Ag$ *at 27 MeV/nucleon are described. Primary masses of the fragments can then be calculated. The excitation energy partition between the two fragments is derived from the number of particles evaporated by each fragment, and thermalization times are deduced. Finally linear momentum, mass and Z balances are presented.*

A large experimental effort has been devoted, in the last decade, to the investigation of nucleus-nucleus collisions at intermediate energies (15-100 MeV/u) providing probes for the competition between mean-field dynamics, dominant at lower energies, and nucleon-nucleon collisions known from the relativistic domain [1,5]. At first glance in this transitional regime, even around the Fermi energy, the experimental situation was found rather complex and very plausible arguments stated that higher energy should lead to new reaction aspects. In this context the necessity of rather complete exclusive experiments was important to tentatively determine the intervening processes.

In this contribution we present an overview of violent collisions involved at 27 MeV/u for the Ar + Ag system. Through exclusive measurements between heavy residues (HR) and light charged particles (LCP) or intermediate-mass-fragments (IMF), the main characteristics of the mechanisms were derived namely fusion and deeply inelastic collisions preceded and/or accompanied by a preequilibrium emission. Thermalization is discussed through excitation energies involved. Finally, balances of linear momentum, mass and atomic number are also presented.

A 1090 MeV Ar beam delivered by GANIL bombarded a 220 $\mu g/cm^2$ $^{nat}$Ag target. LCP and IMF were detected and identified using a sectorized detection system [6]. Each sector subtended a solid angle of 3.2 msr. The angular range 5-80° with respect to the beam axis was explored in two positions. Coincident heavy residues were detected with nine silicon detectors

[1]Permanent address: Hahn-Meitner-Institut, Glienicker Strasse, D-1000 Berlin 39
[2]Permanent address: IPEN, Apt. 1687, Lima, Peru

of 600 $mm^2$ active area located at 40 cm from the target; five detectors were positioned on the opposite side of the beam axis (in-plane) from $10^o$ to $50^o$ , and four in a perpendicular plane containing the beam axis (out-of-plane) from $10^o$ to $40^o$ . The HR masses were deduced from the time of flight measured relative to the beam bunches and the experimental velocity threshold was 0.4 cm/ns.

# 1 COINCIDENCES BETWEEN HR AND IMF: DYNAMICS OF COLLISIONS

## 1.1 Experimental results.

Exclusive velocity spectra of IMF, examples of which are shown on the left part of Fig. 1 have shapes completely similar to those observed in single measurements [7] over the whole angular range studied. Coincident IMF and HR are strongly coplanar; in-plane angular distributions of HR for different Z values of coincident IMF exhibit a bell shape and the value of the most probables angle $< \theta_{HR} >$ is shifted towards larger angles when the atomic number (or the mass) of the detected IMF increases, as observed for binary processes. The connection of these observations indicates that the intervening mechanisms are essentially binary [8].

Following these observations, coincidence data between HR and IMF were analyzed assuming a two-body splitting of a system recoiling at $0^o$ , and that the average velocity and direction of the fragments are not modified by their deexcitation. Moreover, to get rid of large evaporation effects, due to the strong dissipation involved, only mean quantities were used; for each $Z_{IMF} > 4$ and at each value of $\theta_{IMF}$ , the most probable angle of the correlated heavy partner $< \theta_{HR} >$ was determined and the associated mean velocities $< v_{IMF} >$ and $< v_{HR} >$ were deduced. The relative velocities $v_{rel}$ between the two partners were then derived as well as the recoil velocity of the system before splitting $v_{recoil}$ .

At a given angle $\theta_{IMF}$ , both $v_{rel}$ and $v_{recoil}$ were found independent of $Z_{IMF}$ indicating the dominant common origin of all IMF detected. The various results are summarized in Fig. 2. The plotted values are non-weighted averages over the different $Z_{IMF}$ , with standard deviations represented by bars. The top part of the figure shows the relative velocity between HR and IMF which clearly exhibits a damping of the relative motion which increases with the emission angle of the light partner $\theta_{IMF}$ . Complete damping occurs for $\theta_{IMF} > 30^o$ . This situation fully resembles the observations in the case of deeply inelastic collisions at low energies. The features of these DIC differ, however, in some aspects from those observed at low energies for the same system [9]; firstly, all the Z of light partners studied here are much lower than the projectile Z. At low energies such Z had been associated with completely relaxed events, whatever the detection angle. Secondly, incompletely damped collisions are observed over a large angular range, well beyond the grazing angle ($7^o$ ). As the incident

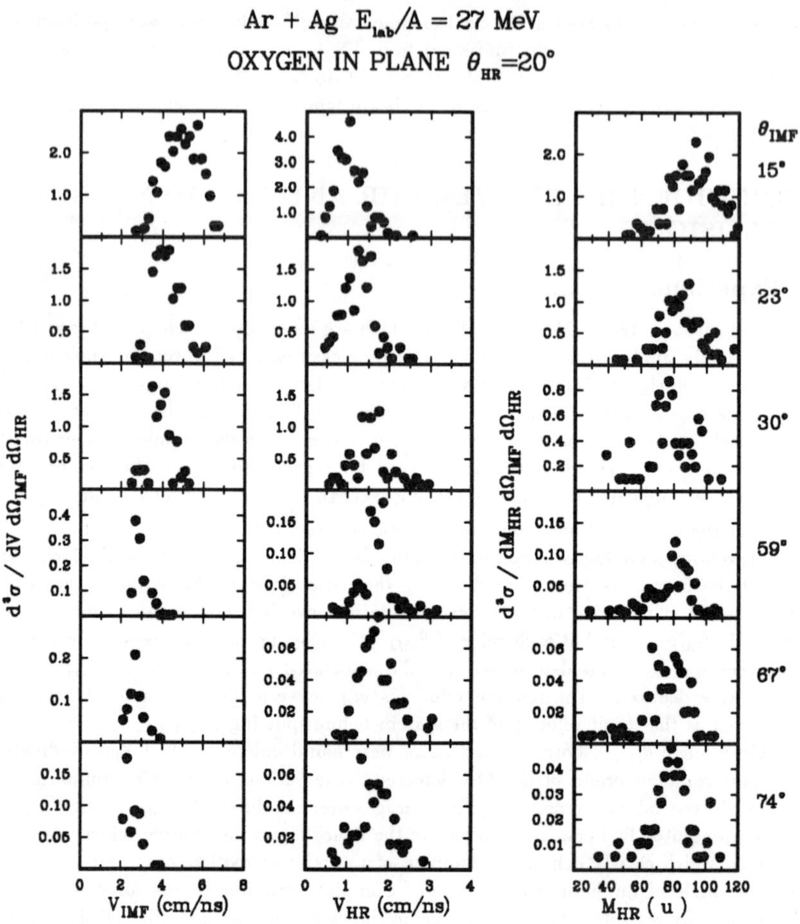


Figure 1: Example of the evolution of the velocity spectra (IMF and HR) and mass spectra (HR) with the emission angle of the IMF, for $Z_{IMF} = 8$. The HR is detected at $20^o$ from the beam, which is close to the most probable correlation angle of the HR. All spectra are normalized with respect to one another.

velocity is large there is an enormous amount of energy to be dissipated, which probably induces large angular fluctuations. Consequently, the final light partners studied here are the remnants of very excited primary fragments and final low Z are dominant [8]. From the assignment of mechanism made just before the decrease of $v_{rel}$ when $\theta_{IMF}$ increases should be correlated with increasing excitation energies. Experimentally this fact can be seen directly from the evolution of the average final mass $< M_{HR} >$ of HR with $\theta_{IMF}$ , which reflects the large increase of energy deposition (Fig. 1).

For $\theta_{IMF}$ larger than $30^o$ , the relative velocity of the two fragments is close to 3 cm/ns which corresponds roughly to the Coulomb repulsion of the two fragments, provided they are spherical, and indicates that the initial relative motion is completely damped. When a complete damping of the initial relative motion is observed experimentally, an IMF can be regarded as the light partner of a DIC but also as evaporated from a fused system as well as a very asymmetric fission product. As we shall see later, theoretical support and considerations on IMF emission properties are needed to conclude on this last point.

The bottom part of Fig. 2 shows how the recoil velocity of the system before splitting evolves with $\theta_{IMF}$ . This velocity can be compared with the c.m. velocity of the total system, denoted $v_{FMT}$ in the figure, and the deviations observed from this value are taken as reflecting the occurence of preequilibrium emission and its amount. The gentle evolution of $v_{recoil}$ with $\theta_{IMF}$ , which moves from about $v_{FMT}$ at $7^o$ to $\sim 0.8\ v_{FMT}$ at large angles, reveals the appearance and the increase of preequilibrium emission with the degree of dissipation in the collision and consequently with the decrease of the impact parameter.

## 1.2 Comparison with a semi-classical calculation.

A semi-classical calculation, based on the Landau-Vlasov equation was shown to reproduce some dynamical aspects of nuclear reactions at intermediate energies [10]. In this approach, the interplay between mean field features and two-body collisions is taken into account and the dissipative mechanisms responsible for the transfer of translational energy into intrinsic excitation energy are expected to be described by the respective roles of these two components. It was therefore appealing to compare the results of such a calculation with our experimental results and to eventually obtain more precise information.

Calculations ( using a Zamick force) were performed for several values of the impact parameter $b$; similar results were obtained using the effective finite-range Gogny force [11,12]. The evolution of the density distribution as a function of time is displayed in Fig. 3 for $b = 0, 4, 5, 5, 7$ *and* $10\ fm$ [13]. The mechanisms that are predicted for violent collisions are the same as those known at lower energies, namely fusion and deeply inelastic collisions. An important difference however is *the abundant preequilibrium emission which accompanies both types of reactions*; the number of preequilibrium particles strongly increases when the impact

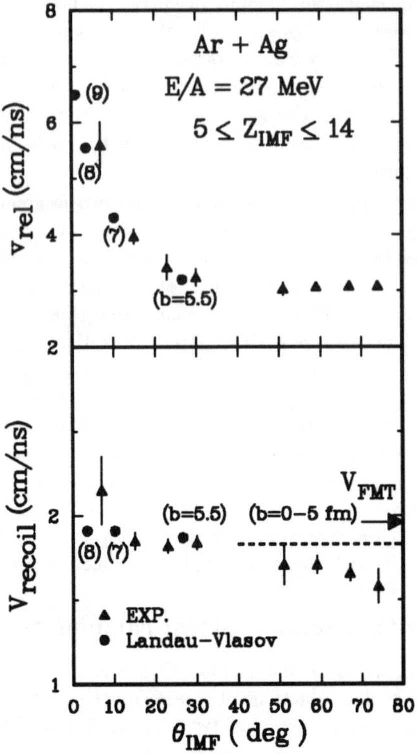


Figure 2: Evolution of the relative velocity of the two partners (HR and IMF) and of the recoil velocity of the system before splitting as a function of the laboratory emission angle of the IMF. Black points and dashed line are the results of semi-classical calculations (impact parameters are indicated within parenthesis).

parameter decreases. Fusion, or more exactly incomplete fusion, occurs for central collisions (0-5 fm): the two nuclei merge together, several nucleons are rapidly ejected at the beginning of the collision, both forward and backward; their momentum anisotropy, $\eta = \sum(2v_z^2 - v_x^2 - v_y^2)$ (where $z$ is the beam direction and $x$ the impact parameter axis), is large when the emission begins and vanishes around $t = 120\ fm/c\ (4.10^{-22}s)$, indicating that the subsequent nucleon emission is isotropic, as in an evaporation process. At the end of the collision only one heavy fragment subsists whose mass is between those of the target and of the compound nucleus. Its velocity, in correlation with the momentum anisotropy of the emitted nucleons, decreases until 120 fm/c and remains constant afterwards. The linear momentum missing in the heavy residue has been carried away by the preequilibrium nucleons.

For larger impact parameters, fusion no longer occurs. Again the two incident nuclei form a compact shape entity, where however two centers can still be distinguished. Fast nucleons are ejected as in central collisions. But after a time depending on the impact parameter the system re-separates into two fragments close to the target and the projectile; the separation time is $\sim 5 - 8.10^{-22}s$ for impact parameters between 5 and 6 fm, and twice shorter when the impact parameter is equal to 9 fm. The relative velocity (at infinity) of the two fragments was calculated. An excellent agreement with the experimental data is observed (Fig. 2). From the calculation we learn that while this damping occurs the system rotates and even for large impact parameters (9 fm ) the fragments are emitted *at negative angles*, reflecting the persistence of the nuclear mean field influence. All these features are reminiscent of deeply inelastic collisions as they are known between 5 and 10 MeV/nucleon. Once again it should be underlined that around 30 MeV/nucleon, this process is preceded (for the more central collisions) or accompanied (for larger impact parameters) by preequilibrium emission.

From the experiment it is impossible to decide whether one observes, for $\theta_{IMF} > 30^o$ , completely damped DIC or fusion followed by evaporation or asymmetric fission. Here the calculation helps to make a choice : a dynamical splitting as expected for completely damped DIC is not observed ; it is therefore reasonable to attribute the events with $\theta_{IMF} > 30°$ to fusion-fission/evaporation processes and thus to link them to incomplete fusion processes ($b < 5.5$ fm) given by the calculation.

From the momentum anisotropy $\eta$ of emitted particles, as previously mentioned, it was possible to determine the duration of the preequilibrium emission stage for central collisions : $4.10^{-22}$ sec. For DIC it was difficult to apply the same criterion to separate the preequilibrium emission, as probably no isotropy can be expected in the emission from a very deformed system. For the sake of comparison, however, the velocity of the system before splitting was calculated at the same time for all impact parameters up to b = 8 fm and it is compared in the bottom part of Fig. 2 to the same quantity deduced experimentally. Experiment and calculation agree on one point: the amount of preequilibrium emission is found to increase when the impact parameter decreases. Moreover the calculation indicates that more and more

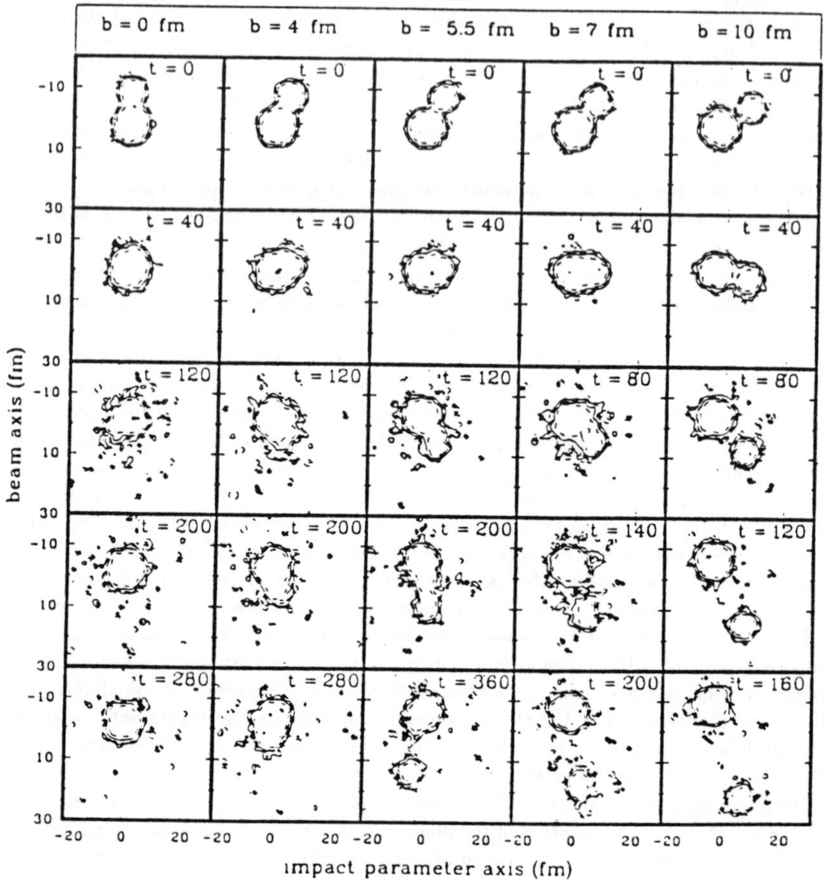


Figure 3: $^{40}Ar +^{108} Ag$ collisions at 27 MeV/nucleon calculated with the Landau-Vlasov equation for several impact parameters. The z (initial velocity) axis is vertical, and the impact parameter axis (x) horizontal. The lines represent equal density levels in the (x,z) plane. Times are expressed in fm/c (30 fm/c = $10^{-22}$ sec).

preequilibrium nucleons are emitted when we scan from peripheral to central collisions, while the average laboratory kinetic energy removed per nucleon is remaining roughly constant around 22 MeV.

Concerning excitation energies involved, at the present time, due to the sampling of phase space, the energy partition in DIC cannot be derived. By correction for energy removed by preequilibrium emission only the total excitation energy can be estimated. Fig. 4 shows the values derived in a schematic two step picture of the process. The energy removed by the preequilibrium stage was calculated assuming not only proton and neutron emission, but also complex particle emission as experimentally observed (see next section). The impact parameter range 0-8 fm corresponds to a value of 1.8 *b* for the HR cross-section which is close to the experimental one.

# 2 THE HOT NUCLEI PRODUCED

## 2.1 Coincidences between HR and IMF.

Primary masses and total excitation energies for DIC and fusion reactions can be deduced from the experimentally determined kinematic quantities of the two partners, by incorporating the results from semi-classical calculations for preequilibrium emission (number of nucleons $A_{PE}$ and energy removed $E_{PE}$ , as a function of $v_{recoil}$ ).

Fig. 5 illustrates the details of the results for two IMF detection angles which select DIC ( $\theta_{IMF} = 15^o$ ) or fusion reactions ( $\theta_{IMF} = 67^o$ ); the measured missing mass $\Delta$A, which is associated to evaporated particles, and the extracted excitation energy are given as a function of the detected $Z_{IMF}$ . Firstly, depending on $Z_{IMF}$ , total excitation energies are found to vary between 250 and 500 MeV. To make the symbols for the two quantities in Fig. 5 coincide, one has to assume that 10-11 MeV of excitation energy are connected with one missing mass unit. Since the removal of 11 MeV per evaporated nucleon is a reasonable assumption, the two compared independent experimental quantities are compatible with each other in the sense of the underlying reaction scenario (preequilibrium emission, binary splitting, evaporation). Secondly we observe a strong difference of behaviour between IMF from DIC and fusion. The evolution of $\Delta$A or E* with $Z_{IMF}$ is found pronounced for DIC (low Z associated with the largest excitation energies), which indicates rather a narrow primary fragment mass distribution and sustains the hypothesis of DIC. On the contrary for emission of an IMF from fused nuclei $\Delta$A or E* is found rather constant, whatever $Z_{IMF}$ (the Q + $B_{IMF}$ happens to be approximately constant over the $Z_{IMF}$ range studied).

Finally, information concerning the energy partition in DIC [14] was derived from the primary masses of IMF calculated by applying conservation of perpendicular momentum :

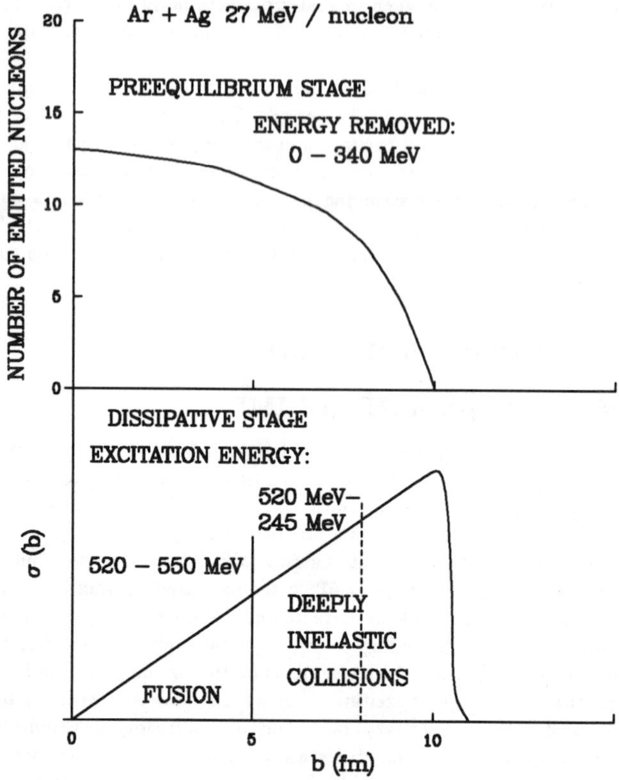


Figure 4: Schematic representation of the evolution of the mechanisms involved in the $^{40}Ar + ^{108}Ag$ reaction in the framework given by semi-classical calculations. Top : number of preequilibrium nucleons as a function of the impact parameter ; the energy so removed is indicated for b = 10 to 0 fm. It contributes to decrease the total excitation energy (bottom part) of the HR (fusion) or of both DIC fragments.

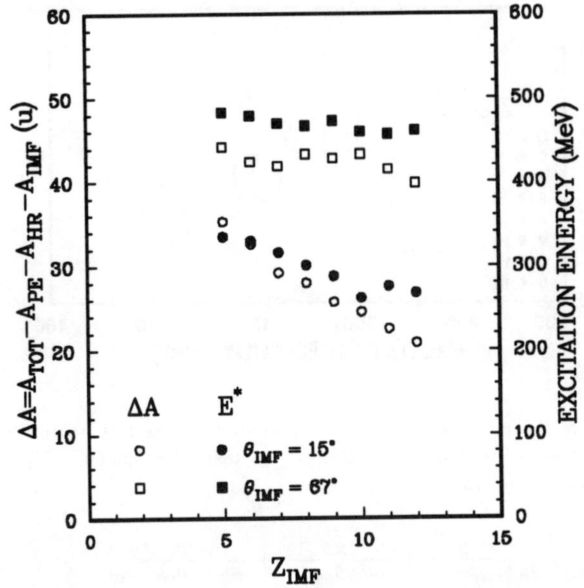


Figure 5: Measured missing mass (except $A_{PE}$ , see text) and excitation energy deduced from kinematics of HR-IMF coincidences as a function of $Z_{IMF}$ , for two IMF detection angles : i) $15^o$ which selects mainly DIC collisions and ii) $67^o$ which selects a fusion-evaporation process.

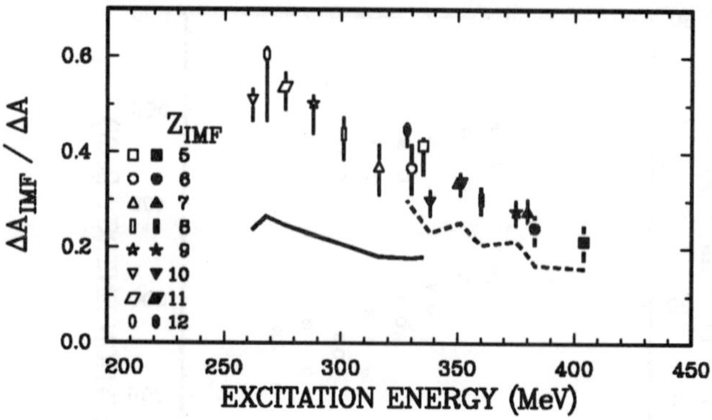


Figure 6: Ratio of the number of nucleons evaporated by the IMF to the total number of evaporated nucleons deduced from kinematics (see text), for different $Z_{IMF}$ . Open (full) points refer to $\theta_{IMF} = 15^o$ ($30^o$ ).

$$A'_{IMF} = \frac{148 - A_{PE}}{\sin\theta_{IMF}} \times \left(\frac{< v_{IMF} >}{\sin < \theta_{HR} >} + \frac{< v_{HR} >}{\sin < \theta_{IMF} >}\right)^{-1} \times < v_{HR} >$$

Note that, firstly, the $A'_{IMF}$ were accurately deduced for $\theta_{IMF} \geq 15^o$ (no bias due to angular distributions) and from experimental average quantities (see section 1) and do not suffer from the uncertainties discussed in ref. [15] and secondly that preequilibrium emission affects the values of the total excitation energy but not its repartition. The evaporated mass $\Delta A_{IMF}$ then was defined as $A'_{IMF}$ - (2 $Z_{IMF}$ + 0.5) and normalized to the total evaporated mass $\Delta$A. Fig. 6 shows the evolution of this quantity for different degrees of relaxation characterized by two detection angles, $\theta_{IMF}$ = 15 and $30^o$ , and for different $Z_{IMF}$ , as a function of the total excitation energy. The values observed, which reflect the energy partition provided that evaporated nucleons from IMF and HR carry away in average the same energy (which is confirmed by an evaporation code [16], are found to evolve from energy equipartition for total excitation energy around 270 MeV to approximate equilibrium for total excitation energy larger than 350 MeV.

Such an evolution has been observed at lower energy in the last few years within the excitation energy range 50-200 MeV [17]: a few $10^{-21}$ s are necessary to approach thermal equilibrium in asymmetric systems. At 27 MeV per nucleon, from the direct correspondence between $\theta_{IMF}$ and the impact parameter obtained from semi-classical calculations, we can deduce interaction times in the range $5-10.10^{-22}$ s. This shortening of the duration necessary to approach thermal equilibrium in DIC has to be connected with the large enhancement of two-body collisions which reduce the energy relaxation time. The reduction of a factor 2-3 (for similar systems in the same impact parameter range) observed between 10 [17] and 27 MeV per nucleon bombarding energy well agrees with calculations of ref. [18].

## 2.2 Coincidences between HR and LCP: evaporation from the HR.

For the analysis of the coincidences between the HR and the LCP, four regions in the velocity-mass plane of the HR were defined, which select "symmetric fission" ($45 \leq A_{HR} \leq 60$) and different HR ($A_{HR} > 60$) parallel velocity bins, $V_{HR//}$ , which are mainly connected with the violence of the collision and with the impact parameter ; edges are 0.6, 1, 1.6 and 2.2 cm/nsec corresponding respectively to median values $V_{HR//}/V_{cm}$ = 0.41, 0.66 and 0.97 for the three bins [8].

The main goal of this part of the experiment was to isolate the evaporation component of the HR which was taken as the largest isotropic component in the reference frame of a source moving at $0^o$ with the average parallel HR velocity. However, we just have shown before that part of the HR results from DIC. Then when dealing with binary reactions the assumption that the HR recoils at $0^o$ is questionable. Nevertheless in this case the maximum of the HR angular distribution $d\sigma/d\Omega$ is situated at an angle smaller than $10^o$ , so the treatment we applied should correctly take into account the HR evaporation.

The evaporative component was determined by using only the most backward detection angles ($55^o$ to $80^o$ ) with respect to the beam. The LCP spectra were fitted using a parametrization :

$$\frac{d^2\sigma}{d\Omega dE} = k(E-B) \times \exp\left[-E\left(\frac{\sin^2\theta}{T_{in}} + \frac{\cos^2\theta}{T_{out}}\right)\right]$$

which takes into account the enhancement of particle emission perpendicularly to the spin axis of the emitter and the increase of this anisotropy with energy and mass of the particle. In the formula $\theta$ is the emission angle with respect to the nucleus spin in the emitting nucleus frame, and $T_{in}$ and $T_{out}$ are free parameters. It can be shown that the $T_{out}$ value derived can be viewed as a good approximation to the true average temperature [8]. In Fig. 7, we present, as an example, the laboratory velocity spectra of protons and alpha particles detected in coincidence with HR of selected velocities ($V_{HR//}/V_{cm}$ = 0.97) emitted at $10^o$ from the beam direction. At $59.2^o$ the spectra have been rather well characterized by the parametrization (solid curves in the figures). The emission barrier B was taken from [19], and $T_{in}$ and the amplitude factor k were deduced from the in-plane spectra ; fixing these values $T_{out}$ was deduced from the out-of-plane spectra. At $15.2^o$ the anisotropic component which populates

| $\theta_{HR}$ | | 10° | | | 20° | |
|---|---|---|---|---|---|---|
| $V_{HR//}/V_{cm}$ | 0.41 | 0.66 | 0.97 | 0.41 | 0.66 | 0.97 |
| | | | PROTONS | | | |
| $T_{in}$ (MeV) | 4.4 | 5.1 | 5.3 | 4.8 | 5.1 | 5.5 |
| $T_{out}$ (MeV) | 4.1 | 4.8 | 5.0 | 4.1 | 4.5 | 4.7 |
| R | 1.2 | 1.2 | 1.2 | 1.6 | 1.4 | 1.5 |
| $M_{iso}$ | 1.5± 0.2 | 2.3± 0.3 | 3.2± 0.4 | 1.6± 0.3 | 2.2± 0.4 | 2.7± 0.4 |
| | | | ALPHA PARTICLES | | | |
| $T_{in}$ (MeV) | 5.8 | 6.4 | 6.6 | 7.0 | 8.2 | 9.0 |
| $T_{out}$ (MeV) | 5.2 | 5.8 | 5.8 | 5.6 | 6.2 | 6.4 |
| R | 1.6 | 1.5 | 1.7 | 2.4 | 2.9 | 3.5 |
| $M_{iso}$ | 0.9± 0.2 | 1.4± 0.2 | 1.9± 0.3 | 1.2± 0.2 | 1.6± 0.3 | 2.0± 0.4 |

Table 1: Parameters ($T_{in}$ , $T_{out}$ , R), and multiplicities derived from the parametrization used to disentangle isotropic (iso) and anisotropic components (see text).

the high velocity part of the spectra is clearly observed. We observe also an additional yield of particles at low velocities which suggests that either the simple sharp cutoff, in transmission coefficients, implied by the parametrization is inadequate or that the barriers fluctuate.

Table I summarizes, for the three velocity bins of heavy residues detected at 10° and 20° with respect to the beam, the values derived for the free parameters $T_{in}$ and $T_{out}$ , the anisotropy R (in-plane/90° out-of-plane) and the multiplicity $M_{iso}$ of protons and alpha particles evaporated from the HR deduced by integrating the above equation.

For both protons and alpha particles it is seen that $T_{out}$ and $M_{iso}$ increase with HR velocities reflecting the increase of excitation energy. Moreover, we clearly observe the different influence of excitation energy and angular momentum on the decay of protons and alpha particles as expected from statistical calculations. $T_{out}$ is found larger for alpha particles, reflecting a more favorable decay probability ratio $\Gamma_\alpha/\Gamma_P$ for the $\alpha$ particles at the higher excitation energies. The difference $T_{in}$ - $T_{out}$ and the in-plane/out-of-plane anisotropy R both of which reflect the angular momentum effects are larger for alpha particles. Note also that these quantities increase with the HR detection angle, which is due to a better definition of the reaction plane and consequently to a better definition of the HR spin vector.

The excitation energies of heavy residues which can be inferred from the values of $T_{out}$ are in good agreement with those deduced in the previous subsection.

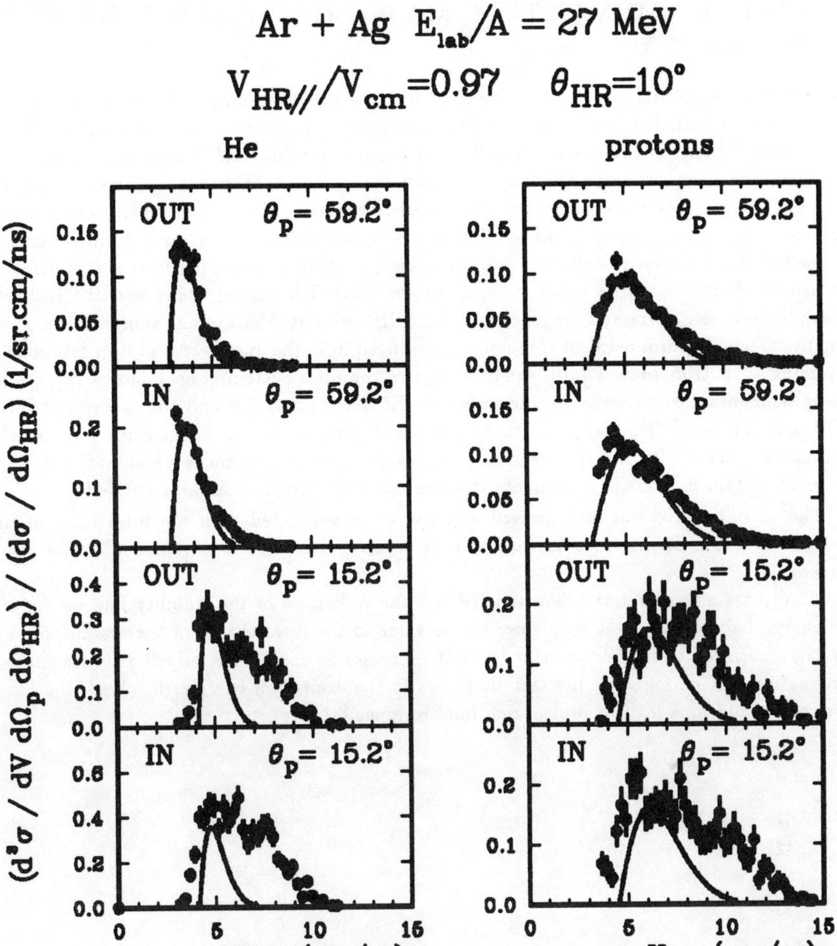


Figure 7: Examples of LCP (protons and alpha particles) velocity spectra in coincidence with HR detected at 10° from the beam. The full lines correspond to the isotropic component fixed at large angles (see text).

## 3 LINEAR MOMENTUM, MASS AND ATOMIC NUMBER BALANCES

Considering the results presented in the previous sections concerning IMF and LCP detected in coincidence with HR, one is tempted to derive the balances of some mean values. As we used time-of-flight techniques for HR, Z identification for IMF and A and Z identifications for LCP, the most accurate quantity we could derive is the linear momentum. However, we need also information on neutron emission. Fortunately neutron multiplicity distribution from a rather similar system, Ar + Rh, has been measured at the same incident energy by means of a neutron calorimeter [20]. The higher neutron multiplicities were associated, by means of the measured cross-sections, to the three HR velocity bins and the deduced multiplicities were corrected for efficiency using HR velocity and average temperature $T_{out}$ . Finally, preequilibrium neutron emission was deduced from the anisotropic proton component assuming a neutron/proton ratio equal to that of the total system. Fig. 8 indicates parallel linear momenta carried away by the different HR detected at $10^o$ and the associated IMF, LCP and neutrons. The total has to be compared with the linear momentum value of the beam 9078 MeV/c. Within the error bars we observe a good agreement which indicates that we really obtain in average a complete description of the violent collisions studied.

The mass balance was also derived and has to be compared with the total mass of the system 147-149. Taking into account the large error bars, we also notice a rather good agreement.

Finally assuming that the detected HR has the A/Z ratio of the stability line we deduce the charge balance. $Z_{TOT}$ is very close to the value of the total charge of the system, 65, for the bin associated to the lowest velocities but decreases for higher HR velocities; a reasonable explanation can be found in the fact that, due to the excitation energies involved, the most dissipative collisions lead to proton rich final residues [21].

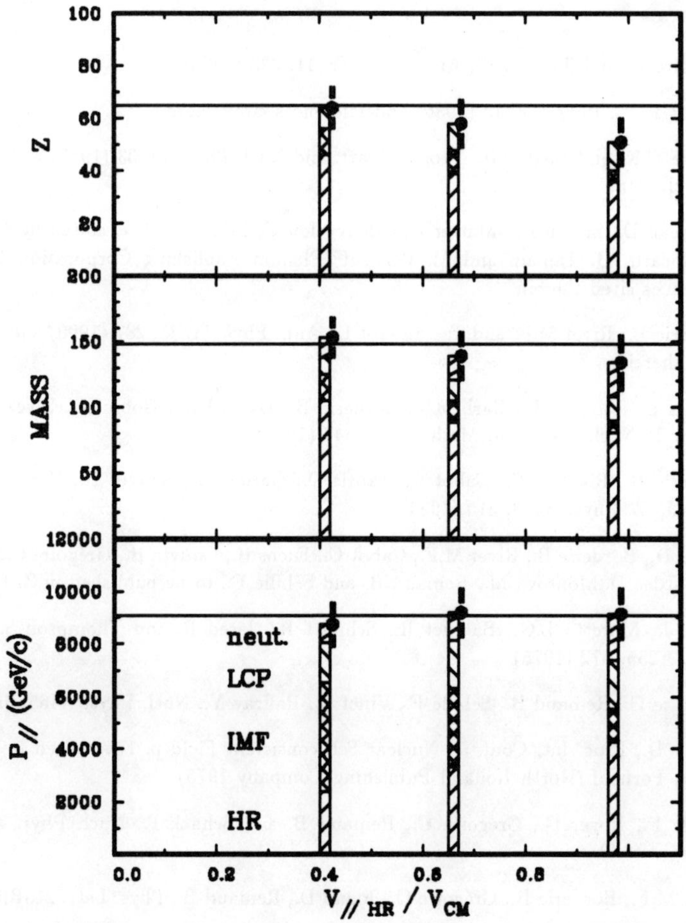


Figure 8: Linear momentum, mass and Z balances. The contributions due to HR, IMF, LCP and neutrons are indicated.

## References


[1] Grégoire C. and Tamain B., Ann. Phys. Fr. 11, 323 (1986)

[2] Dayras R., J. Phys 47c, 13 (1986) and references cited therein.

[3] Gelbke C.K. and Boal D.H., Prog. in Part. and Nucl. Phys. 19, 33 (1987) and references cited therein.

[4] Guerreau D., in Nuclear Matter and Heavy Ion Collisions, p. 187, edited by M. Soyeur, H. Flocard, B. Tamain and M. Porneuf (Plenum Publishing Corporation 1989) and references cited therein.

[5] Borderie B., Rivet M.F. and Tassan-Got L., Ann. Phys. Fr. 15, 287 (1990) and references cited therein

[6] Gardes D., Monnet F., Barbey S., Borderie B., Dumail M., Gobbi A., Rivet M.F. and Volkov P. Nucl. Instr. and Meth. A247, 347 (1986)

[7] Borderie B., Rivet M.F., Cabot C., Fabris D., Gardes D., Gauvin H., Hanappe F. and Peter J., Z. Phys. A318, 315 (1984)

[8] Jouan D., Borderie B., Rivet M.F., Cabot C., Fuchs H., Gauvin H., Grégoire C., Hanappe F., Gardes D.,Montoya M., Remaud B. and Sébille F., to be published in Z. Phys.

[9] Galin J., Moretto L.G., Babinet R., Schmitt R., Jared R. and Thompson S.G., Nucl. Phys. A255, 472 (1975)

[10] Grégoire C., Remaud B., Sébille F., Vinet L., Raffray Y., Nucl. Phys. A465, 317 (1987)

[11] Gogny D., Proc. Int. Conf. on Nuclear Self-consistent Field p. 176, edited by G. Ripka and P. Porneuf (North Holland Publishing Company 1975)

[12] Sebille F., Royer G., Grégoire C., Remaud B. and Schuck P., Nucl. Phys. A501, 137 (1989)

[13] Rivet M.F., Borderie B., Grégoire C., Jouan D., Remaud B., Phys. Lett. 215B,55 (1988)

[14] Borderie B., Rivet M.F., Cabot C., Fuchs H., Gardes D., Hanappe F., Jouan D., and Montoya M., Z. Phys. (in press)

[15] Toke J., Schroder W.U. and Huizenga J.R., Phys. Rev. C40, 1577 (1989)

[16] Tassan-Got L. and Stephan C., Nucl. Phys. in press

[17] Wile J.L., Datta S.S., Schroder W.U., Huizenga J.R., Toke J., and De Souza R.T., Phys. Rev. C39, 1845 (1989) and references cited therein

[18] Toepffer C. and Cheuk-Yin Wong, Phys. Rev. C25, 1019 (1982)

[19] Alexander J.M., Guerreau D. and Vaz L.C., Z. Phys. A305, 313 (1982)

[20] Guerreau D. and Galin J., private communication of unpublished results.

[21] Hubert F., Del Moral R., Dufour J.P., Emmermann H., Fleury A., Poinot C., Pravikoff M.S., Delagrange H. and Lleres A., Nucl. Phys. A456, 535 (1986)

# PROTON AND SUBTHRESHOLD PION PRODUCTION CORRELATED WITH LIGHT FRAGMENTS : A STATISTICAL PROCESS ?

J.L. LAVILLE[+], A. BADALA[++], R. BARBERA[++], G.BIZARD[+], D. DURAND[+], A. PALMERI[++], G.S. PAPPALARDO[++] and F. RIGGI[++]

+ *Laboratoire de Physique Corpusculaire - LA34-ISMRA-IN2P3-CNRS-Université de Caen - 14032 CAEN Cedex(France)*

++ *Istituto Nazionale di Fisica Nucleare, Corso Italia 57, 95129 Catania (Italy)*

## ABSTRACT

Coincident data between charged pions or protons and elements of $Z \leq 8$ detected over a large solid angle are presented for reactions induced by 94 MeV/u $^{16}O$ on Al,Ni,Au targets. Charged pion and particle emission is shown to occur via the formation and decay of a highly excited reaction zone, the characteristics of which (velocity, temperature, number of constituant nucleons) are deduced from large angle particle-particle kinematical correlations.

## I) INTRODUCTION

The main interest to study subthreshold pion production in nuclear reactions stands in the fact that pions are created in the interaction zone. It is commonly admitted at that time that it mainly occurs via the elementary interaction between quasi-free nucleons above a laboratory energy of ≅ 150 MeV/u (réf. 1,2). Reversely, the observation of sizeable pion yields at such a low energy as 20 MeV/u (réf. 3,4) shows that pion production requires a large cooperation of the nuclear system. In the "intermediate region" set about 100 MeV/u one can expect this process to happen through various mechanisms implying a more or less important number of the nucleons, with various impact parameters and complex multi-body final states.In this paper, we present quasi-exclusive results from a study of pion and particle production induced by $^{16}O$ projectile of 94 MeV/u on 3 targets (Al,Ni,Au).

## II) THE CATANE-GANIL-LPC EXCLUSIVE EXPERIMENT

### a) Why an exclusive experiment ?

The expected complexity of many particle final states produced with pions requires obviously an exclusive experimental approach. Nevertheless we wish to point out the three following reasons :

1) To disentangle the various reaction mechanisms and/or to distinguish between various models.

2) To evaluate the impact parameter range of the pionic reaction, the first step beeing to select central and peripheral collisions.

3) To provide particle-particle kinematical correlations very usefull to get quantitative informations about the interacting sub-system (source size etc...)

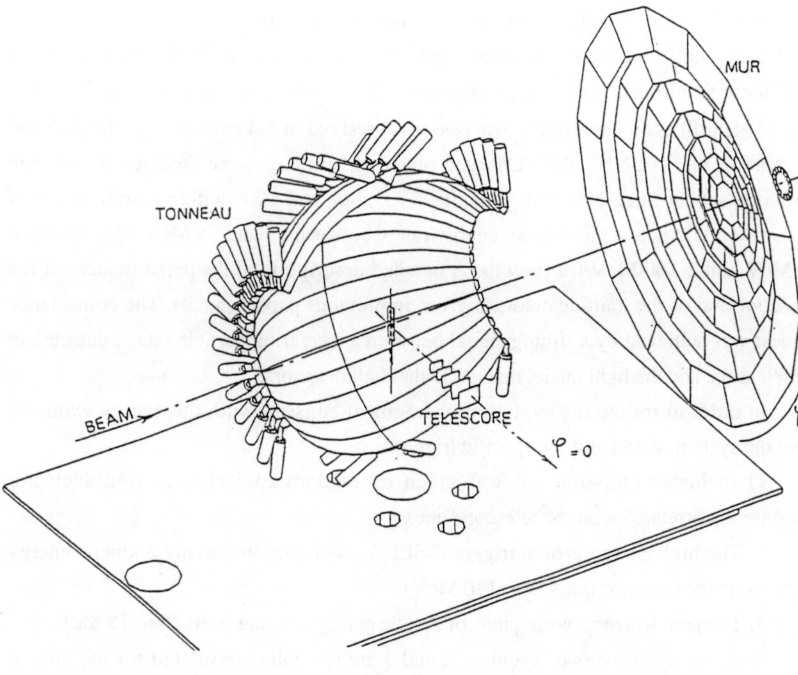


*Fig. 1 :*
*Experimental set-up*

### b) Experimental considerations

To produce a pion is to extract an amount of energy of at least 140 MeV, which corresponds to the pion rest mass. Hence, the proposed "method" consists in comparing fragment emission with pion production to fragment emission triggered by protons the kinetic energy of which is higher than 140 MeV. We studied charged pion and proton production in coincidence with light fragments in reactions induced by 94 MeV/u $^{16}O$ projectiles on Al, Ni, Au targets. Grazing angles are $\theta_g = 0.7, 1.2$ and 3° respectively for the three systems.

An aim of the experiment was to study both peripheral collisions characterized by projectile-like fragment emission and central ones giving rise to high particle multiplicity at large angle. The use of two large area multidetectors installed in the vacuum chamber "NAUTILUS" at GANIL (see fig. 1) allowed an almost exclusive detection of light fragments ($1 \leq Z \leq 8$). The forward one ("MUR" i.e. Plastic Wall) consists in 8 concentric rings of scintillators covering the whole azimuthal acceptance for polar angles between 1.5 and 30°. The second multidetector ("TONNEAU" i.e. Plastic Barrel) is made of 36 slates, displayed in two sets : a forward one $30° < \theta < 90°$ and a backward one ($90° < \theta < 150°$) with azimuthal acceptance between 90° and 270°.

This whole detection system rests upon the same $\Delta E$ time-of-flight device allowing for Z identification and velocity measurement. Its overall solid angle is about $2\pi$ sr, but clear Z identification requested a low velocity threshold of 5.4 cm/ns in the "MUR", and 4 cm/ns in the "TONNEAU". Charged pions and protons were detected by a range telescope set at three laboratory angles : 70°, 90° and 120° within a solid angle of 58 msr. It could detect pions in an energy range between 25 and 75 MeV ; and between 40 MeV and $\geq$ 160 MeV for protons. A detailed description of the performances of the telescope and of the multidetectors is given in previous papers.(5,6,9). The coincidence request was achieved by a timing signal between a triggering particle (start) detected in the telescope and the light nuclei marked in the multidetectors.

In order to follow the evolution of fragment emission with the energy extracted from the system, we used three specific triggers :

1) Inclusive emission ; it was given by random coincidences (multidetector response uncorrelated with the telescope one).

2) The high energy proton trigger ("HEP") : with protons having a kinetic energy higher than the pion rest mass (Ep>140 MeV).

3) The pion trigger : with pions of kinetic energy ranging from 25 to 75 MeV.

Note that the classes of events n°2 and 3 are specially considered for the sake of comparison which is the leading clue of this experiment.

About 23000 pions ($\pi^+$ and $\pi^-$) in coincidence with $\cong 10^5$ light nuclei (mainly H and He, up to = 85 %) have been observed. This set includes $\cong$ 2000 significant coincidences between pions and fragments of Z>2, which were observed for the first time in subthreshold pion production.

## III) EXPERIMENTAL RESULTS AND DATA ANALYSIS

### a) Comparaison of "π" and "p" events

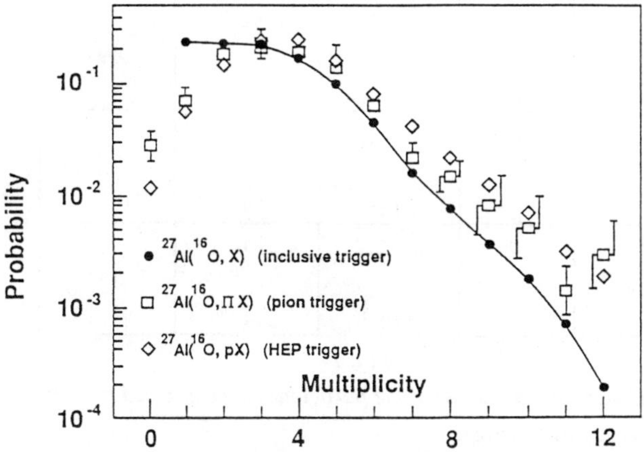


*Fig. 2 : Multiplicity distribution of charged particles in coincidence with different triggers*

In fig. 2, we see that the fragment multiplicity distribution evolves clearly with the trigger energy : the low multiplicity contribution, dominant for "inclusive" events, vanishes for the HEP or pion trigger. Such a trend characterizes the transition from peripheral "soft" collisions to central "violent" collisions .

Moreover pion and proton induced multiplicities show very similar shapes : both pion and HEP production are favoured in central collisions giving rise to a large emission of particles.

### b) Velocity contour plot for coincident π*He and p*He events : evidence of two sources contribution :

The search for sources of particle emission is a convenient way to characterize the reaction process(es). The contour plots of fig. 3 show for π*He (3a) and for p*He events (3b) from the $^{16}O$+Al system the specific topology of the participant/spectator scenario : a projectile-like source with a velocity slightly lower than the beam one, and a medium participant source. The target-like component is not seen because of poor statistics at backward angles. Table I shows the velocity and "temperature" parameters obtained by a standard moving source analysis. Fitted velocities of the medium source are very close to the participant model calculated ones. We shall now characterize the "participant" zone from typical angular correlations.

| System | | $T_S$ | $\beta_S$ | $\beta_{part}$ |
|---|---|---|---|---|
| Target | Trigger | MeV | | |
| Al | π | 20±1. | .18±.01 | .191 |
| | p | 20.±.7 | .17 " | |
| Ni | π | 21±1. | .16 " | .165 |
| | p | 24±.7 | .16 " | |
| Au | π | 25±1.7 | .14 " | .125 |
| | p | 25±.8 | .14 " | |

*TABLE I*

*Temperature and velocity (in c units) of the participant zone deduced from He velocity spectra for various targets and triggers*

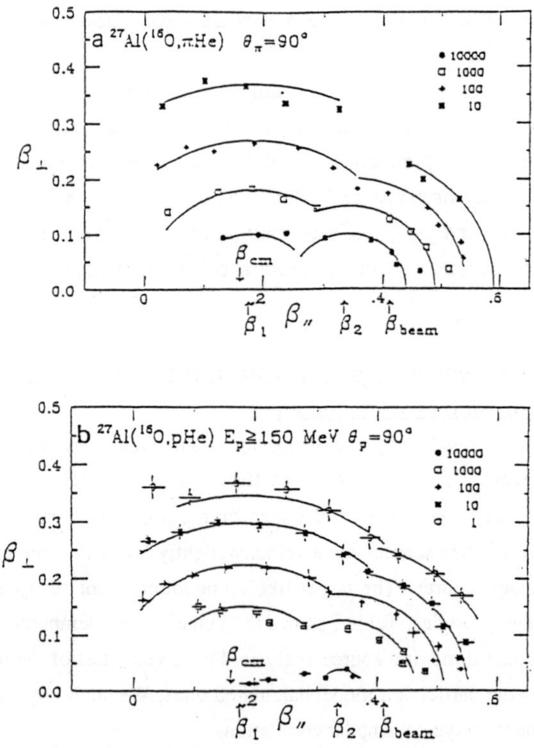


*Fig. 3 : He velocity contour plot of the invariant cross section for π-triggered (3a) and p-triggered (3b) events*

c) <u>Azimuthal distributions</u> : <u>kinematical signature of a hot source</u>

Typical Z=2 distributions for a polar angle $\theta = 23°$ and the 3 systems are displayed in fig.4 for p triggered correlations. They all exhibit a peak at $\phi = 180°$, i.e. on the opposite direction (see fig.1) of the trigger with respect to the beam axis. This $\phi$ peaking may be related to the hot source emission for the following reasons :

1) The $\phi$ peaking begins above $\theta_{He}=20°$, in the angular domain where the hot source emission is expected to be dominant, (see fig.3).

2) It appears to be due to a recoil effect since the triggering proton carries out a large momentum transverse (up to 600 MeV/c) which has to be balanced on the other side by the recoil of the residual system so favouring fragment emission around $\phi = 180°$. These typical $\phi$ correlations are very interesting since they show that the source recoils as a whole. We shall see that such a picture is confirmed by a "fireball" model calculation leading to an estimate of the mass number of this nuclear subsystem. Since this calculation has been developped in ref. 5, we limit ourselves to recall its main features : two reaction stages are considered ; a primary formation one according to the classical fireball geometry, and a sequential decay one described by the Weisskopf evaporation theory. For the light system $^{16}O$+Al, we got an overall agreement with the fragments experimental observables (angular and velocity spectra) in the range where the contribution of the particle decay of the hot zone is expected to dominate.

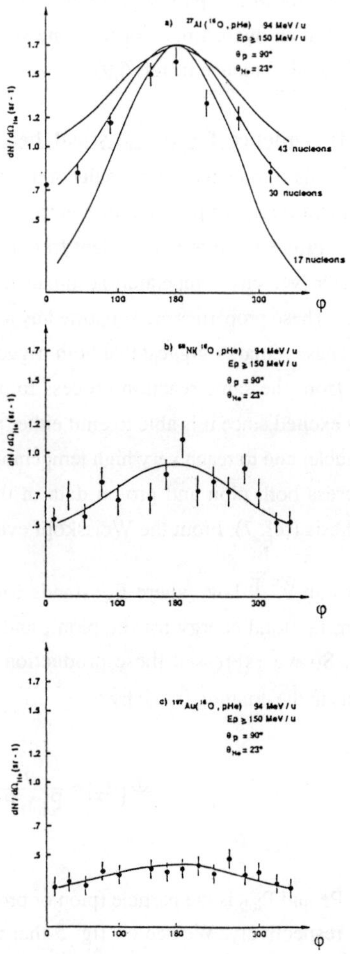


*Fig. 4 : Azimuthal distributions of p-triggered He particles for the $^{16}O$+Al (4a), $^{16}O$+Ni (4b) and $^{16}O$+Au (4c) systems. The full curves represent the statistical model calculation. This calculation is normalized to the data by a factor of .62 in the Ni case and of .43 in the Au case.*

The good agreement observed in fig. 4 between the statistical calculation and the $\phi$ distributions corresponds to given numbers of participant nucleons : 30, 35 and 50 nucleons for the $^{16}O$+Al, Ni, Au systems respectively. These numbers are obtained from the model without any fitting procedure. The different curves on fig. 4a, implying various mass numbers of the zone, show that the shape of the $\phi$ spectrum is very sensitive to its size. Of course, it is fully reliable only for the light $^{16}O$+Al system ; various effects, like energy dissipation or Pauli blocking, (ref. 6, 9) which are not accounted for in that purely geometrical model might be important with heavier targets and lead to an overestimate of the zone size. Including intermediate energy corrections slightly tend to reduce it (ref. 6,9).

**d) Evidence for a statistical behaviour of $\pi$ and p emission**

Let us sum up the main results of the present analysis :

1) Similarity of pion and proton triggered observables.

2) Pion and proton coincident fragments are emitted by the same source since the parameter (velocity, temperature) estimate leads to similar values in both cases.

3) These properties are supported by a statistical model calculation.

These remarks suggest that both triggering particles and coincident light fragments araise from the same reaction process implying the formation of a hot zone which is highly excited since it is able to emit either pions or very energetic protons together with light nuclei and to reach very high temperatures. A way to confirm such an assumption is to express both pion and proton data in the same framework of the statistical decay hypothesis (ref. 7). From the Weisskopf evaporation theory, the particle production rate follows an $e^{-\frac{E_x}{T}}$ law, where $E_x$ stands for the energy extracted from the interacting system, i.e. total energy for the pion ; and kinetic energy for the proton in the source frame. So we expressed these production rates $\phi$ ($E_x$) from the experimental cross-sections in the emitting frame by :

$$\Phi\left(E_x\right) = \frac{1}{P_{lab}\,P^*}\left[\frac{d^2\sigma}{d\Omega dE_x^*}\right]_{\pi,P}$$

where P* and $P_{lab}$ is the particle (pion or proton) momentum in the source and laboratory frame respectively. We see on fig. 5 that the $\Phi$ ($E_x$) data points agree with the $e^{-E/T}$ dependance which is observed whatever the particle and target ; it is of course confirmed by the complete model calculation : the proton data agrees fairly well within a factor of 2,

2, but the pion ones show an increasing discrepancy with the target mass, though the slopes remain in all cases well reproduced. For the sake of clarity, the calculation is displayed on fig. 5 for the Au target case only. This target effect reveals the dramatic importance of the final state absorption of the pion in the residual nuclei. It can be roughly accounted for by a standard absorption factor in e-(R/λ) where λ ≡ 5 Fm is the mean free path of the pion in the nuclear medium and R the size of the residual nuclear matter in which it might be absorbed.

Let us notice here that this problem of pion absorption has already been pointed out (ref. 4,8), and could be responsible of important deviations from the primary distributions.

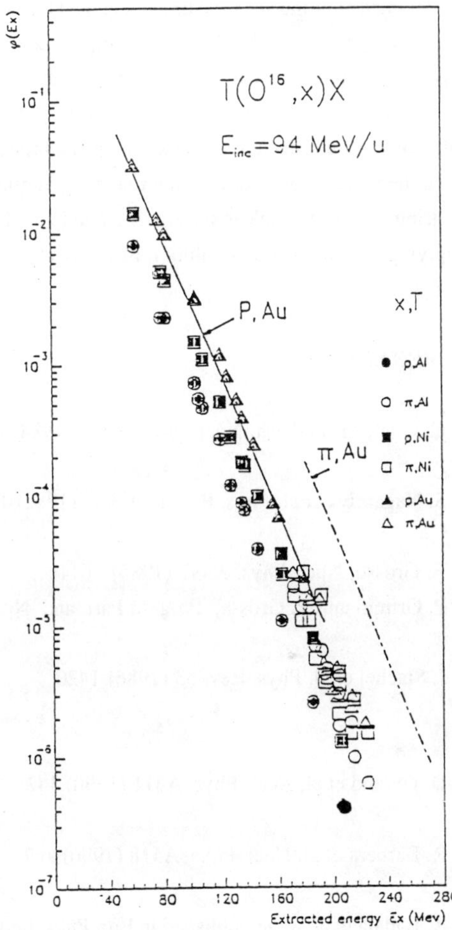


*Fig. 5 : "Production rates" (see text) of pions and protons for the three targets as a function of the extracted energy. The full (dashed) curve is the statistical model calculation for proton (pion) emission (Au case only)*

## IV - CONCLUSIONS

Exclusive pion/proton-fragments measurements have provided a relevant tool to describe both the pion and proton data in the same framework :

1) A standard fireball model assuming statistical particle emission is able to give a coherent description of both pion and proton triggered events. Nevertheless, one has probably to account for important absorption and dissipative effects.

2) Specific p*He azimuthal correlations provided an original method to estimate the size (i.e. mass number) of the decaying hot zone.

3) From π/p-fragment correlations ranging over a wide solid angle, we could derive the parameters of the hot zone producing the pions and particles.

Pion production by the hot participant zone has also been observed by an other experimental program using a different apparatus and a different analysis (ref.10).

From our viewpoint, the conceipt of pion production in an "individual" NN collision does not make sense at such a low subthreshold energy, since the extra energy has to be pumped out from the internal fermi motion implying de facto some collective contribution of the nuclear system. A suitable treatement needs a dynamical calculation implying the space-time evolution of the reaction, as it is done in a BNV framework (ref. 9).

**REFERENCES**

1) C. Guet, M. Prakash, Nucl. Phys. A428 (1984), 119c

2) S. Nagamiya et al, Phys. Rev. Lett. 48, (1982) 1780

3) E. Grosse, Nucl. Phys. A447 (1985), 611c
P. Grimm and E. Grosse, Prog. in Part. and Nucl. Phys. 15 (1985)

4) J. Stachel et al, Phys. Rev. 33 (1986) 1420

5) D. Durand et al, Nucl. Phys. A511 (1990) 442

6) R. Barbera et al, Nucl. Phys. A518 (1990) 767

7) A. Badala et al, to be published in Eur. Phys. Lett.

8) W. Cassing et al, Phys. Rep. 188 (1990) 369

9) A. Badala et al, Phys. Rev. C 43 (1991) 190

10) D. Lebrun et al, Phys. Lett. B223 (1989) 139, Z. Phys. A335 (1990) 73

# PROJECTILE BREAKUP AND NUCLEAR ASTROPHYSICS*

H. UTSUNOMIYA

Institut de Physique Nucléaire, 91406 Orsay Cedex, France

Texas *A&M* University, Cyclotron Institute
College Station, TX 77843, USA

**Abstract.** A new experimental approach to radiative capture processes via projectile breakup is discussed in relation to primordial and stellar nucleosyntheses. The spectrograph technique which plays a complementary role to the conventional telescope method has been devised to detect two particles emitted in extremely close proximity. This technique seems to be indispensable in the astrophysical application. Projectile breakup reactions of $^7Li$ studied with the spectrograph technique along with a recent measurement of $^{16}O$ breakup at 30 MeV/nucleon are reported.

## I. Coulomb breakup - a new approach to radiative capture reactions

Direct measurements of radiative capture cross sections have been limited at energies that are far above the astrophysical energy. This is because it is impossible to replicate such astrophysical environments as big bang expansion or interior of massive stars in laboratory. As a result, many nuclear reaction rates, the basic input for theoretical calculations of nucleosyntheses, are only poorly known. An innovative method recently proposed by Baur, Bertulani and Rebel[1,2] is based upon Coulomb breakup which may play an inverse role to radiative capture processes. If the validity of this method is proved, a fruitful branch of heavy ion physics would be formed in the future. Several attempts to break up nuclear species including both stable[3,4,5] and unstable[6,7] nuclei are underway along this line.

## II. Spectrograph method

Preceding to the proposal of the new approach, a few experimental works had been conducted by Shotter *et al.*[8] They studied $^7Li$ breakup reactions with an arrangement of detector telescopes. These earlier works might have stimulated the appearance of the theoretical proposal. Experimentally, however, there was a technical problem in measuring two particle correlations. The conventional telescope method suffered from a detection threshold to relative kinetic energies in two particle correlation measurements. The detection threshold stems from a finite angle between a pair of detector telescopes. The threshold is fatal in this method and cannot be eliminated regardless of the number of detector telescopes and of their arrangement.

* Supported by the U.S. Department of Energy under Contract No. DE-FG05-86ER40256

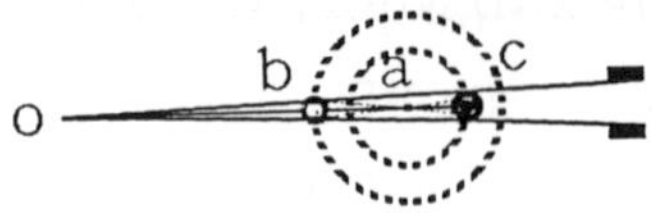


Figure 1: Unique coincidence phase space in the spectrograph measurement.

In order to eliminate the detection threshold, it is obviously necessary to detect two particles at the *same* angle. This unique coincidence geometry is depicted in Fig. 1. It is not impossible to accomplish the unique detection geometry with the telescope method. For example, using a multi-element detector telescope[9], one can measure correlations between two particles that have different range energies. However, this is not generally useful. A new technique that utilizes a magnetic spectrograph[10] was devised to measure two particles emitted in extremely close proximity. A similar technique was independently developed by Rebel[11] *et al.*

Some of the advantages of the spectrograph method over the telescope method are: (1) it permits measurements at small angles including zero degrees; (2) it provides the best resolving power at small relative kinetic energies ($\varepsilon$); (3) it eliminates the interference from elastic scattering, the origin of intense background to correlation measurements and of radiation damage to detectors; (4) it allows neutron-charged particle correlation measurements if the spectrograph has a straight-through port. Particle correlation measurements with the spectrograph method can be classified into two groups, depending upon m/q values of the two particles. The two types of measurements are schematically shown in Fig. 2. When they have different m/q values (type I), it is possible to completely eliminate the detection threshold because two particles with small $\varepsilon$ are focussed at different positions along the focal plane. A broad range magnetic spectrograph, such as an Enge split pole[12], is crucial to this type of measurements. In contrast, when they have identical m/q values (type II), the detection threshold is not completely eliminated though it can be sufficiently lowered. Large dispersion spectrographs are well-suited to this type of study. The type II measurements were previously made to detect two protons from (d,$^2He$) and ($^3He$,$^2He$) reactions[13,14,15].

The first experiment of type I was made to study $^7Li$ breakup reactions[3,4,16] in conjunction with nuclear astrophysics. Very recently, a type II measurement was carried out to study $^{16}O$ breakup reactions at 30 MeV/nucleon. These studies are reported below.

## III. Projectile breakup of $^7Li$

The $^7Li$ is one of the most controversial nuclear elements between two different models of the early Universe[17,18,19,20,21,22,23,24]. The standard big bang model[17] limits the universal baryon density such a way that the universe is far open unless there is a dark matter. In contrast, inhomogeneous big bang models based upon the density fluctuation triggered by the quark-hadron phase transition may overproduce $^7Li$ though they could close the universe[18,19,23]. Three independent measurements[25,26,27] of the $t(\alpha,\gamma)^7Li$ reaction rate, however, have ended up with a discrepancy by a factor of nearly two at zero relative energy.

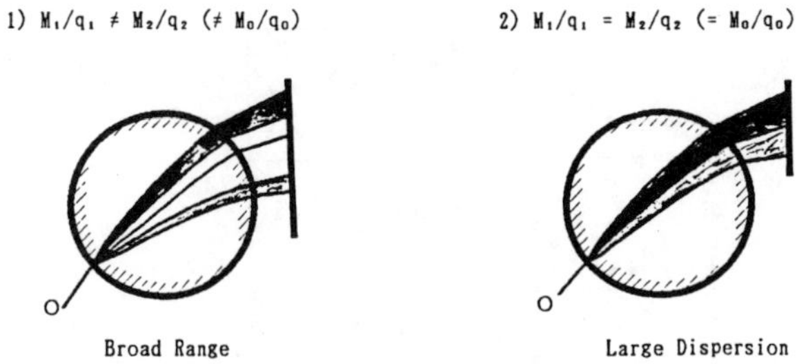


Figure 2: Two types of spectrograph measurements.

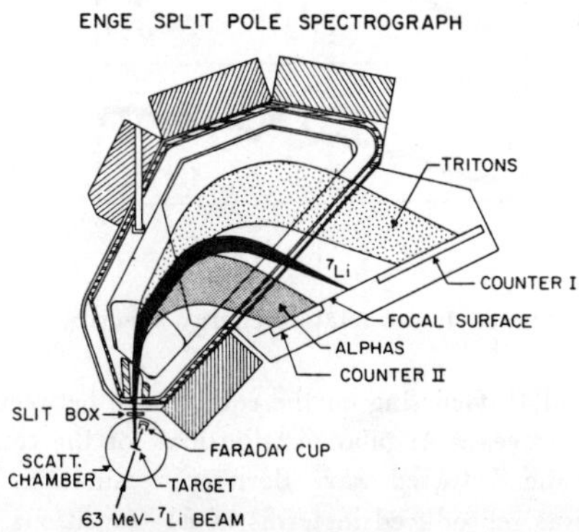


Figure 3: Schematic of the experimental set-up.

Two pieces of $^7Li$ breakup experiments were done, one at the bombarding energy of 63 MeV at the Cyclotron Institute of Texas *A&M* University and the other at 42 MeV in the Oak Ridge National Laboratory. The experimental set-up used at Texas *A&M* is illustrated in Fig. 3. A similar set-up was used for the Oak Ridge experiment. Breakup pairs of $\alpha$'s and tritons entered an Enge split pole spectrograph, were analysed according to their magnetic rigidity and were focussed at different positions along the focal plane. Two single-resistive-wire proportional counters mounted at appropriate locations along the focal plane were used to detect $\alpha$'s and tritons in coincidence. The elastic scattering impinged upon the dead space between the two focal plane detectors, causing no interference to the coincidence measurement.

The $^7Li$ breakup reactions were studied systematically. The systematic study included the 2 bomarding energies, 5 targets ranging from $^{27}Al$ to $^{208}Pb$ (five different intensities of the Coulomb field) and many detection angles (different impact parameters). Several of the measured relative energy spectra are shown in Fig. 4. For the first time, $\alpha$-$t$ correlations were measured, without detection threshold, at energies down to zero. One can see that intense $\alpha$-$t$ correlations arize from *continuum* states just above the $\alpha$-$t$ threshold energy. The continuum is considered to be the same as intermediate states populated in the radiative capture $t(\alpha,\gamma)^7Li$ process.

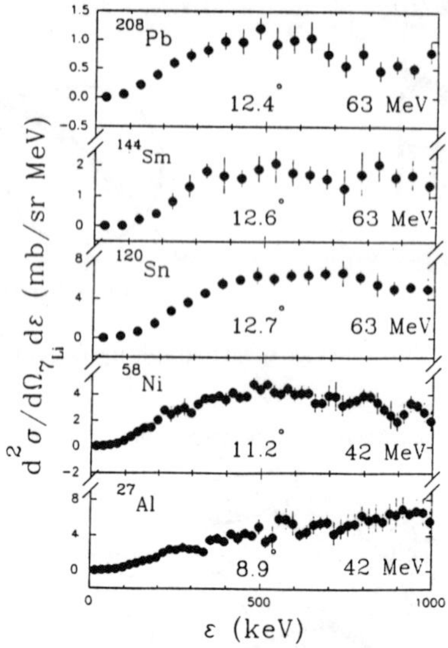


Figure 4: $\alpha$-$t$ relative energy spectra.

The data were examined[4], focussing on the equivalence between the breakup reaction and the radiative capture process. Angular distributions for the continuum were analysed within the framework of the distorted wave Born approximation. The gross feature of the angular distribution was reproduced in terms of E1 transitions. The nuclear breakup contribution ranges from 5% to 30% at forward angles inside the grazing angle. Although the Coulomb breakup seems to dominate at specific forward angles, the nuclear contribution cannot be entirely ignored at the present bombarding energies. The effect of the post Coulomb acceleration was also studied by examining the location of the spectral minima corresponding to the cross section at zero relative energy. No strong indication of the effect was observed.

An attempt was made to **obtain** the reduced transition probablity, $B(E1;\varepsilon)$, by dividing the measured cross section by the theoretical E1 Coulomb excitation function[16]. The deduced $B(E1;\varepsilon)$ are shown in Fig. 5. It turned out that despite of the possible nuclear-Coulomb interference, the $\varepsilon$ dependence of the reduced transition probability is nearly universal, independent of the two bombarding energies, the five targets and the detection angles. Motivated by this finding, a DWBA calculation was performed to study the energy dependence of the Coulomb and nuclear excitation functions, assuming that the continuum results from one-step E1 transitions from the ground state of $^7Li$. The DWBA calculation introduced the concept of *parallelism* in the two excitation functions. However, the parallelism is not confirmed by another DWBA analysis[28]. Further theoretical analyses are necessary. Such analyses should address, in addition to the parallelism, the effects of continuum-continuum and/or discrete-continuum channel coupling and of final state interactions.

The universal reduced transition probablities allowed us to extract astrophysical S-factors. After normalizing them to the direct measurements at $\varepsilon$ = 500 keV, the extracted astrophysical S-factors are plotted in Fig. 6. It was concluded that the present breakup data

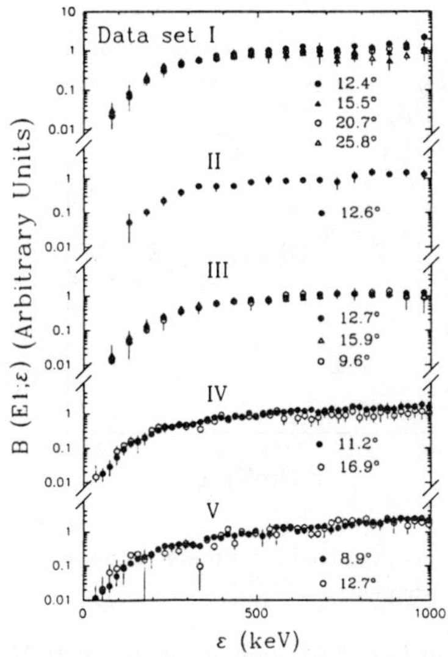


Figure 5: Reduced transition probability. The target and the bombarding energy for the data sets I - V are: (I) $^{208}Pb$, 63 MeV; (II) $^{144}Sm$, 63 MeV; (III) $^{120}Sn$, 63 MeV; (IV) $^{58}Ni$, 42 MeV; (V) $^{27}Al$, 42 MeV.

support the recent direct measurement of Schröder *et al.*, which exhibits a significant energy dependence at small relative energies. However, the result seems to contradict the most recent theoretical calculations[29,30,31,32,33] which are rather consistent with the data of Griffiths *et al.* Another direct measurement of the $t(\alpha,\gamma)^7Li$ reaction rate will be necessary.

## IV. Projectile breakup of $^{16}O$

It is very interesting to direct the present research idea at other nuclear elements of astrophysical interest. Very recently, we studied $^{16}O$ breakup reactions at 30 MeV/nucleon in relation to the stellar helium burning process, i.e., $^{12}C(\alpha,\gamma)^{16}O$. The reaction rate at astrophysical energies around 300 keV is a key for stellar evolution which determines the internal structure of massive stars and, therefore, is a key for mechanisms of the supernova explosion. In spite of many efforts[34,35,36,37] for the last two decades, experimentally this reaction rate remains unknown at relative energies below 1 MeV. The current situation is very unsatisfactory, providing a serious challenge to experimentalists.

For this experiment, we have developed a new technology with a plastic phoswich for heavy ion identification in magnetic spectrograph measurements[38]. When combined with the p/q selection by the spectrograph, the plastic phoswich was found to provide a complete identification for light heavy ions at energies of some tens of MeV/nucleon. This technology offers the simplest method for HI identification in magnetic spectrograph measurements, eliminating thick ionization chambers from the focal plane detector complex. A further test

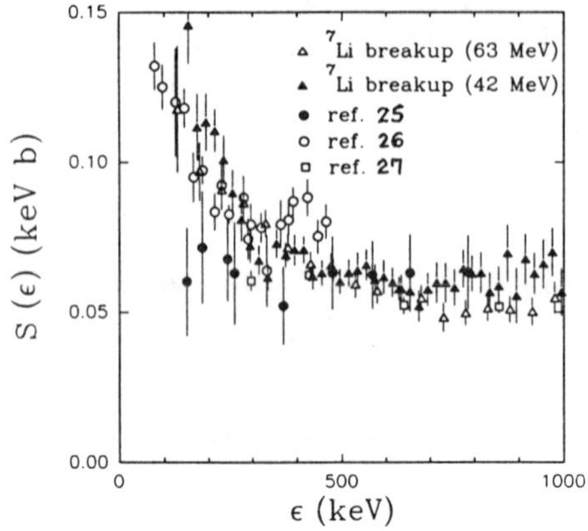


Figure 6: Astrophysical S-factors extracted from the breakup data in comparison with those deduced from the direct measurements.

is underway with a 30 MeV/nucleon-$^{40}Ar$ beam at the Texas *A&M* Cyclotron Institute.

Two 40 cm single-resistive-wire proportional counters backed by plastic phoswiches were mounted along the focal plane of the Enge split pole magnetic spectrograph. The plastic phoswich is 40 cm long, consisting of a 400 $\mu$m BC400 fast plastic scintillator and a BC444 slow plastic scintillator of 25 mm thickness. The scintillation light was read out with a RCA S83010E 38mm diameter photomultiplier tube mounted on one end of the crystal. The dead space between the two focal plane detectors was 12 cm long. The associated detection threshold to $\alpha$-$^{12}C$ relative kinetic energies was 200 keV. The dynamic range for the relative energy spanned from the threshold energy up to 6 MeV.

Figure 7 shows a 2-dimensional plot of $\alpha$-$^{12}C$ correlations along with an energy sum, $E_\alpha + E_{^{12}C}$, spectrum in reactions with a $^{208}Pb$ target. Reaction mechanisms involved in the $\alpha$-$^{12}C$ correlation are classified into three groups. The large peak in the energy sum spectrum is identified as elastic breakup, in which both $^{12}C$ and $^{208}Pb$ are in their ground states. The small peak next to the elastic breakup peak resulted from inelastic breakup, in which the $^{12}C$ was excited in the $2^+$ state at 4.44 MeV. The broad bump with large negative Q-values originated from *deep inelastic* breakup, in which one of the $^{12}C$ and the $^{208}Pb$, (probably both), underwent substantial excitation. In the 2-d plot, kinematical loci corresponding to the elastic and inelastic breakup are indicated by the solid and dashed lines, respectively.

The elastic breakup needs to be examined from astrophysical point of view. Many resonant states of $^{16}O$ were populated. However, it seems that the continuum below $\varepsilon = 2$ MeV was not strongly excited. The final result must await the completion of the off-line analysis which is in progress.

## V. Summary - toward the future

A test of the innovative approach to radiative capture processes was made for the $^7Li$

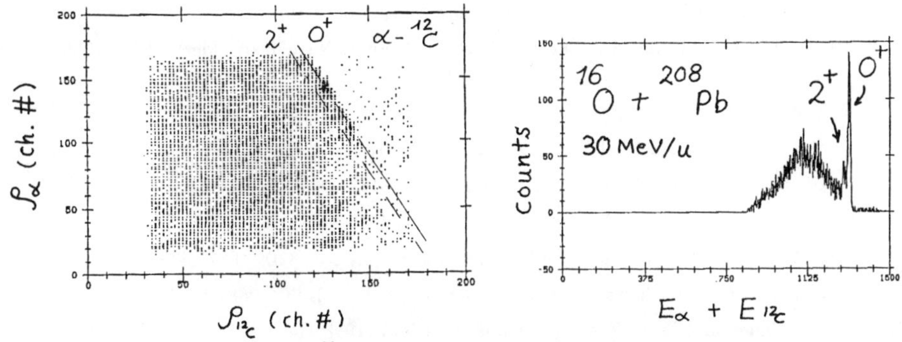


Figure 7: Scatter plot of $\alpha$-$^{12}C$ correlations and an energy sum ($E_\alpha$ + $E_{12_C}$) spectrum.

system. The test has shed new light on the $t(\alpha,\gamma)^7Li$ reaction rate. Very recently, an attempt was made of extending the test to the $^{16}O$ system using breakup reactions at 30 MeV/nucleon. A proposal to study $^{16}O$ breakup at a higher bombarding energy is also underway[39]. One must tackle an elaborate task of clarifying reaction mechanisms involved in projectile breakup reactions on the way to the goal of the innovative approach. It is expected that those mentioned above along with the on-going experiments with unstable beams[6,7] produce fruitful results in the next few years. This field may occupy a unique position in nuclear physics since it could newly bridge between nuclear physics and astrophysics.

# References

[1] H. Rebel, Workshop on Nuclear Reaction Cross Sections of Astrophysical Interest, Kernforschungszentrum Karlsruhe (Feb. 1985), unpublished.

[2] G. Baur, C.A. Bertulani and H. Rebel, Nucl. Phys. **A458** (1986) 188

[3] H. Utsunomiya *et al.*, Phys. Lett. **211B** (1988) 24

[4] H. Utsunomiya *et al.*, Nucl. Phys. **A511** (1990) 379

[5] J. Kiener, H.J. Gils, H. Rebel and G. Baur, Z. Phys. **A332** (1989) 359

[6] T. Motobayashi, *private communications*

[7] P. Aguer, *private communications*

[8] A.C. Shotter *et al.*, Phys. Rev. Lett. **46** (1981) 12; ibid. **53** (1984) 1539 and references therein

[9] D.L. Disdier *et al.*, Phys. Rev. Lett. **27** (1971) 1391

[10] H. Utsunomiya, Y.-W. Lui and R.P. Schmitt, Nucl. Instr. Meth. **A278** (1989) 744

[11] H. Rebel and D.K. Srivastava, KfK Report 4761 (August, 1990)

[12] J.E. Spencer and H.A. Enge, Nucl. Instr. Meth. **49** (1967) 181

[13] E.H.L. Aarts, A.M. van den Berg and R.J. de Meijier, Nucl. Instr. Meth. **176** (1980) 605

[14] C. Ellegard *et al.*, Phys. Rev. Lett. **59** (1987) 974

[15] T. Motobayashi *et al.*, Nucl. Phys. **A481** (1988) 207

[16] H. Utsunomiya *et al.*, Phys. Rev. Lett. **65** (1990) 847

[17] D.N. Schramm and R.V. Wagoner, Ann. Rev. Nucl. Part. Sci. **27** (1977) 37

[18] C.R. Alcock, G.M. Fuller and G.J. Mathews, Astrophys. J. **320** (1987) 439

[19] G.M. Fuller, G.J. Mathews and C.R. Alcock, Phys. Rev. D**37** (1988) 1380

[20] R.A. Malaney and W.A. Fowler, Astrophys. J. **333** (1988) 14

[21] N. Terasawa and K. Sato, Phys. Rev. D**39** (1989) 2893

[22] H. Kurki-Suonio and R.A. Matzner, Phys. Rev. D**39** (1989) 1046

[23] G.J. Mathews, B.S. Meyer, C.R. Alcock and G.M. Fuller, Astrophys. J. **358** (1990) 36

[24] C.R. Alcock, D.S. Dearborn, G.M. Fuller, G.J. Mathews and B.S. Meyer, Phys. Rev. Lett. **64** (1990) 2607

[25] G.M. Griffiths *et al.*, Can. J. Phys. **39** (1961) 1397

[26] U. Schröder *et al.*, Phys. Lett. **192B** (1987) 55

[27] S. Burzynski *et al.*, Nucl. Phys. **A473** (1987) 179

[28] J. Kiener *et al.*, preprint, 1991

[29] T. Kajino and A. Arima, Phys. Rev. Lett. **52** (1984) 739; T. Kajino, Nucl. Phys. **A460** (1986) 559

[30] K. Langanke, Nucl. Phys. **A457** (1986) 351

[31] T. Mertelmeier and H.M. Hofmann, Nucl. Phys. **A459** (1986) 387

[32] T. Kajino, G.F. Bertsch and K.-I. Kubo, Phys. Rev. C**37** (1988) 512

[33] T. Kajino G.J. Mathews and K. Ikeda, Phys. Rev. C**40** (1989) 525

[34] B.W. Filippone, Ann. Rev. Nucl. Par. Sci. **36** (1986) 717 and references therein

[35] R.M. Kremer *et al.*, Phys. Rev. Lett. **60** (1988) 1475

[36] X. Ji, B.M. Filippone, J. Humblet and S.E. Koonin, Phys. Rev. C**41** (1990) 1736

[37] F.C. Barker and T. Kajino, preprint to appear in Australian J. Phys. (june, 1991)

[38] H. Utsunomiya and Y.-W. Lui, Nucl. Instr. Meth. **A294** (1990) 674

[39] T. Motobayashi *et al.*, '91 RIKEN proposal

# SYSTEMATICS OF THE POWER LAW PARAMETER AND THE ROLE OF ANGULAR MOMENTUM IN THE EMISSION OF INTERMEDIATE MASS FRAGMENTS

W. TRAUTMANN, U. MILKAU, U. LYNEN
GSI Darmstadt, D-6100 Darmstadt, Germany
J. POCHODZALLA
Universität Frankfurt, D-6000 Frankfurt, Germany

ABSTRACT

The element yield distributions of intermediate mass fragments from proton and heavy-ion induced reactions show a universal behaviour as a function of the bombarding energy. It is suggested that this observation is linked to the role of angular momentum in the emission process.

It has become customary to fit power law functions to the element (or mass) yield distributions of intermediate mass fragments (IMF's) from proton and heavy-ion induced reactions on nuclear targets [1]. This was originally motivated by the idea [2] that a critical disassembly of nuclear matter in a liquid-gas phase transition should lead to power law distributions, i.e. elemental cross sections $\sigma(Z) \propto Z^{-\tau}$, with a critical value of $\tau \simeq 2.3$. While the observed yield distributions can also be explained by other models, the power law parameterizations were found useful for characterizing the fragment emission in nuclear reactions.

In fig. 1 we present a compilation of power law parameters $\tau$ for proton and heavy-ion induced reactions on Ag, Xe, Au, Pb, Th, and U targets. They were either directly taken from the literature or, in order to ensure consistency, obtained from fits to the reported cross sections [2-28]. The fits included IMF's with atomic number $Z \geq 3$ except for Be as $^8$Be is not stable and the measured Be cross sections are generally below the trend. Yields from projectile fragmentation were excluded, and data suffering from high detection thresholds were not considered for this compilation.

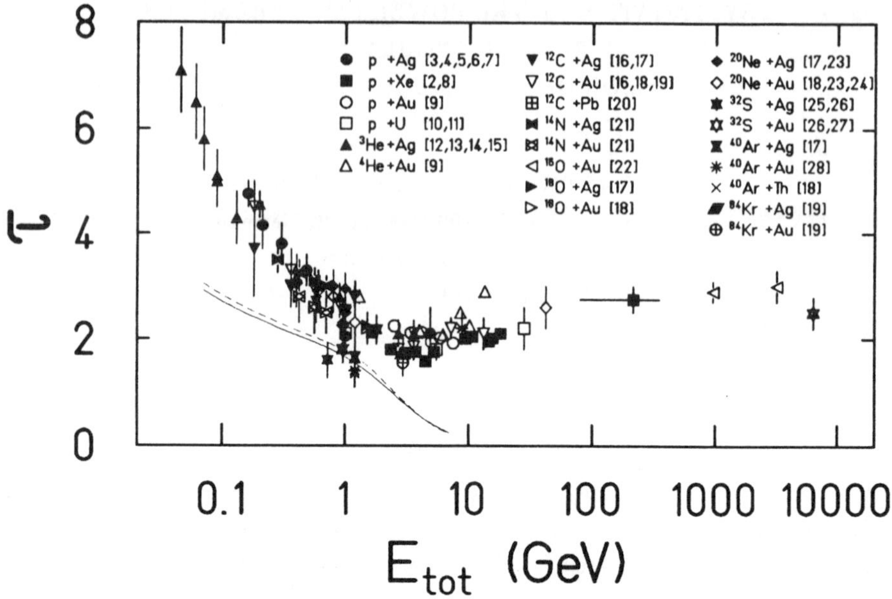


Figure 1: Systematics of the $\tau$ parameter. The meaning of the symbols and the references are indicated. The lines are the results of the calculations explained in the text for Ag (full lines) and Au (dashed lines) targets.

It turns out that the $\tau$ parameters exhibit a universal behaviour when plotted as a function of the total projectile energy $E_{tot}$. At low energies $\tau$ is large, values up to $\tau = 7$ were extracted from the low energy $^3$He + $^{nat}$Ag data [13]. With increasing energy $\tau$ monotonically decreases to a minimum $\tau \simeq 2.0$ at $E_{tot} \cong 2$ GeV to 5 GeV. Beyond this energy $\tau$ levels off or perhaps rises very slowly at the highest bombarding energies. In the chosen representation no significant dependence on the projectile nor on the target is observed. In some cases explanations for somewhat unusual $\tau$ parameters may exist. So, e.g. in the $^{40}$Ar + $^{197}$Au reaction at $E_{tot} = 1.2$ GeV, dissipative processes may cause an enhancement of the cross section for $Z \approx Z_{proj}$ and thus be responsible for $\tau = 1.4$ measured in this case [28].

The invariance of $\tau$ at $E_{tot} > 10$ GeV is consistent with the concept of limiting fragmentation. The production cross sections for IMF's were found to saturate at projectile energies near 5 to 10 GeV in a variety of reactions [4,8,29,30]. When all the individual IMF cross sections $\sigma(Z)$ have reached their limiting values $\tau$ will remain constant. The $\tau$ parameters observed in that region are between 2 and 3. Most multifragmentation models are able to reproduce such values but similar $\tau$ parameters also follow from more conventional approaches. The mass yield distributions measured in the p + Xe reaction at $E_p$ = 80 - 350 GeV were nearly quantitatively reproduced with a statistical model that takes the separation energies of the excited primary fragments and their subsequent decays into account [31].

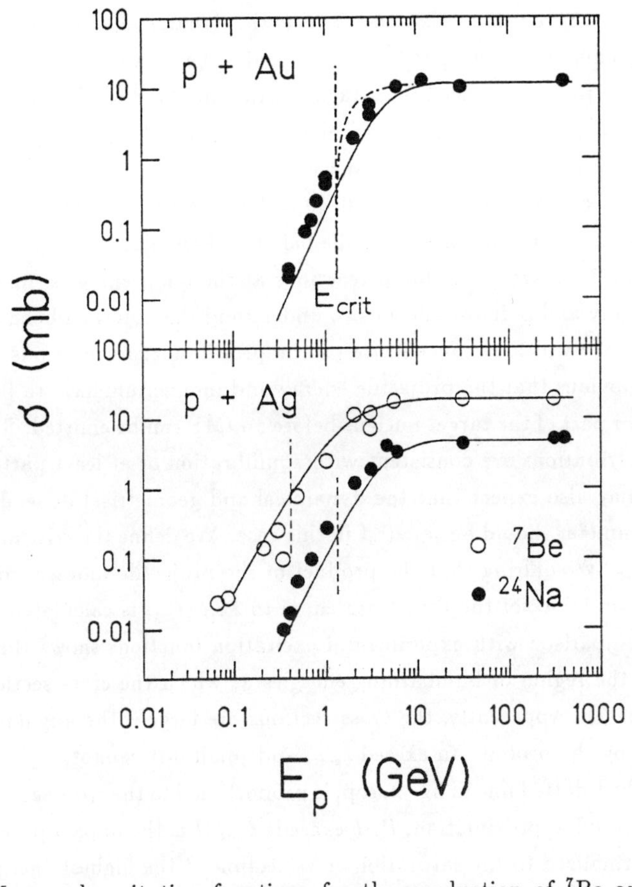


Figure 2: Measured excitation functions for the production of $^7$Be and $^{24}$Na in reactions of protons with Au and Ag targets. The data are from refs.[4,30,36]. The vertical dashed lines mark $E_{crit}$, the dashed-dotted line (top) gives the result of the sharp-cutoff and the full lines give the results of the smooth-cutoff calculations explained in the text.

At projectile energies near the lower end of the range considered here, it was found that the IMF emission is associated with a specific partition of the angular momentum residing in the system [32]. A physical interpretation was given by applying the concept of critical angular momentum, first introduced to explain the excitation functions of complete fusion reactions [33,34], to the dinuclear configuration at the time when the fragment separates. The picture is the following: The relative angular momentum $\ell$ of an IMF with given mass and charge with respect to the emitting system is *critical* if the sum of the nuclear, Coulomb, and centrifugal forces vanishes at a critical distance. This is the dynamically most favourable situation for the emission of this particular fragment. The emission will be suppressed for $\ell > \ell_{crit}$ due to the enhanced centrifugal barrier and for $\ell < \ell_{crit}$ due to the resulting attractive net force at the critical distance. Applied with the parameterization of Wilczynski [34] this concept was successful in explaining the angular momentum distributions in heavy-ion reactions at $10MeV \lesssim E/A \lesssim 20MeV$ [32,35].

In this contribution we show that the balance of angular momentum, found to be so important in heavy-ion induced reactions at the lower energies, may be relevant more generally and perhaps allow us to understand the universal behaviour of the $\tau$ parameter. We will first concentrate on the proton-nucleus reactions because here it is most obvious that the projectile energy and momentum have to be distributed over a major part of the target nucleus before an IMF can be emitted. The measured angular distributions are consistent with equilibration or at least partial equilibration. We may also expect that the dynamical and geometrical dependences on the impact parameter should be smallest in this case. We define the critical bombarding energy $E_{crit}$ by requiring that the product of the projectile momentum $P_p$ and the radius $R_t = 1.2A^{\frac{1}{3}}$ of the target are equal to $\ell_{crit}$ ($\ell_{crit}$ is calculated according to [34]). A comparison with experimental excitation functions shows that $E_{crit}$ coincides with the region of bombarding energies at which the cross sections rise very steeply (fig. 2). Apparently, the cross sections are large if the angular momentum brought in by the proton can exceed $\ell_{crit}$, and small if it cannot.

The dashed-dotted line in fig. 2, top, is proportional to the cross section for which, in a sharp-cutoff approximation, $P_p \cdot b$ exceeds $\ell_{crit}$ ($b$ is the impact parameter). The curve is normalized to the saturation cross section at the highest energies at which this condition is valid for virtually all partial waves. This prescription reproduces the dominant features of the measured excitation functions.

The agreement can be further improved if a smooth-cutoff is chosen and some obvious modifications are added. The full lines given in fig. 2 were obtained by adjusting the parameter that determines the width of the cutoff region. In addition, realistic density distributions for the target nucleus, a mean-free path of 1.5 fm for the incident proton, and a second parameter describing the incomplete transfer of linear momentum were used. However, it is not intended here to reproduce the experimental excitation functions with high accuracy but rather to demonstrate that the concept of critical angular momentum is sufficient to arrive at a qualitative understanding.

The critical angular momentum is a function of the fragment mass and charge. It is larger for larger fragments. The excitation functions of $^{7}$Be and $^{24}$Na fragments from p + Ag reactions reflect that behaviour (fig. 2, bottom). This implies that the yield of larger fragments will be suppressed if the angular momentum residing in the system at the time of IMF emission is not sufficiently large. As the projectile energy and momentum decrease and, presumably, the transferred angular momentum decreases, this suppression increases which leads to an increase of $\tau$. The thin lines shown in fig. 1 were obtained with the smooth-cutoff prescription and with the same parameters as used for the excitation functions in fig. 2. They demonstrate that the deviation of $\tau$ from its limiting value may be qualitatively understood as an angular momentum effect. As $\ell_{crit}$ depends very little on the target (cf. fig. 2) this interpretation is consistent with the observed invariance of $\tau$ with respect to the target nucleus (fig. 1).

## REFERENCES


1. see, e.g., W.G. Lynch, Ann. Rev. Nucl. Part. Sci. 37, 493 (1987).
2. A.S. Hirsch et al., Phys. Rev. C29, 508 (1984).
3. G.D. Westfall et al., Phys. Rev. C17, 1368 (1978).
4. R.E.L. Green and R.G. Korteling, Phys. Rev. C22, 1594 (1980).
5. R.E.L. Green, R.G. Korteling, and K.P. Jackson, Phys. Rev. C29, 1806 (1984).
6. L.N. Andronenko et al., Phys. Lett. B174, 18 (1986).
7. S.J. Yennello et al.,Phys. Rev. C41, 79 (1990).
8. N.T. Porile et al., Phys. Rev. C39, 1914 (1989).
9. V.V. Avdejchikov et al., Sov. J. Nucl. Phys. 48, 1043 (1988).
10. A.M. Poskanzer, G.W. Butler, and E.K. Hyde, Phys. Rev. C3, 882 (1971).
11. L.P. Remsberg and D.G. Perry, Phys. Rev. Lett. 35, 361 (1975).

12. L.G. Sobotka et al., Phys. Rev. Lett. 51, 2187 (1983).

13. M.A. McMahan et al., Phys. Rev. Lett. 54, 1995 (1985).

14. K. Kwiatkowski et al., Phys. Lett. B171, 41 (1986).

15. S.J. Yennello et al.,Phys. Lett. B246, 26 (1990).

16. C.B. Chitwood et al., Phys. Lett. B131, 289 (1983).

17. R. Trockel, Ph.D. thesis, Universität Heidelberg, Report No. GSI-87-17, 1987 (unpublished).

18. R. Trockel et al., Phys. Rev. C39, 729 (1989).

19. U. Milkau et al., to be published.

20. B. Grabež et al.,Phys. Rev. C34, 170 (1986).

21. D.E. Fields et al., Phys. Lett. B220, 356 (1989).

22. B. Berthier et al., Phys. Lett. B193, 417 (1987).

23. N.H. Papadakis et al., Phys. Lett. B240, 317 (1990).

24. A.I. Warwick et al., Phys. Rev. C27, 1083 (1983).

25. D.J. Fields et al., Phys. Rev C34, 536 (1986).

26. A. Lleres et al., Phys. Lett. B185, 336 (1987).

27. R. Boisgard et al., Z. Phys. A338, 243 (1991).

28. U. Milkau et al., preprint GSI-91-08, 1991.

29. K. Aleklett, L. Sihver, and W. Loveland, Phys. Lett. B197, 34 (1987).

30. S.B. Kaufman and E.P. Steinberg, Phys. Rev. C22, 167 (1980).

31. D.J. Fields et al., Phys. Lett. B187, 257 (1987).

32. W. Dünnweber et al., Phys. Rev. Lett. 52, 1405 (1984).

33. R. Bass, Phys. Lett. 47B, 139 (1973).

34. J. Wilczyński, Nucl. Phys. A216, 386 (1973).

35. Z. Majka et al., Phys. Rev. C40, 2124 (1989).

36. G.D. Cole and N.T. Porile, Phys. Rev. C24, 2038 (1981).

# V. Ultra Relativistic Heavy-Ion Beams

# DILEPTON PRODUCTION WITH PROTON AND ION BEAMS AT THE CERN-SPS

D. CONTARDO*

**Abstract**

The NA38 collaboration has measured the two muon decays of the $\rho$, $\omega$, $\phi$ and $J/\psi$ vector mesons produced in 200 GeV per nucleon proton-ion and ion-ion interactions. For the latter reactions, the $\phi$ and $J/\psi$ yields relative to the muon-pairs produced in the mass continuum, are found to strongly depend on the transverse energy released in the collisions. Such effects were predicted to occur in case of Quark-Gluon Plasma formation, but explanations by hadronic processes are not ruled out. The NA34-3 experiment has upgraded its apparatus for the detection of low mass dimuons. A new experiment, NA45, is dedicated to the measurement of electron-pairs.

* I.P.N. Lyon, CNRS-IN2P3, 43 Bd du 11 Novembre 1918, F-69622 Villeurbanne Cedex, France

## 1. Introduction

In high energy nucleus-nucleus reactions, among all predicted signals of quark-gluon plasma formation, it is expected a thermal production of dileptons [1], a Debye screening of the $J/\psi$ formation [2] and an increase of the $\phi$ yield as a consequence of the general strangeness production enhancement [3]. Recent measurements of electron pairs produced by p-Be reactions at 4.5 GeV/c [4] have shown the possible opening of a two pion annihilation channel at 300 MeV/$c^2$, this phenomenon could provide a new signature for a phase transition.

Since muons, filtered by a hadron absorber, provide a highly efficient and selective trigger, dimuon spectroscopy is a unique tool to measure such rare processes. However, the combinatorial background, originating from decay muons of the pions or kaons copiously produced in the nuclear collisions, is a limitation to the achievable resolution and statistical accuracy in the measurement of low mass dimuons. An experimental alternative consists in detecting dielectrons. In this case, in spite of a reduced luminosity, the combinatorial background, which arises mainly from Dalitz decays and conversion pairs, is expected to be one order of magnitude smaller. Due to the minimization of the radiation length on the electron trajectories, a better resolution is also foreseen.

In the 200 GeV per nucleon ion beam program at the CERN-SPS facility, the NA38 experiment is dedicated to the dimuon measurement with the NA10 spectrometer, complemented with an electromagnetic calorimeter and a segmented active target. In the NA34-3 experiment an adequate hadron dump is now set up upstream the HELIOS dimuon spectrometer. Finally, the NA45 collaboration has built a double-solenoid electron spectrometer based on ring-image Čerenkov techniques.

## 2. The dielectron measurement with the NA45 detector

In this experiment, electron pairs are measured between 0.1 and 3.0 GeV/c$^2$. The acceptance of the apparatus covers the rapidity range $2 < y_{lab} < 3$ and the expected mass resolution is about 4-5% at the $\rho$, $\omega$ and $\phi$ resonances. The experimental set-up consists of two ring-image Čerenkov located on each side of the magnet. The Čerenkov thresholds are chosen sufficiently high such that essentially only electrons create rings. The electron momentum is obtained from its change in direction as measured directly in the RICH detectors. For each pair produced in the interactions, the correlated charged particle multiplicity is measured by silicon detectors. The first data were registered in August 1990, they show a good identification of rings in the Čerenkov counters, but it is probably too early to extract physical results.

### 3. The dimuon data from NA34-3 Experiment

The HELIOS spectrometer used in this experiment, covers the forward rapidity range $3.5 < y_{lab} < 6.0$. It measures the continuum muon pairs in the mass region between 0.2 and 2.0 GeV/c$^2$ and the $\rho$, $\omega$ and $\phi$ vector mesons. A new hadron dump made of alumina optimizes the mass resolution at the resonances and silicon ring detectors measure the charged particle multiplicity of the collisions. In spite of a large background originating from the dump, preliminary results indicate a good mass resolution of about 10% at the $\phi$ resonance. The prospect for physics is thus encouraging.

### 4. The NA38 dimuon experiment

This experiment measures the mass and transverse momentum of dimuons emitted in the $2.8 < y_{lab} < 4.0$ rapidity acceptance of the spectrometer [5]. The data acquisition is triggered on two tracks of any sign, pointing at the target, and surviving complete absorption of hadrons, in two hadron dumps located upstream and downstream the toroïdal magnet. This geometry implies a cut on individual muons with a longitudinal momentum $P_{L_\mu} < 4.0$ GeV/c. In order to handle the high event rates, up to 600 triggers during the 4s burst, a second level trigger rejects pairs with both muons having a transverse momentum $P_{T_\mu} < 0.26$ GeV/c. The mass resolution at the $J/\psi$ ($\phi$) is about 300 (210) MeV/c$^2$ (FWHM).

The electromagnetic calorimeter is made of scintillating fibres embedded in lead. It is 12 cm long and covers the pseudorapidity range $1.7 < \eta_{lab} < 4.1$. The neutral transverse energy released in the collisions ($E_T$) is estimated from the measured energy flow after calibration and correction for the charged hadron contribution (40%).

The p-Cu, p-U, O-Cu, O-U and S-U reactions were studied in five different periods with 200 GeV per nucleon beams. The maximum available intensities up to $\sim 10^7$ particles/sec were used. When analysing the data, events with two and only two reconstructed muon tracks, pointing at the target, are selected. In order to prevent any pile-up in the transverse energy measurement, a beam hodoscope is used to select events with only one incoming particle within a time window of $\pm$ 20 nsec, while the segmented active target (10 to 17 subtargets, representing a 20% total interaction length, surrounded by 24 ring scintillators) rejects events showing a secondary interaction of a projectile fragment. These two cuts are not applied for the analysis performed independently of $E_T$, which is the case for the proton data and for the cross-section estimations.

Most of the ion beam events at masses below 3.0 GeV/c$^2$ are pairs of pion or kaon decay muons. This background in the sample of opposite sign pairs is estimated to amount to $2\sqrt{N_{\mu^+\mu^+}N_{\mu^-\mu^-}}$ from the number of registered like sign pairs. Provided that acceptances of each of the muons are identical, this formula is valid bin by bin in any dimuon distribution. To fulfill this condition in the mass region below 1.7 GeV/c$^2$, an additional rejection of pairs, where one of the muons has $P_{T_\mu} < 0.7$ or $P_{L_\mu} < 9.5$ GeV/c, is needed. This reduces the kinematical domain for the remaining $\rho$, $\omega$ and $\phi$ to $P_T > 1.1$ GeV/c.

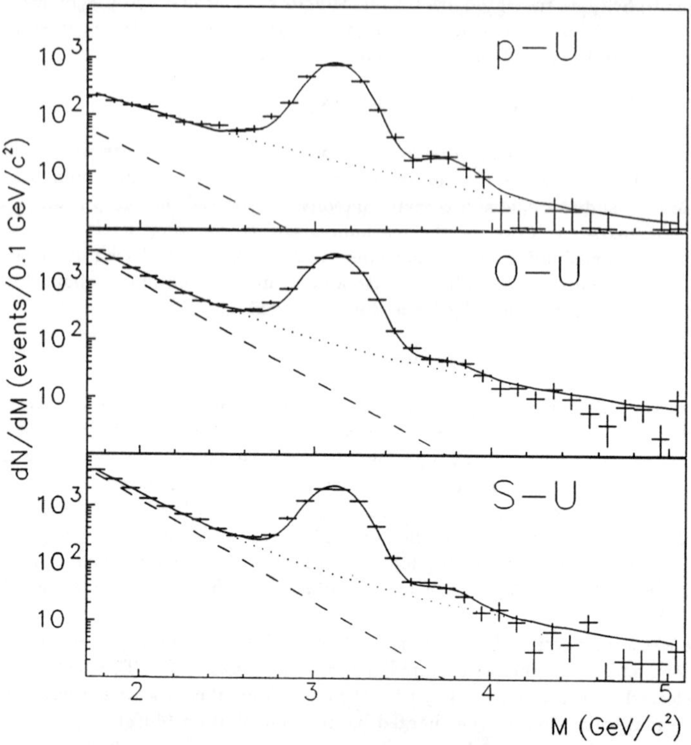


**Fig. 1.** *Opposite sign muon-pair mass distributions measured in p-U, O-U and S-U interactions. The full lines are the results of the fit explained in the text. The backgrounds are represented by the dashed lines. The dotted lines are the sum of the continuum and background curves.*

### 5. NA38 : The $J/\psi$ suppression

Presently, only the results for muon-pairs with $M_{\mu\mu} > 1.7$ GeV/c$^2$ are corrected for the acceptance of the apparatus ($\simeq 10\%$ for $J/\psi$ with Feynman variable $X_F > 0$). Figure 1. shows the mass distributions of opposite sign pairs, obtained in various reactions. They were fitted to a sum of the signal, modified by the acceptance and smearing effects of the apparatus, plus the background previously defined. The signal is itself the $J/\psi$ and $\psi'$ amplitudes, at their fixed mass, added to a continuum term behaving as $const./M^{A_c}$. The deformation in the detector is obtained from a Monte-Carlo simulation. The background is estimated by a simultaneous fit to the measured $\mu^+\mu^+$ and $\mu^-\mu^-$ mass distributions. This analysis was performed for different $E_T$ event samples of similar statistics.

| Reaction | $E_T$ (GeV) | $\varepsilon$ (GeV/fm$^3$) | $A_c$ | R |
|---|---|---|---|---|
| p-Cu | all $E_T$ | | $5.11 \pm 0.17$ | $2.00 \pm 0.09$ |
| p-U | all $E_T$ | | $5.46 \pm 0.20$ | $1.56 \pm 0.08$ |
| O-Cu | $< 13$ | 1.0 | $6.23 \pm 0.35$ | $1.40 \pm 0.11$ |
| | $13 - 21$ | 1.1 | $6.20 \pm 0.37$ | $1.36 \pm 0.11$ |
| | $21 - 30$ | 1.3 | $5.91 \pm 0.30$ | $1.10 \pm 0.08$ |
| | $\geq 30$ | 1.6 | $5.98 \pm 0.44$ | $1.16 \pm 0.11$ |
| O-U | $< 20$ | 0.9 | $5.73 \pm 0.23$ | $1.23 \pm 0.07$ |
| | $20 - 31$ | 1.3 | $5.77 \pm 0.29$ | $1.23 \pm 0.08$ |
| | $31 - 39$ | 1.6 | $6.15 \pm 0.32$ | $1.02 \pm 0.07$ |
| | $39 - 45$ | 1.8 | $6.31 \pm 0.36$ | $0.92 \pm 0.07$ |
| | $45 - 52$ | 2.1 | $5.57 \pm 0.35$ | $1.02 \pm 0.09$ |
| | $\geq 52$ | 2.5 | $5.72 \pm 0.30$ | $0.71 \pm 0.06$ |
| S-U | $< 34$ | 1.2 | $5.56 \pm 0.19$ | $1.72 \pm 0.09$ |
| | $34 - 49$ | 1.5 | $5.67 \pm 0.22$ | $1.02 \pm 0.06$ |
| | $49 - 62$ | 1.6 | $6.07 \pm 0.27$ | $1.03 \pm 0.06$ |
| | $62 - 71$ | 1.8 | $5.97 \pm 0.32$ | $1.12 \pm 0.09$ |
| | $71 - 78$ | 2.0 | $5.84 \pm 0.33$ | $0.90 \pm 0.07$ |
| | $\geq 78$ | 2.3 | $5.52 \pm 0.33$ | $1.07 \pm 0.10$ |

**Table 1 :** *Fitted values of $A_c$ and R for p-Cu, p-U, O-Cu, O-U and S-U interactions as a function of $E_T$ ($\epsilon$). The quoted errors are statistical only.*

Results are summarized in table 1. We report the shape parameters $A_c$ of the mass continuum and the ratios R of the $J/\psi$ cross-sections divided by the continuum ones estimated for muon-pairs with a mass between 1.7 and 2.7 GeV/c$^2$. All the dimuons have a Feynman's variable in the range $0 < X_F < 0.5$. The $\psi'$ contribution which is found to be too small to allow any significant conclusions, has been excluded. A 40% decrease of R is observed between the extreme $E_T$ bins for both the O-U and S-U interactions.

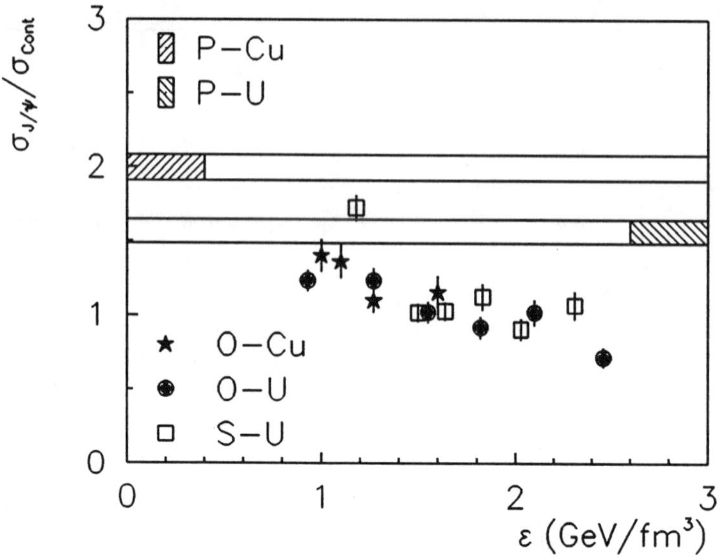


**Fig. 2.** *J/ψ to continuum cross-section ratios as a function of the Bjorken's energy density for p-Cu, p-U, O-Cu, O-U and S-U reactions.*

In order to compare the effects observed in the various reactions, the different mean values of the $E_T$ bins have been related to the Bjorken's energy density which governs the QGP formation, namely : $\epsilon = 3<E_T> / \Delta\eta A_\perp \tau_0$ (table 1). $\Delta\eta = 2.4$ is the pseudorapidity interval in which $E_T$ is measured, $\tau_0$ = 1fm is the QGP formation time, and $A_\perp$ is the mean transverse overlap area of the colliding nuclei evaluated using a geometrical model [6]. The ratios R, obtained in the ion-ion reactions, are plotted on fig. 2 with respect to $\epsilon$ which appears to give a good common representation of the data. However, since the transverse size of the interaction region inside the target nucleus , and thus $\epsilon$, is ill defined for incident protons, the p-Cu and p-U results are reported as horizontal dashed area. It has to be noted that, according to the QGP model, the energy density achieved in the most central O-U and S-U collisions, should be sufficient to create it. Nevertheless, the trend of the data is not well enough defined to observe a kink, at some $\epsilon$ threshold, which could be the cleanest signature of a change of regime from hadronic final state absorption of the $J/\psi$ to Debye screening in the QGP.

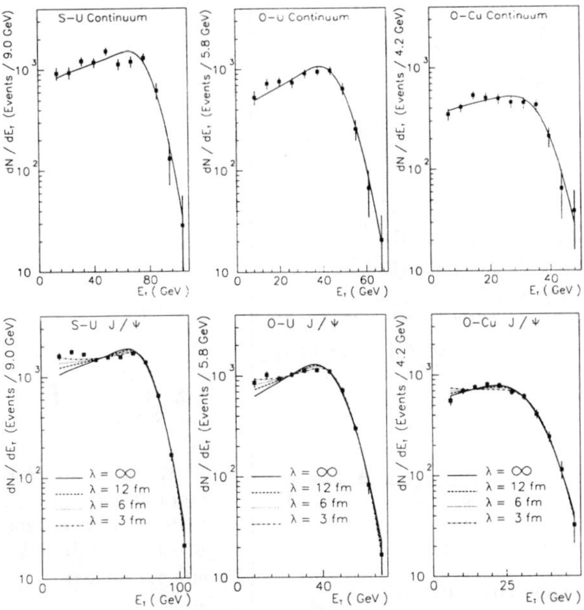


**Fig. 3.** *Neutral transverse energy distributions associated to the continuum muon pairs ($1.4 < M_{\mu\mu} < 2.7$ $GeV/c^2$) and to the $J/\psi$ ($2.7 < M_{\mu\mu} < 3.5$ $GeV/c^2$). The curves are the results of the analysis described in the text, for different $J/\psi$ constant mean free paths ($\lambda$) in the nuclear matter.*

An enhancement of the continuum production rate, as a function of $\epsilon$, is unlikely to account for the observed decrease of R. A first argument against this interpretation is given by the mass shape parameter $A_c$ which is found to be a constant as a function of $E_T$ or of the reaction. The small increase, if any, between the proton and ion induced reactions could be due to their different isospins. Even more convincing, is the analysis performed on the separate $J/\psi$ and continuum $E_T$ distributions [6]. In the framework of a geometrical model of the ion-ion collisions, which accounts well for the $E_T$ inclusive interaction distributions, it has been shown that the continuum $E_T$ spectra are also well reproduced when assuming for the cross-section an $A_{p.proj.} A_{p.targ.}$ dependence, where $A_{p.proj.}$ ($A_{p.targ.}$) is the number of projectile (target) nucleons entering the collision at a given impact parameter. Such a feature indicates that our continuum muon pairs mainly originate from the Drell-Yan mechanism. In hadron experiments, these muon-pairs have been measured , above the $\psi'$ mass, to be produced proportionnally to the target atomic number. Moreover, this process is not expected to be affected by the QGP formation [2]. On the contrary, for the $J/\psi$, an additive absorption term is necessary to account for the data. However, it is difficult to determine quantitatively its value and its dependence with $E_T$ ($\epsilon$). Results of this analysis are reported on fig. 3.

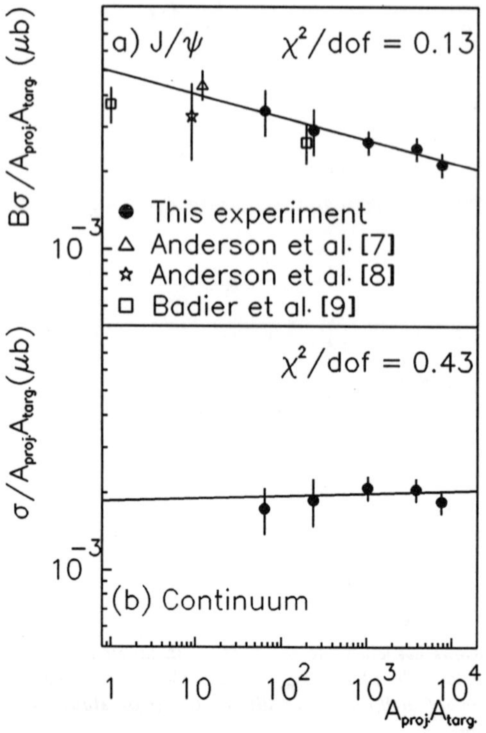


**Fig. 4.** *p-Cu, p-U, O-Cu, O-U and S-U cross-sections (a) for $J/\psi$ (b) for continuum muon-pairs with $1.7 < M_{\mu\mu} < 2.7$ GeV/c$^2$. Lines are the fits to the NA38 data. Systematic errors are included. B is the $J/\psi$ branching ratio into two muons.*

Also of interest for the comparison with available hadron data, the $E_T$ integrated $J/\psi$ and continuum cross-sections have been studied as a function of the product of the projectile and target atomic numbers. Fig. 4 shows the preliminary estimated values with their uncertainties, including the systematic errors. The results are given for muon-pairs with $0 < X_F < 0.5$ according to the apparatus acceptance range. The $J/\psi$ cross-sections measured in other proton experiments, at the same c.m. energy $\sqrt{s}$=19.4 GeV [9], and at 20.6 and 16.8 GeV [7,8], are reported. A good agreement is observed between these values and the $(5.0 \pm 1.5)(A_{proj.}A_{targ.})^{(0.91\pm0.04)}$ behaviour estimated from the NA38 data. The continuum production is found to be compatible with a $(1.9 \pm 0.6)(A_{proj.}A_{targ.})^{(1.01\pm0.04)}$ dependence, in agreement with the $E_T$ analysis, and as expected if it is dominantly of Drell-Yan type. However, it has to be noted that effects, such as the proton versus neutron composition or the quark and gluon densities inside the nucleus are not estimated when comparing the reactions with free protons or with ions. The measured cross-section dependence on the reaction, thus reflects both the initial and final state mechanisms. Anyhow, they indicate once again, a global lack of nuclear effects on the continuum yield, and a suppression in the $J/\psi$ production. In case of the latter, since high energy density effects are washed out by the integration over $E_T$, a better accuracy would be necessary to probe a stronger absorption when going from hadron to nuclear reactions.

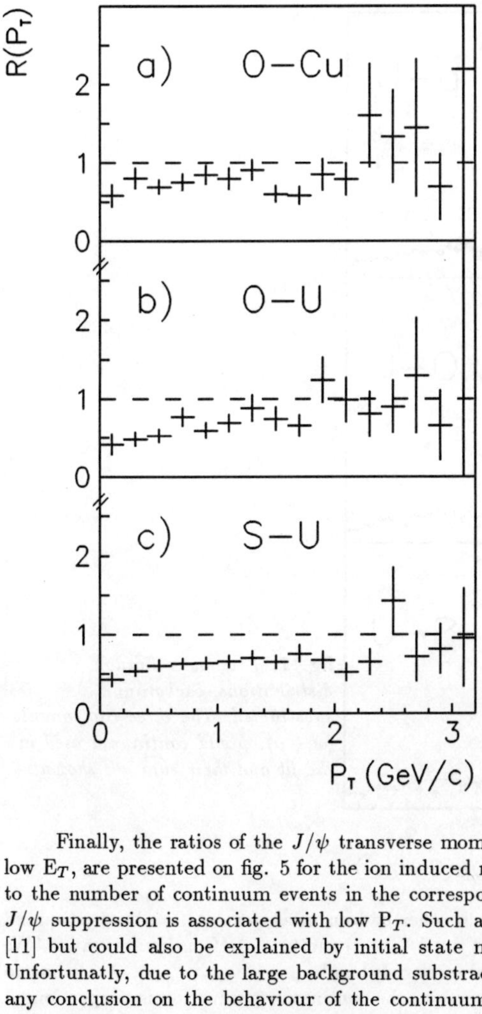


**Fig. 5.** *Ratios of the $J/\psi$ transverse momentum distributions at high and low $E_T$. The data are normalized to the number of continuum in the corresponding $E_T$ bin.*

Finally, the ratios of the $J/\psi$ transverse momentum ($P_T$) distributions, at high and low $E_T$, are presented on fig. 5 for the ion induced reactions [10]. The data are normalized to the number of continuum events in the corresponding $E_T$ bin. They indicate that the $J/\psi$ suppression is associated with low $P_T$. Such a feature is expected from QGP models [11] but could also be explained by initial state multiple scattering of the partons [12]. Unfortunatly, due to the large background substraction, uncertainties prevent us to draw any conclusion on the behaviour of the continuum muon pairs, when plotting the same ratio as a function of $P_T$ and $E_T$. Nevertheless, the mean square transverse momentum $<P_T{}^2>$ can be studied with respect to the energy density (fig. 6). It indicates for the various nuclear reactions a similar large broadening of the $J/\psi$ distributions as $\epsilon$ increases, but a rather flat behaviour for the continuum data. The $<P_T{}^2>$ for small $\epsilon$ values are compatible with the proton data. Progress in this peculiar field is expected from improvement of the statistics at high $P_T$ and in the continuum region.

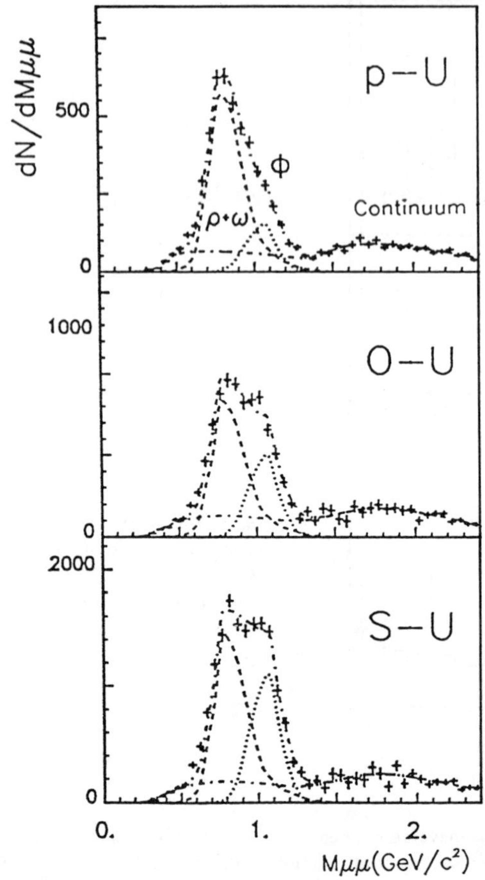


**fig. 7.** *Low mass dimuon distributions, background substracted. The three components ($\omega + \rho$), $\phi$ and continuum used in the fit and their sum are shown.*

In order to compare these results to other hadron measurements, we have estimated the $A^{\Delta\alpha}_{proj.}$ atomic number dependence of the $\phi/(\omega+\rho)$ and $(\omega+\rho)/(\text{Continuum})$ ratios. We find, $\alpha_\phi - \alpha_{\omega+\rho} = 0.32 \pm 0.05$ (statistical error only) and $\alpha_{\omega+\rho} \simeq \alpha_{continuum}$. According to the results presented in the preceding section, the continuum production is proportionnal to $A_{proj.}$. We thus observe a similar behaviour of the $\rho$ and $\omega$ resonances, and on the other hand a strong enhancement of the $\phi$ when going from p-U to O-U and S-U reactions. Although the kinematical domains were not exactly identical (here $P_T(\rho,\omega,\phi) > 1.1$ GeV/c), none of the previous hadron beam experiments [13] has measured such a $\phi/(\omega+\rho)$ enhancement as a function of the target atomic number. New experimental data on this striking feature with improved resolutions and $P_T$ acceptances are expected soon. As in the $J/\psi$ case, alternative explanations to the QGP formation have been proposed in the assumption of a dense hadron gas creation [14].

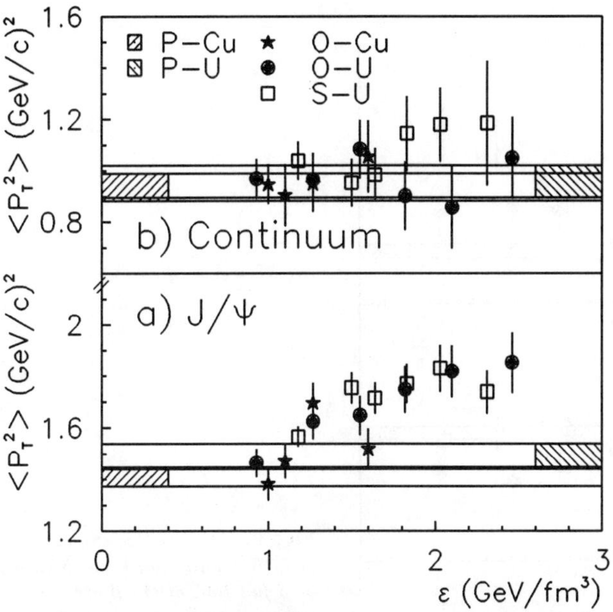


**fig. 6.** *$<P_T{}^2>$ as a function of the Bjorken's energy density reached in the various reactions (a) for $J/\psi$ and (b) for continuum muon-pairs ($1.7 < M_{\mu\mu} < 2.7$ $GeV/c^2$).*

### 6. NA38 : The $\phi$ enhancement

Figure 7 shows the mass spectra, background substracted, obtained in p-U, O-U and S-U reactions for muon-pairs with $M_{\mu\mu} < 2.4$ GeV/c$^2$. As we are interested here in the $\rho$, $\omega$ and $\phi$ production, the specific cuts mentionned in section 4 are applied to the data.

The distributions have been parametrized as the three resonances, with shapes determined according to the apparatus resolution, added to a phenomenological continuum. Amplitudes of the $(\omega+\rho)$, $\phi$ and continuum are the fitted values, assuming the same $\omega$ and $\rho$ production. Results are reported on fig. 8 where the $\phi/(\omega+\rho)$, $(\omega+\rho)/(\text{Continuum})$ and $\phi/(\text{continuum})$ ratios, are plotted with respect to the energy density previously defined. The continuum in the latter ratio, is not the fitted one under the resonances, but is estimated between 1.7 and 2.4 GeV/c$^2$ with similar cuts as for the $J/\psi$ analysis, providing thus a common reference. The $\phi/(\omega+\rho)$ ratio is observed to increase by a factor 3 between the p-U and the most central S-U collisions. On the contrary, the $(\omega+\rho)/(\text{Continuum})$ values do not depend on $\epsilon$ for the ion induced reactions and are compatible with the proton measurement.

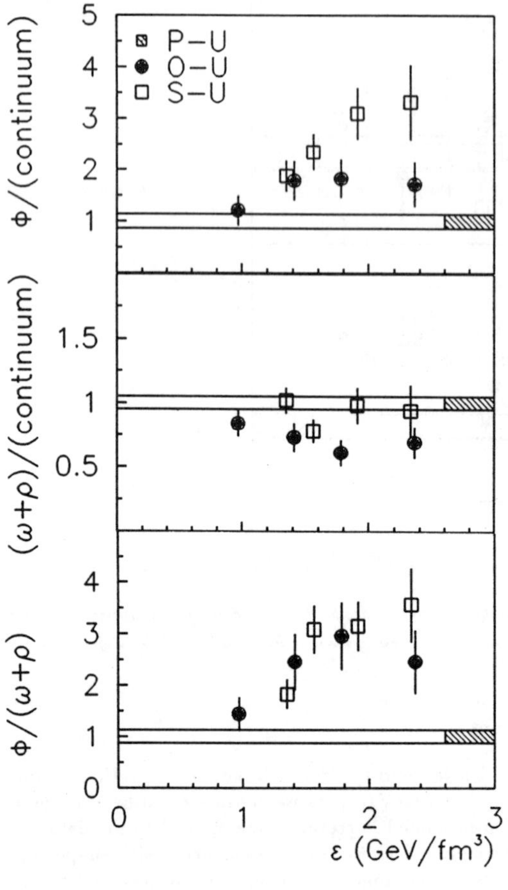


**Fig. 8.** *ratios, $\phi/(continuum)$, $(\omega+\rho)/(continuum)$ and $\phi/(\omega+\rho)$ as a function of the Bjorken's energy density. Numbers of continuum muon-pairs are estimated between 1.7 and 2.4 GeV/$c^2$. Values are normalized to the p-U data and errors are statistical only.*

## 7. Conclusion

The NA38 results have shown the interest of measuring lepton-pairs produced in high energy nuclear collisions. Besides the study of high density matter, either quark-gluon plasma or dense hadron gas, these results have revived interest in the description of the vector meson production and absorption inside the nucleus.

The first CERN exploratory ion beam program is now extended to new beam periods and to new experiments. Three of them will be dedicated to the dilepton measurement.

**References**

[1] - K.Kajantie et al., Phys.Rev. D 34 (1986) 811.
[2] - T.Matsui and H.Satz, Phys.Lett. B 178 (1986) 416.
[3] - J.Rafelski and B.Mueller, Phys.Rev.Lett. 48 (1982) 1066.
A.Shor, Phys.Rev.Lett. 54 (1985) 1112.
[4] - G.Roche et al., Phys.Rev.Lett. 61 (1988) 1069.
[5] - C.Baglin et al., Phys.Lett. B 220 (1989) 471.
C.Baglin et al., Phys.Lett. B 255 (1991) 459.
[6] - C.Baglin et al., Phys.Lett. B 251 (1990) 472.
[7] - K.J.Anderson et al., Phys.Rev.Lett. 42 (1979) 944.
[8] - K.J.Anderson et al., Phys.Rev.Lett. 37 (1976) 799.
[9] - J.Badier et al., Z.Phys. C 20 (1983) 101.
[10] - C.Baglin et al., Phys.Lett. B 251 (1990) 465.
C.Baglin et al., submitted to Phys.Lett. B.
O.Drapier, thesis, Université de Lyon (1990), unpublished.
[11] - F.Karsch and R.Petronzio, Phys.Lett. B 212 (1988) 255.
M.C.Chu and T.Matsui, Phys.Rev. D 37 (1988) 1851.
J.P.Blaizot and J.Y.Ollitrault, Phys.Rev. D 39 (1989) 232.
[12] - S. Gavin and M. Gyulassy, Phys.Lett. B 214 (1988) 241.
J. Hűfner, Y. Kurihara and H.J. Pirner, Phys.Lett. B 215 (1988) 218.
J.P. Blaizot and J.Y. Ollitrault, Phys.Lett. B 217 (1989) 392.
[13] - M. Binkley et al., Phys.Rev.Lett. 37 (1976) 571.
J.G. Branson et al., Phys.Rev.Lett. 38 (1977) 1334.
J. Badier et al., Phys.Lett. B 122 (1983) 441.
R. Bailey et al., Z.Phys. C 22 (1984) 125.
[14] - P.Koch U.Heinz and J.Pišut, Phys.Lett. B 203 (1990) 149.
P.Koch, U.Heinz and J.Pišut, Z.Phys. C 47 (1990) 477.

# A DILEPTON SPECTROMETER PROJECT FOR SIS

L. MERITET

Université Blaise Pascal
IN2P3 - CNRS
Laboratoire de Physique Copusculaire
de Clermont–Ferrand
F–63177 AUBIERE CEDEX FRANCE

## Abstract

We present an experimental project to study $e^+e^-$ pairs production in nucleus–nucleus collisions at GSI. Dileptons appear to be a clean probe to investigate in medium properties of nuclear form factors, baryon excitations ($\Delta$) and mesons ($\pi$–dispersion relation ; $\rho$–form factor) at high nuclear matter densities. The design of the detector system will be based on the experience obtained with the DLS–spectrometer at LBL and other dilepton detector systems in operation at high energy facilities. The project includes a fast electron trigger provided by a ring imaging Cherenkov detector with a segmented photomultiplier array.

## Introduction

Dilepton production in proton–nucleus collisions below 10 GeV has been first observed by the DLS Collaboration at the LBL Bevalac [1]. Since then, more results have been obtained with the DLS in p + Be collisions at 1.0, 2.1 and 4.9 GeV beam kinetic energies [2], and in nucleus–nucleus collisions, for the Ca + Ca reaction at 1.0 and 2.0 GeV/A [3]. Analyses of Nb + Nb, p + p and p + d measurements are presently in progress. Dilepton production experiments in the energy range from one to a few GeV per nucleon are shown to be feasible.

Following the work of C. Gale and J. Kapusta [4], who first expressed the idea of using dileptons as a probe of pion dynamics in nuclear matter, many theorists have performed calculations to interpret the DLS results [a complete list of references is not presented here]. Several points of disagreement between the experimental data and the theoretical predictions, or even between the theoretical models themselves, are found, and thus more work is needed.

Advantages of dileptons compared to hadronic probes are summarized as follows: (i) they are a penetrating probe due to their electromagnetic interaction and can carry information out of the nucleus without distortion; (ii) their coupling with other particles is well known even though there may be problems with electromagnetic form factors; (iii) they are mainly produced in the first stage of the collision, in hot and dense nuclear matter, which makes possible to access nuclear density effects. Their main disadvantage is the low production rate (about $10^{-4}$ $e^+e^-$ per NN collision due to $\alpha^2$).

Gammas would present the same advantages and their production rate would be higher by a factor of 100, but they present a severe experimental difficulty with the huge background due to $\pi^0$ decay.

Both experimental and theoretical issues were recently discussed at several Workshops (GSI, Saturne [5], Clermont-Fd).

The spectrometer project has to meet two essential requirements. Firstly, its acceptance has to cover a wide range of mass M (from 100 MeV to 1 GeV, typically) and transverse momentum $p_t$ (from 0 to about 800 MeV/c). Let us notice that the DLS acceptance is limited; in particular at low mass (100 – 300 MeV), the $p_t$ range is very narrow. Secondly, high statistics is desirable, which is possible with the luminosity expected at SIS, assuming a fast electron trigger with adequate background rejection power. This last requirement can be achieved with a gas Cherenkov detector and a photomultiplier array readout capable of reconstructing the Cherenkov ring at the trigger level.

## Dilepton production at intermediate energies

Let us recall we are considering the energy range from 1 to a few GeV per nucleon, therefore far from the field where the Drell Yan process applies. In this intermediate energy range, the various dilepton production processes are essentially:

- $\pi^0$ Dalitz decay, which actually is a background (contribution only for dilepton invariant mass lower than the $\pi^0$ mass) (fig. 1a).

- $\eta$ Dalitz decay (fig. 1a).
- $\Delta$ resonance Dalitz decay (fig. 1b), which raises difficulties of analysis considering the three bodies decay but may contain some interesting data.
- $N^*$ resonance Dalitz decay found to give a negligible contribution.
- pn bremsstrahlung (fig. 1c) which was initially considered as a background. It is nevertheless necessary to understand correctly this process, which concerns mainly nucleon–nucleon and nucleon–nucleus experiments. This process may contain information on the nuclear electromagnetic form factors and the test of VDM, but it would be very difficult to isolate such a process in nucleus–nucleus collisions.
- $\pi$N annihilation–bremsstrahlung (fig. 1d).
- $\rho$, $\omega$, $\phi$ decay, very interesting to study the characteristics of these vector mesons in the hadronic matter. High luminosities are needed.
- $\pi^+\pi^-$ annihilation (fig. 1e) on which bears a dominant interest now. For $|q|$ (q being the virtual photon 4–momentum) of about a few hundreds of MeV, photons couple to hadrons through vector mesons (VDM assumption). So, the annihilation $\pi^+\pi^- \to e^+e^-$ proceeds dominantly through the rho meson which decays into a virtual photon (fig. 1e). Thus, measurement of the dilepton gives us a direct access to pion dynamics in nuclear matter. The dispersion relation links the pion energy $\omega(\vec{k})$ to its momentum $\vec{k}$, and, in the particular case where the two pions have opposite momentum in the nuclear matter c.m. frame, the two electrons $e^+$ and $e^-$ emerge "back to back" and their invariant mass M is equal to $2\omega(\vec{k})$. The dispersion relation evaluation is essentially theoretical. C. Gale and J. Kapusta [4] point out the great sensitivity of the invariant mass spectrum to nuclear density, mostly close to threshold, and show that the $\pi^+\pi^-$ annihilation rate is a rapidly increasing function of temperature. Figure 2 schematizes the nuclear density effect. These early calculations have stimulated interest among theorists and much work has been performed but the medium effect on the process (pion propagator and vertex) is still an open question [6,11].

The DLS date in p + Be at 2.1 and 4.9 GeV/A (fig. 3) suggest that $\pi^+\pi^-$ annihilation is an important dilepton source and the structure near twice the pion mass cannot be explained without including medium effects (though the issue is not yet settled). The Ca + Ca results (fig. 4) suffer from low statistics and more calculations about the elementary processes, particularly bremsstrahlung, is needed.

## Spectrometer project

### General aspects

Though the "back to back" pairs kinematical range is particularly interesting, it appears that it is necessary to largely extend measurements to test the models with cuts on various kinematical variables. At SIS energies, we can limit ourselves to rapidity about zero in the collision c.m. frame, but we are thinking of measuring invariant masses from 50 – 100 MeV to 1 GeV and transverse momenta from 0 to 500 – 800 MeV/c. So, as a first step of thinking, we have considered a double–arm type spectrometer, symmetrical about the beam direction, including in each arm a magnet set in a vertical position, with a relatively small $\Delta\theta$ horizontal aperture but an adequately large vertical aperture to accept $\Delta\phi$ angles for two electrons between 90° and 180°. This configuration gives a symmetrical $e^+e^-$ detection, decouples the magnet bending and the diffusion angle, and allows a better transverse momentum acceptance.

With a gas threshold Cherenkov detector, we can typically obtain hadron rejection power of a few $10^{-3}$, limitations being essentially due to scintillation and "knock–on" electrons. Scintillation, often ignored for small optical aperture detectors (due to the directional property of the Cherenkov light compared to the isotropic nature of scintillation), may be a nuisance in our case where the mirror aperture is large. A single detector does not allow to get the needed rejection power and the DLS uses two of these (the magnet sweeps away "knock–on" electrons between the two).

It is clear that if we measure the Cherenkov ring, we introduce a much stronger constraint than with a simple counting of photons and electron identification is considerably improved. Thus, scintillation effect is strongly reduced. But "knock–on" electrons remain and measurement of their direction allows to reject them. Moreover, Cherenkov ring reconstruction improves the false pair background. These false pairs are due to uncorrelated electrons coming directly from two $\pi^0$ Dalitz decays, or from the conversion of the associated photons. This background can be experimentally evaluated by "like–sign" pairs, sample and subtracted from the "opposite–sign" pairs, but statistical accuracy is reduced by the subtraction. It is thus better to reject these false pairs as much as possible. For this, we can try to recognize the presence of two electrons in one arm, which identifies a small opening angle Dalitz or conversion pair.

Amplitude measurements can select events with two electrons in one detector (as it is done with DLS) but good events are also lost due to bad resolution. Reconstruction of two Cherenkov rings should be more efficient. At 1 GeV/c, the $\pi^0$ Dalitz pair mean opening angle is 25° and particle distribution in the laboratory is not very much peaked forward; it will not be possible to reject many Dalitz false pairs, except by using a mirror larger than the magnet acceptance. On the other hand, conversion pairs are weakly opened and their rejection should be favored.

Furthermore, to get acquisition rates ten times and more higher than for the DLS, a fast and selective trigger is needed; the Cherenkov detector we propose is described in the next section.

The spectrometer will be completed with a tracking system in the magnet (for reason of size), scintillator hodoscopes contributing to the trigger, and an outer counter (which could be and electromagnetic calorimeter) in order to get the needed hadron rejection power of $10^{-4} - 10^{-5}$.

Figure 5 is a scheme of a horizontal section of one arm of the spectrometer.

### Cherenkov counter

Except for electrons produced by $\pi^0$ decay, dielectron production cross section is very small. The $e^+e^-/\pi^+\pi^-$ rate is typically about $10^{-5}$ which determines the needed level of hadron rejection power. In order to include the Cherenkov counter signal into the fast trigger logic, we propose to measure Cherenkov rings with a segmented photomultipliers array. The photomultipliers main characteristics are: small diameter, good single photoelectron response and U.V. extended sensitivity. The assumed gas is isobutane. A simulation has shown that we should be able to reconstruct circles with a sufficient number of points (from 7 to 10 average according to tested configurations) and spatial resolution on the center of about 2 mm (for a ring radius of about 3 cm).

## Conclusion

Study of dileptons produced in dense and hot nuclear matter should give access to medium effects on particle propagation and interactions. In view of detailed measurements with cuts on kinematical variables, and despite the low yield compared to hadron, we propose a spectrometer able to get good statistics owing to a fast and selective trigger and high luminosity as expected at SIS. One major component of the spectrometer is the Cherenkov ring reconstructing detector using a finely segmented photomultiplier array.

A group of people willing to contribute to such a program has been found. It presently includes scientists from four institutions: The starting collaboration is quite open and more people would certainly be very welcome (and are actually needed).

## References

[1] G. Roche et al. (DLS Collaboration), Phys. Rev. Lett. **61** (1988) 1069

[2] C. Naudet et al. (DLS Collaboration), Phys. Rev. Lett. **62** (1989) 2652

[3] G. Roche et al. (DLS Collaboration), Phys. Lett. **B226** (1989) 228

[4] C. Gale and J. Kapusta, Phys. Rev. **C35** (1987) 2107

[5] "Dilepton production in nucleon–nucleon and nucleon–nucleus collisions at relativistic energies", 4th Journées des Théoriciens, Nov. 22–23, 1990, Laboratoire National Saturne, LNS/Ph/91-03

[6] L.H. Xia et al., Nucl. Phys. **A485** (1988) 721

[7] C.L. Korpa and S. Pratt, Phys. Rev. Lett. **64** (1990) 1502

[8] C.M. Ko et al., Phys. Lett. **B231** (1989) 16

[9] C.L. Korpa et al., Phys. Lett. **B246** (1990) 333

[10] M. Hermann et al., see ref. [5]

[11] G. Chanfray et al., see ref. [5]

[12] L. Xiong et al., Phys. Rev. **C41** (1990) 1355

[13] J. Kapusta and P. Lichard, Phys. Rev. **C40** (1989) 1574

[14] G. Wolf et al., Nucl. Phys. **A517** (1990) 615; see also ref. [5]

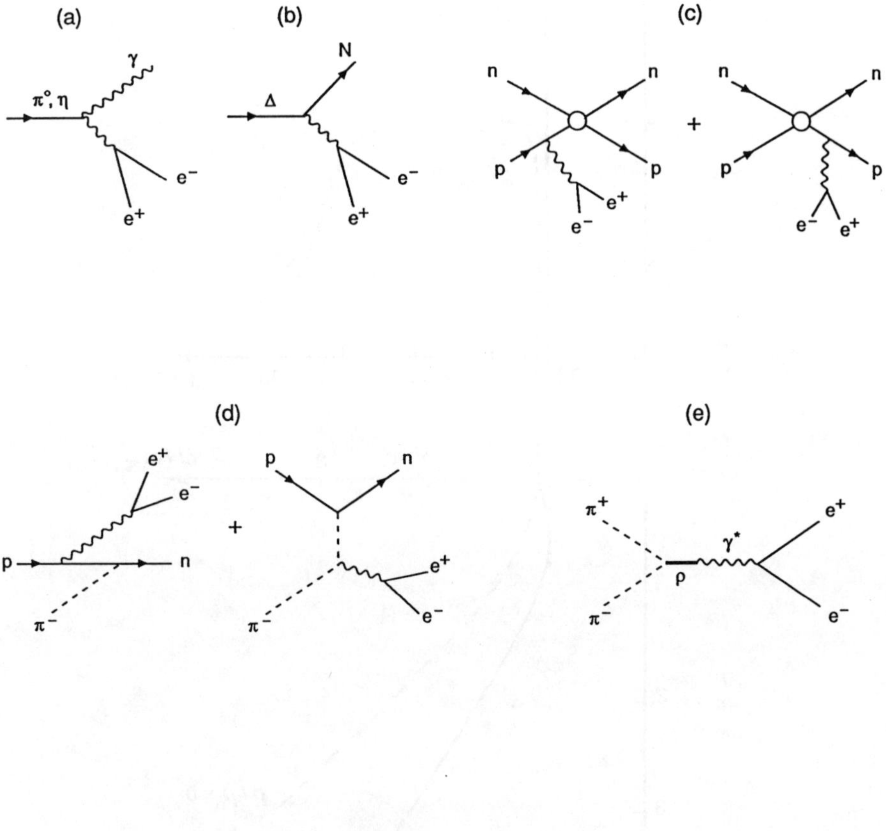


**Figure 1** – examples of dilepton production processes :

(a) – $\pi^0, \eta$ Dalitz decay
(b) – $\Delta$ Dalitz decay
(c) – pn bremsstrahlung from the external proton lines
(d) – $\pi$N annihilation – bremsstrahlung
(e) – $\pi^+\pi^-$ annihilation through the rho meson.

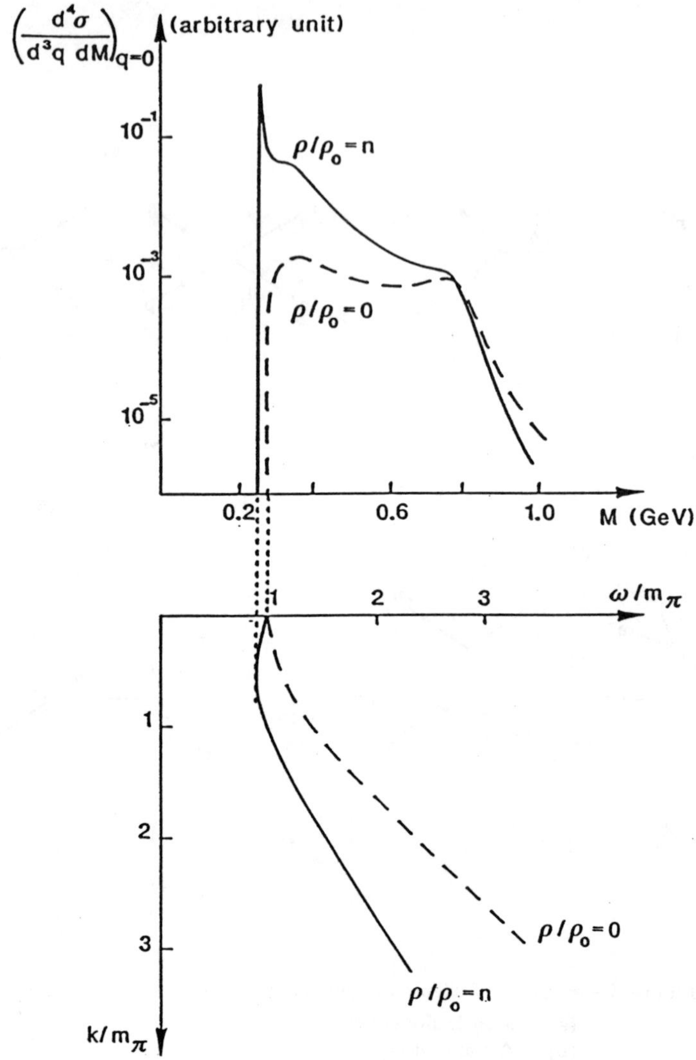


**Figure 2** – schematization of direct relation between the dilepton invariant mass spectrum and the pion dispersion fonction.

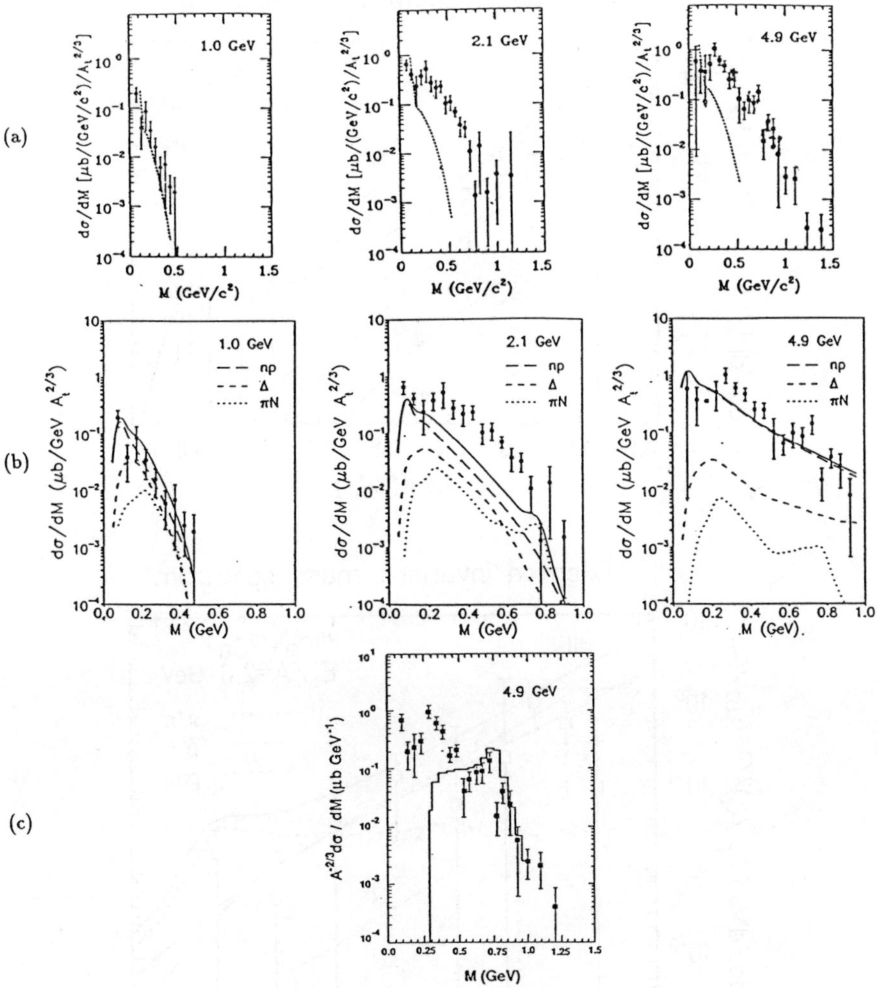


**Figure 3** – the invariant mass spectra for p – Be at three beam energies (1.0, 2.1 and 4.9 GeV/A). Data of ref [1, 2]. The various lines shown in each figure are estimated contributions of the main dilepton sources.

(a) – estimated contribution due to wide angle Dalitz decay [2]
(b) – theoretical productions from the pn bremsstrahlung, the $\Delta$ decay, and the $\pi$N interaction [12]
(c) – the histogram is the result of $\pi^+\pi^-$ annihilation model [13]

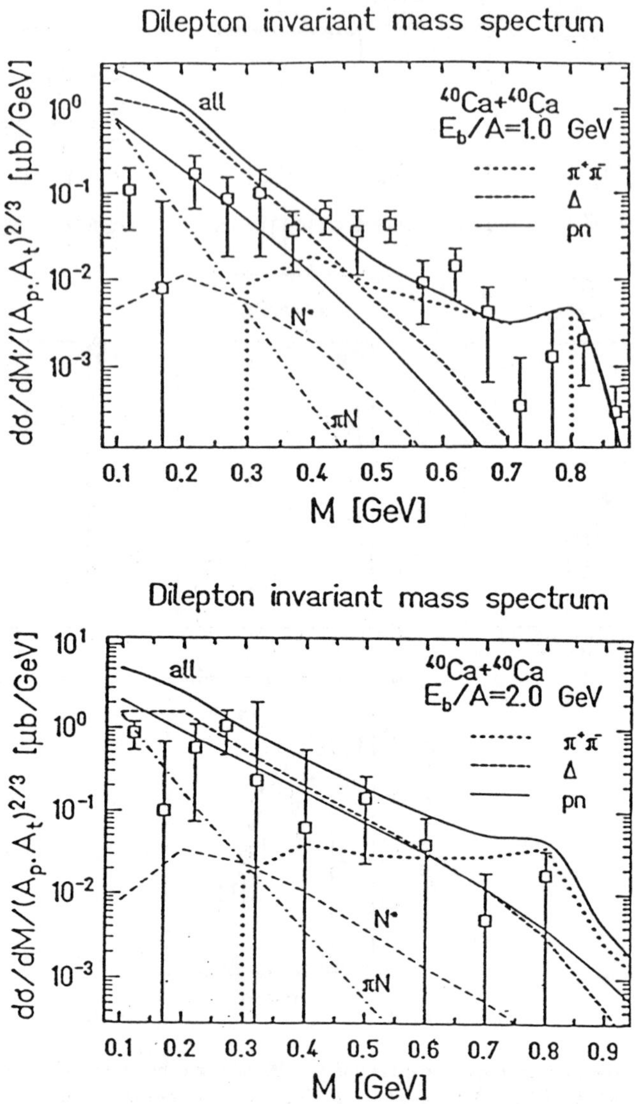


**Figure 4** – the invariant mass spectra for Ca – Ca at two beam energies (1.0 and 2.0 GeV/A). Data of ref. [3].
The various lines shown in each figure are theoretical productions from the pn bremsstrahlung, the $\Delta$ decay, the N* decay, the $\pi$N interraction and the $\pi^+\pi^-$ annihilation [14].

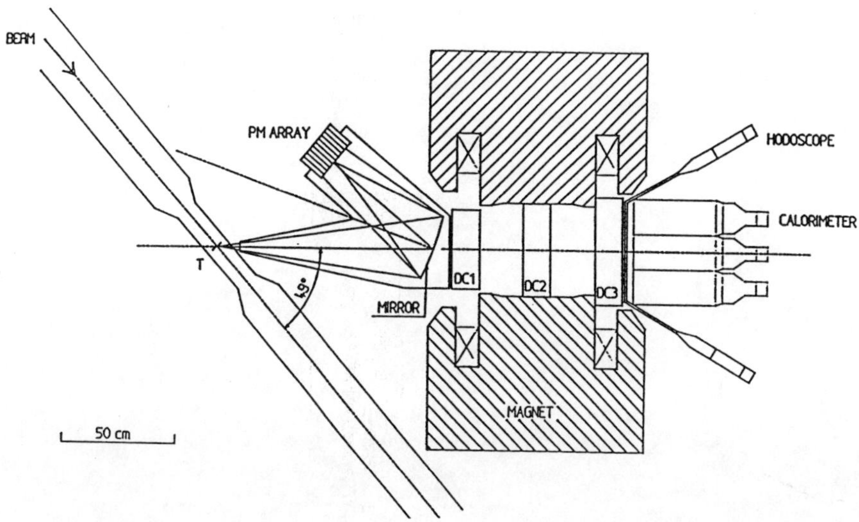


**Figure 5** – scheme of a horizontal section of one arm of the spectrometer.

# VI. Future Equipment in Heavy-Ion Domain

# AGOR

## A SUPERCONDUCTING CYCLOTRON FOR LIGTH AND HEAVY IONS PLANS FOR EXPERIMENTAL FACILITIES AND PHYSICS PROGRAM

S. Galès

Institut de Physique Nucléaire, 91406 Orsay Cedex France
K.V.I. Groningen, Zernikelaan 25, 9747 AA Groningen, The Netherlands

Abstract
The construction of the K600 superconducting cyclotron AGOR, a joint undertaking of the KVI Groningen and the Institut de Physique Nucléaire at Orsay, has reached the stage where the assembly of major subsystems is underway. Field measurements are scheduled to start in the fall of this year, beam tests should start at Orsay by the end of 1992 before AGOR final installation at Groningen.
The beam guiding system, the location and equipments of the main experimental areas are currently being designed. Taking advantage of the broad range of ions and energies that AGOR will made available (from 200 MeV protons to 100 MeV/A $\alpha$ down to 6 MeV/A Pb ions), the first ideas about the physics research to be done will be presented.

## I - INTRODUCTION

The construction of the K600 superconducting AGOR, a joint undertaking of the KVI Groningen and the IPN Orsay, is underway.

Details of the design have been published elsewhere[1]. They will be briefly recall in section II with emphasis on the novel features of the accelerator and on the status of the subsystems.

In sec III a description of the general lay-out of the beam guiding system and experimental areas with the specific apparatus which are to day considered will be given.

The uniqueness of the AGOR accelerator which is able to accelerate protons as well as heavy ions from the low to the intermediate energy regime greatly enlarges the research domain accessible to the nuclear physicists.

As an example, the high resolution polarized proton and deuteron beams at 200 MeV are very well suited to the study of the spin or spin-isospin response of nuclei, throught the excitation of specific states and/or the exploration of the nuclear continuum[2].

On the other hand, inelastic, charge exchange or transfer reactions using both light and heavy ions in a broad energy range (30 to 100 MeV/nucleon) are the best tools to explore new collective and single particle degree of freedom at rather high excitation energies[3].

The general trend is to investigate the decay characteristics of these high-lying states embedded in a large continuum through exclusive experiments using multidetecteurs systems (for proton, neutron and $\gamma$ rays) in coïncidence with a magnetic spectrometer[4-6].

The observation of high-energy photons in nucleus-nucleus collisions has also attracted considerable attention in the recent years[7]. The produced photons may reveal an undisturbed signal of the phase of the colliding system in the nuclear collision. With the AGOR beams these processes producing hard photons can be studied both in the bare proton-neutron, proton-proton system as well as in nucleon-nucleus and nucleus-nucleus systems. Few examples of the above mentionned experiments will be described in the last section of this paper.

## II - CONSTRUCTION OF THE AGOR CYCLOTRON

The AGOR cyclotron will be a compact cyclotron with a three-fold magnetic symmetry and three RF electrodes for the acceleration of light and heavy ions. The main field is generated by a set of two superconducting coils housed in a split cryostat. The central field strength ranges from 1.7 to 4.1 T. The major components of the machine (yoke, poles, superconducting coils, RF structure, and axial injection line) are shown in fig. 1.

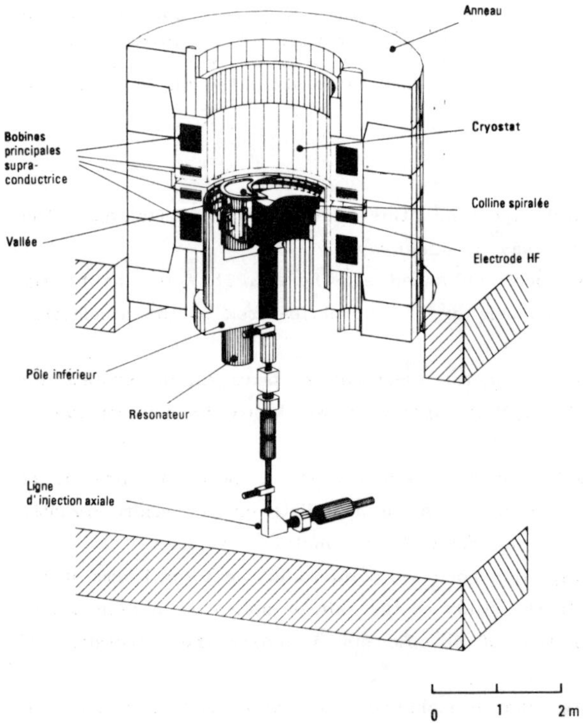


Fig.1 - Simplified cutaway view of the AGOR cyclotron.

The operating diagram (Energy (MeV/nucleon) versus Z/A ratios of the accelerated ions) as well as the operating range (24 to 62 MHz) for the high frequency are shown in fig. 2. The energy range for protons is 127-200 MeV. Heavy ions with Z/A ratio in the range of 0.1 to 0.5 will be accelerated from 6 MeV/A for the lower charge state to 36-95 MeV/A for fully stripped ions.

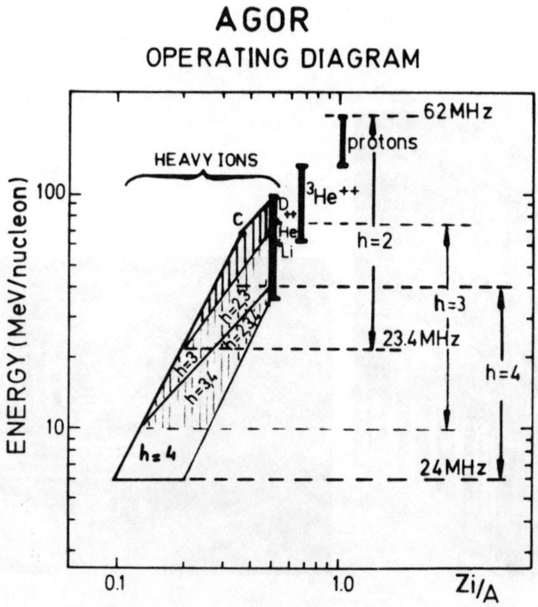


Fig. 2 - Operating diagram of AGOR cyclotron.
Ions energy versus Z/A.

The main cyclotron parameters are listed in table 1.

Table 1 : Cyclotron parameters

| | |
|---|---|
| Bending limit K = 600 | Focusing limit $K_F$= 220 |
| Pole radius 118 cm | Number of sector 3 |
| Min. hill gap 7 cm | Max hill gap 84 cm |
| Main coil ≠ 1 max current density | 4271 A/cm$^2$ |
| " " ≠ 2 " " " | 3270 A/cm$^2$ |
| Central field range | 1.75 - 4.1 T |
| RF frequency range 24-62 MHz | h = 2,3,4 |
| Peak RF Voltage at the center | 85 kV |

**II.1 Magnet structure, trim coils and field mapping equipments**

The 320 t steel structure (that includes the six rings of the cylindrical yoke, the two main poles with lifting and upsetting mechanisms) has been installed at Orsay in July 89. Dimensional checks are all within the specified values and measurements on samples of the saturation magnetization yielded values of M = 2.16 (± 0.01) T, exceeding the goal of 2.14 Tesla. Fig.3 shown the AGOR magnetic circuit installed at Orsay.

Fig. 3 - AGOR magnetic circuit installed in its pit at Orsay (Janv. 91).

Fig. 4 - View of the spiral flutter pole and of the winding of the 15 correction coils.

The 15 room temperature trim coils wound around each flutter pole (see Fig. 4) have been manufactured and delivered in July 90. They have been assembled on the hills and have passed successfully the acceptance tests. The field mapping method for the NSCL K1200 cyclotron[8)] has been selected.

An absolute measurement is made at the center of the machine by a NMR probe and a search coil moving along a radial carbon fiber arm reads the field variation in the median plane. This measurement is repeated every degree by a rotation of the arm. The system is ready for operation and has been used to map the field of the trim coils as test cases.

### II.2 - Superconducting coils and cryostat

The magnet excitation is produced by two pairs of superconducting coils, housed in a "split" cryostat that provides free access in the median plane with the exception of six feedthroughs for support, power and fluid communication. The four coils have been fully vaccum impregnated and thermal stability is improved by direct contact of liquid helium (4°K) at all sides of the winding package. The system is being assembled and tested by the manufacturer (ANSALDO, Genova, Italy) and will be deliver at Orsay during the fall of 1991. Fig. 5 show the assembled cryostat at the manufacturer site.

### II.3 - Helium liquefier

The external cryogenic system consists in :

- a liquefier/refrigerator with two compressors
- a 1000 l liquid helium dewar
- an intermediate 18 $m^3$/15 bar helium storage
- 80°K à 4°K transfer lines

The whole plant is installed at Orsay since the beginning of 1990 and have passed successfully the acceptance tests using a specially built cryostat as a cryogenic load.

The liquid helium plant is specified for the following simultaneous production rates :

- Cooling power at 80 K : 600 W
- Cooling power at 4 K : 50 W
- Liquid helium production : 15 l/h

Fig. 5 - View of the AGOR cryostat during the assembly at the manufacturer site.

The safety margin of this system compared to the expected consumption of the superconducting coils, the cryostat and current leads is between 50 to 100 %.

### II.4 - High frequency system

The high frequency system consist of three compact λ/2 length resonators with continuous and smooth connection to the dees. A full scale model of a resonator was constructed from copper covered plywood in early 1988 (see Fig. 6).

The results of the measurements on the model allows us to forsee a maximum input power of 32 kW per resonator with a nominal accelerating voltage at the center of 85 kV.

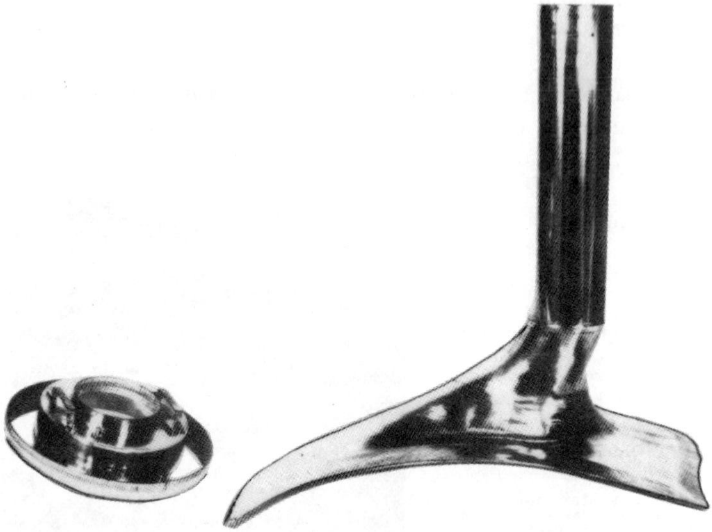

Fig. 6 - Model resonator with its short circuit plate for RF frequency tuning.

A movable short circuit plate permits to fix the length of the resonator and therefore the resonant frequency between 24 and 62 MHz.
Three 70 kW power amplifiers have been delivered by Herfurth GmbH (Hamburg) and are now undergoing long range acceptance tests at Orsay.
An additional original aspect of the AGOR RF system is the design of the vacuum chamber which will at the same time be the RF liner.
The construction of the resonators and of the vacuum chamber is underway and assembly at Orsay should start at the begining of 1992.

### II.5 - Axial injection and central region

The axial injection system is designed to operate with three ion sources : light ions (multi-cups), polarized p and d and ECR for heavy ions.
The injection line lay-out is presented in fig. 7. The first two sections A and B (see fig. 7) are completed and the last part (C) is currently being assembled. A buncher operating at the RF of the cyclotron and common to the 3 harmonic modes h = 2,3,4 will be placed at about 50 cm from the median plane.

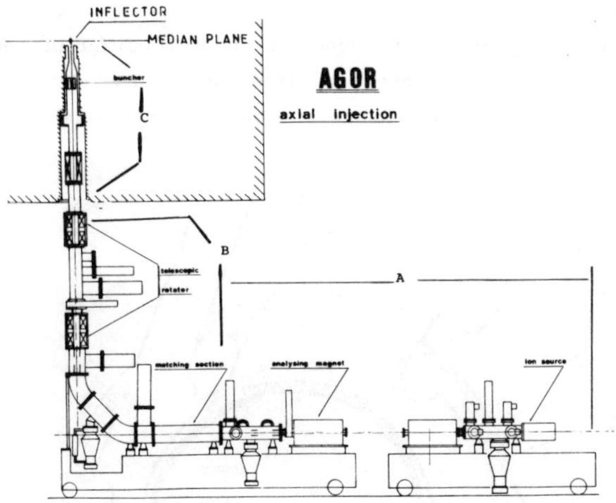


Fig. 7 - Schematic lay-out of the injection line.

A final version of the buncher system is being tested. The central geometry has been designed in conjonction with a spiral inflector. Phase and geometrical focusing have been achieved using a magnetic bump on one hand and posts on the other hand. A CAD view of the central region geometry common to the 3 harmonic modes is displayed in fig. 8.

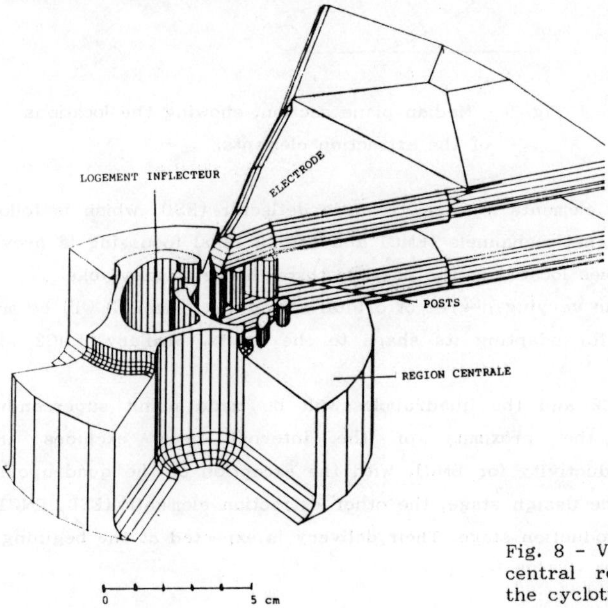


Fig. 8 - View of the central region of the cyclotron AGOR.

### II.6 - Extraction

Beam dynamical calculations are done for about 20 different beams that span the operating range of the cyclotron (see fig. 2). The extraction components of the cyclotron in the median plane section are shown in fig. 9.

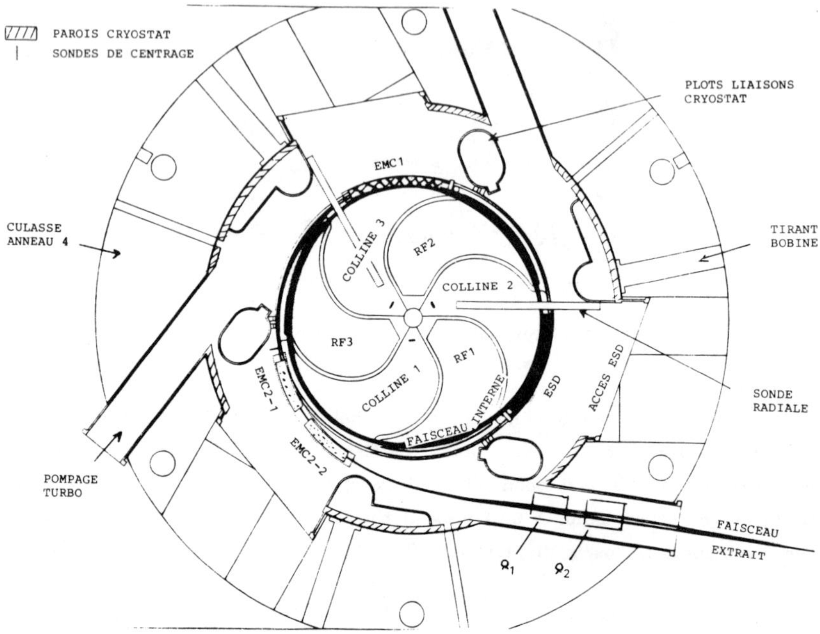


Fig. 9 - Median plane section, showing the locations of the extraction elements.

The first elements is a electrostatic deflector (ESD) which is followed by two electromagnetic channels (EMC1 and EMC2). Final focussing is provided by two quadrupoles located in the passage through the magnet yoke.

Due to the varying degree of orbital scalopping, the ESD will be made in three sections for adapting its shape to the orbits. Similary, EMC2 will have two sections.

Both EMC2 and the quadrupoles will be made using superconducting coils, whereas the proximity of the internal beam excludes the use of supraconductivity for EMC1. With the exception of the quadrupoles which are still in the design stage, the other extraction elements (ESD, EMC1, EMC2) are at the production stage. Their delivery is expected at the begining of 1992.

### II.7 - Vacuum

The poles of the AGOR cyclotron, in contrast to those of most other machines, will be separated for the beam region by a vacuum cover that also serves as RF liner. As a result the surface exposed to vacuum is reduced with a drawback, the complication due to the intricate shape of the liner. The required pressure in the cyclotron has been estimated to be of about $10^{-7}$ torr. This corresponds to a pumping speed for air of 5 000 l/s. Three cryopanels located inside the accelerating electrodes have been built with measured pumping speeds of 1400 l/s for air and 1800 l/s for $H_2$. In addition two turbomolecular pumps are foreseen on the outside of the cyclotron.

### II.8 - Control system

The AGOR control system consists of a network of RT (Real-Time) microVax computers coupled through Ethernet and using VAXeln operating system software. For interfacing to the accelerator equipment, a modular system is used, based on the Bitbus field bus. In close cooperation with industry several modules have been developped among which, a power supply interface, a general positioning interface, a stepmotor and "harp" beam profile monitor interfaces. Several subsystems are already operating under the above mentionned requirements : the injection line, the correction coils, the external cryogenics system. The development of the software necessary for the control console is underway.

### II.9 - Project status

The current status of the project planning is presented in table 2 which shows major milestones both past and future.

<u>Table 2 - AGOR milestones</u>

| | |
|---|---|
| Agreement FOM-IN2P3 | Dec. 1985 |
| AGOR Design Report | Oct. 1986 |
| Formal start construction | May 1987 |
| Civil engineering completed | April 1988 |
| Magnet construction | Jan. 88 - June 89 |
| Main coils and cryostat construction | July 88 - Oct. 91 |
| RF amplifier construction | Jan. 88 - Oct. 89 |
| RF resonator engineering | June 88 - Dec. 89 |
| RF and vacuum chamber construction | Jan. 90 - Feb. 92 |
| Helium liquefier installed | Jan. 90 |
| Cryopumps in use | Feb. 90 |
| Field mapper ready | Dec. 90 |
| Injection line completed | April 91 |
| Mount cryostat in magnet | Oct. 91 |
| Field mapping | Nov. 91 - March 92 |
| Start assembly RF system | March 92 |
| Start beam tests | Dec. 92 |

III - GENERAL DESCRIPTION OF BEAM GUIDING SYSTEM AND EXPERIMENTAL AREAS

The cyclotron AGOR will replace in the same vault, the previous K160 cyclotron in operation at KVI since the beginning of 1970.

A preliminary lay-out of the beam guiding system and of the experimental areas is shown in fig. 10.

Along the extracted beam path, two quadrupoles located inside the yoke are designed to produce a common focusing point for all ions and all energies at about 3m from the cyclotron outside diameter. Just before and after the main analysing magnets, moveable slits are used to adapt the beam emittance and size to the rest of the transport system.

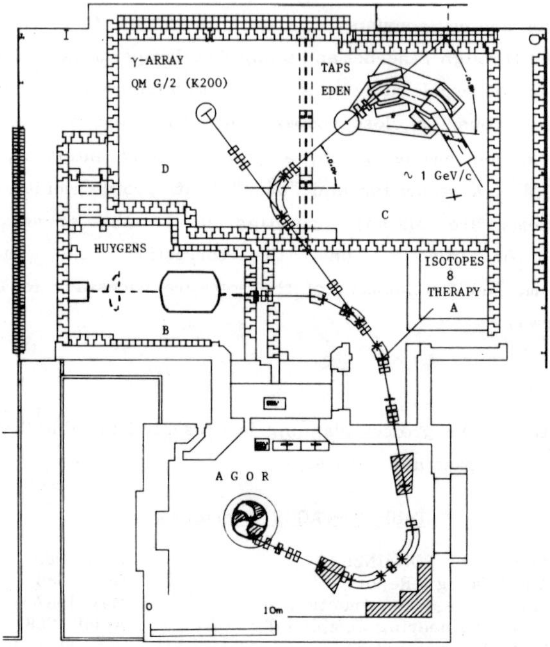


Fig. 10 - Preliminary lay-out of the beam lines and experimental areas associate with the cyclotron AGOR. Local heavy shieldings are indicated by hatched areas.

To date four experimental areas (A to D) are being considered as shown in fig. 10.

A. <u>An Isotope and Therapy station</u>

B. <u>A large scattering chamber named HUYGENSVAT</u>[9]. Its dimensions are of about 3.2 m in diameter and 5.2 m in length. This facility will be built in collaboration between KVI and the Robert Van de Graaff Laboratory at Utrecht. It will be equiped with a $4\pi$ charged particle array consisting of 3 main components :

- The backward wall, contains 64 CsI scintillators with photodiode readout. It will cover a region where evaporated light particles dominate the yield from nucleus-nucleus collision at intermediate energies.
- A "Central Trigger Detector" will most probably be a Time Projection Chamber design including large ionisation chambers and Bragg curve detectors. It will be mainly used to detect intermediate mass fragments from the collision.
- and finally a forward component composed of 30 plastic phoswitch counters for the study of the projectile-like fragments.

The various prototypes of the 4 $\pi$ array and the electronics modules are being tested.

C. A new magnetic spectrometer[10] which fits the high rigidity of the AGOR beams is planned to a one of the main piece of equipment of the facility.

The existing Q3D spectrograph has a mass-energy product of 200 will prohibit the momentum analysis of most reaction products at beam energies higher than 50 MeV/A. The mass energy product for a new spectrograph needs to be at least 500. Giving the qualities of the Q3D, such as momentum resolving power of 10000 and a solid angle of 10 msr, it is intended to built a spectrograph with complementary parameters.

This means a larger momentum bite (≃ 20 %) but with a smaller resolving power (≃ 3000-5000). Two others features are of prime importance in the definition of this new instrument. First, the spectrograph will be use frequently at very small angles including 0°. Secondly, many of the future experiments will be performed using the spectrometer as one arm of a coïncidence set up. The other arm consists of new multidetector arrays (neutron, $\gamma$ rays, etc, ...) and this will requires a rather large free space (1-1.5 m) around the target location. Various combinaison of Quadrupole Dipole elements are being studied with the constraints mentionned above.

The new spectrometer will be in many experiments associated with various detection systems in a coincidence mode. Many of those coïncidence detectors are presently available or under construction. A list and short description of these systems is given below.

*C.1 - The photon spectrometer TAPS*

The photon spectrometer TAPS is a detector for high-energy $\gamma$ rays and is designed for general use to be operated at GANIL, MAMI-B, SIS/ESR and AGOR[11]

TAPS is made of $BAF_2$ scintillators because of their fast response and because $BAF_2$ allows separation of photons and charged particles by pulse shape analysis. Currently the whole set up consists of four blocks, each block containing 64 modules. In the future the number of blocks will be extended to six. For high energy γ-rays, from GEANT simulation based on test experiment a resolution of $\Delta E(fwhm)/E = 0.09E\ (GeV)^{-0.025}$ has been obtained for individual modules. By properly summing the signal of the central detector to its six surrounding neighbours in a test experiment with 186 MeV tagged photon beam, the collaboration measured a resolution of 8 %. Time resolution expected to be of ≤ 200 ps.

A combination of pulse shape analysis and an energy loss signal from thin plastic scintillators placed in front of each $BaF_2$ will be used to discriminate between neutrons and γ-rays in one hand and charged particles on the other hand. Finally a measurement of the flight time can be used to discriminate neutrons from γ-rays.

At a distance of 1 m from the target, the solid angle of the set up with four blocks will be 0.39 sr.

### *C.2 - The Ge-detector setup Dutch-Ball*[12)]

The Dutch-ball is a mechanical structure which has been designed to enable 9 Compton Supressed Ge detectors (CSG) to be placed as close as possible to the target in close-packed geometry.

The structure consists of 3 separate rings each having the possibility of holding three CSG. The distance of each detector to the target is approximately 14 cm.

The CSG's counters has been designed to operate either in a stand alone basis or in combination with the mini-orange or X-ray planar Ge.

A holder for eigth small $BaF_2$ detectors will be supplied for multiplicity filter and fast trigger timing.

For the close-packed geometry, the total solid angle efficiency product will be 1.2 % for γ rays of 1.3 MeV using 9 CSG's detectors, assuming 25 % photo-peak efficiency.

### *C.3 - The lepton-pair spectrometer PEPSI*[13)]

This spectrometer, designed for the detection of electron-position pair, will consist of two concentric parts : the inner one serving as a $4\pi$ mini-orange like structure and the outer one holding thirty phoswich detectors.

The inner part will have a form of regular structure of 20 triangles built from permanent magnets which can focus high energy (E > 5 MeV) electrons. The outer part of the spectrometer will consist of twelve pentagones. Electrons and positrons will be detected by thirty identical phoswich detectors (1mm thick slow scintillator in front of a 65 mm thick fast plastic counter).

The main design parameters of PEPSI are listed below :

- The $\gamma$-ray rejection with the phoswich $\Delta E-E$ telescope was measured to be 97 % with a $^{60}Co$ source.

- The estimated absolute efficiency for the detection of electro-positron pairs at summed energy of 14 MeV is about 1% for uncorrelated pairs.

*C.4 - The neutron detector EDEN*[5)]

EDEN is a joint project between KVI and IPN and it consists of forty identical liquid scintillators specifically suited for the detection of low energy (1-25 MeV) neutrons. EDEN is foreseen to operate at GANIL and at AGOR.

EDEN characteristics are indicated below :

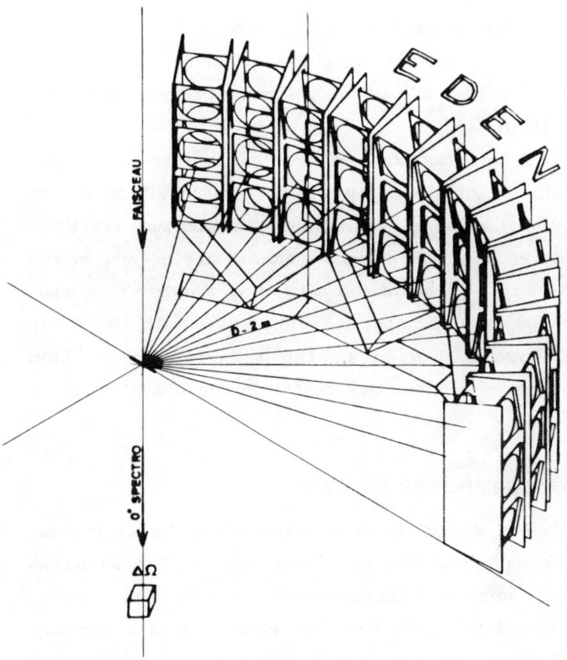


Fig.11 - EDEN used as a second arm in a coincidence experiment with a magnetic spectrometer.

- The forty detectors are made of cylinders with diameter of 20 cm a thickness of 5 cm and filled with the BG-501 scintillator liquid which allows through pulse-shape analysis for an excellent discrimination between neutrons and $\gamma$-rays.

- The detector can be arranged around the target position with a mechanical support which enables easy adjustement of the distance target-detector (see fig. 11).

- The energy will be measured using the time of flight technique where the neutron detector produces the start signal and the spectrograph detector and/or the RF signal from the cyclotron deliver the stop signal. The intrinsic time resolution of a module will be $\Delta t$ (fωhm) $\simeq$ 1 ns.
- The energy resolution depends on the energy of the neutron and on the flight path. Typically one expects for a 1 MeV neutron 30 keV and 300 keV for 5 MeV neutron.

The total solid angle efficiency product depends on the energy of the neutron, on the flight path and on the quantity of light measured with the photomultipliers tubes. Assuming an efficiency of 25 % for 5 MeV neutrons, a distance of 2m between the target and EDEN modules, it is predicted that the solid angle efficiency product will be 80 msr corresponding of 0,6 % of 4 $\pi$.

*D - The QMG/2 spectrograph*

The existing K200 magnetic spectrograph can be relocated in this area and devoted to high resolution reaction studies with light projectiles. It will be quite interesting to install at AGOR new commercially available polarized ion sources which are specified to deliver 100 $\mu$A of polarized protons within an emittance of 100 $\sqrt{E}$ mm.mrad (E in KeV).

This new ion source have to be associated with a beam line of optimum resolution and with the posibility of providing beams with longitudinal and transverse polarisation at the target in addition to the normal (vertical) mode of polarization. To achieved this goal special elements are needed in the beam line, e.g. a combination of two solenoids and a bending magnet. This beam line will serve the existing K200 magnetic spectrograph, where with the study of a focal plane detection system devoted to the measurements of spin observables, AGOR may became a very attractive centre for polarized beams in the nineties.

## IV - SELECTED EXAMPLES OF EXPERIMENTS AT AGOR

The beams which will be delivered by the AGOR cyclotron span the whole mass table region, from protons to lead ions. The incident energies of these beams are also well suited to study nuclear collisions under a wide and varying range of conditions to understand the behaviour and properties of a quantum many-body system in equilibrium (nuclear structure) and non equilibrium (nuclear dynamics).

Among the numerous subfields of nuclear physics which could be investigated by means of the AGOR beams and the planned experimental instruments, a few selected examples of physics experiments will be presented below with emphasis on the specific aspects of the beam nature and energy and/or the associated experimental instruments.

### IV.1 - The spin excitations in nuclei

Polarized protons and deuterons beam with good energy resolution between 100 and 200 MeV are foreseen as one of the unique feature of the accelerator.

With the polarization axis vertical for incoming proton the only spin observable that can be measured in the spectrograph without a second scattering is the analyzing power Ay. Next observable is the spin flip probability $S_{nn}$ which requires a measurement in the focal plane of the polarization of the outgoing particle.

Recently, the $\Delta S = 1$ strength has been localized in several nuclei[14] by measuring $S_{nn}$ which differs from zero only for spin excitations. The energy range between 150 and 400 MeV is optimum for the investigation of spin-isospin or spin excitations in nuclei due to the energy dependance of t-matrix components for free nucleon-nucleon scattering at a momentum transfer $q \simeq 0$[15]. The ratio $t\ \sigma\tau/t\tau$ is maximum as well as the ratio between t - matrix components for $\Delta S = 1$, $\Delta T = 0$ to $\Delta S = 0$, $\Delta T = 0$.

Under such favorable conditions one may cast further light on the quenching mechanisms for spin transitions, on the location of isoscalar $\Delta S = 1$ response using respectively $(\vec{p},p'\gamma)$ and $(\vec{d},d')$ reactions. The usefulness of the $(\vec{d},d')$reaction has been recently demonstrated by M. Morlet et al.[2]. A study of the isoscalar component of the nuclear spin excitation strenght ($\Delta S = 1$, $\Delta T = 0$) has been started using polarized deuteron scattering as a probe. New experimental signatures of $\Delta S = 1$ for $(\vec{d},d')$ as useful as Snn is for $(\vec{p},p')$ were tested on the three lowest natural parity states of $^{12}C$ : $2^+$ at 4.44 MeV, $0^+$ at 7.65 MeV and $3^-$ at 9.64 MeV and on the unatural parity $1^+$ state at 12.71 MeV.

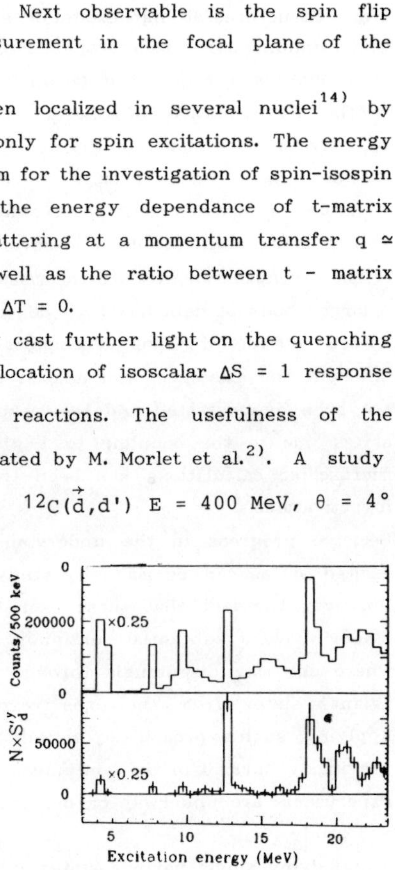


Fig. 12 - Inelastic $(\vec{d},d')$ spectrum measured at a laboratory angle of 4° on $^{12}C$.
(a) Raw spectra 500 KeV bin ,
(b) The same as (a) multiplied by $S'^y_d$ [Taken for fig. 1 of Ref. 2].

Fig. 12a show the inelastic deuteron spectra from the reaction $^{12}C(\vec{d},d')$ at 4° plotted with 500 keV bins. Then this spectra is multiplied by the "reinforced" signature $S'^y_d$ which depends on the vector and tensor analyzing powers of the scattered deuteron and which is equivalent to Snn (spin flip probability) in $(\vec{p},p')$. As demonstrated by the spectrum of fig. 12b, the signature is close to zero for $\Delta S = 0$ levels (natural parity) and large and positive for the 12.7 MeV $\Delta S = 1$ state.

The proposed "reinforced" signatures turn out to be efficient in selecting isoscalar $\Delta S = 1$ spin strength. New experiments using this approach can now be planed to map the isoscalar spin strength, largely unknown, in other nuclei.

**IV.2 - Structures at high excitation energies and exclusive experiments**

Understanding the microscopic structure of collective and single-particle excitations is a major goal of nuclear structure studies. Of special interest is the determination of the contribution to the damping width due to collective surface modes of the nucleus.

The damping of highly excited states is a medium effect and therefore is the signature of how strong is the residual interaction.

This damping manifests itself in several ways of which the shape and the width of the strength function are the most obvious.

A large body of data has been collected already on the gross properties of a limited number of Giant Resonances (GR)[4] and single-particle states (SP)[3]. However, the existence of a number of the initially predicted GR's has so far not been demonstrated and the mechanism responsible of the damping of SP states due to the coupling to 1 phonon, 2 phonons etc... as predicted by microscopic calculations has been the subject of a very limited number of investigations[16,17].

Decisive progress in the understanding of the damping process of highly excited states can be made by studying in detail the decay of these states. Moreover the fact that these excitations are generally observed as "giant" states above a subtantial continuum leads to new generations of experiments where one may disentangle through coincidences the process populating the "giant" states from the ones responsible for the underlying continuum. Applying such approach may as well help to look for unknown high-lying states previously burried in the continuum.

Experiments are underway or planned, using the coïncidence method to locate and study :

- High-lying single-particle states and their decay properties.
- High-lying, high-momentum giant resonances, multiphonons states and their decays.
- Decay of the $M_1$ giant resonance.
- Isovector Monopole and Gamow-Teller Resonances.

Single-particle degree of freedom at high excitation energy

The one-nucleon transfer reactions is a simple method for studying deep-hole states in heavy nuclei and is the only way to populate particle states at high excitation energy in nuclei[3]. Both light and heavy ions transfer at intermediate energies (30-50 MeV/nucleus) are very selective for the population high-spin SP states. The high-lying SP states are observed as broad structures above the particle threshold in transfer reactions such as $(\alpha,{}^3He)^{1,18}$, $({}^{20}Ne,{}^{19}Ne)^{19,20}$ as shown in fig. 13 a-b).

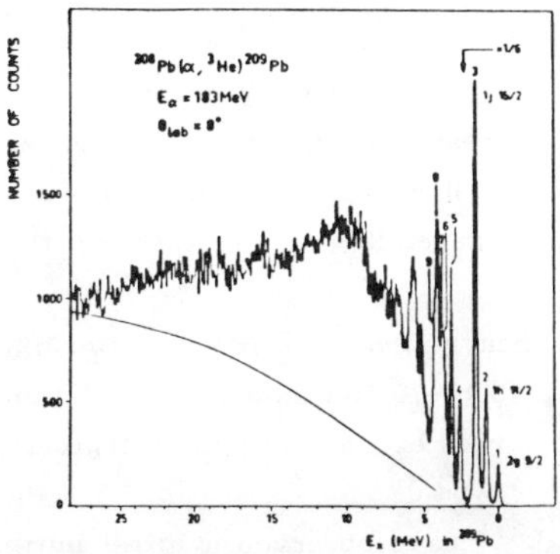


Fig.13-a) $^3$He energy spectrum from $^{208}Pb(\alpha,{}^3He)^{209}Pb$ reaction at 183 MeV. The solid line indicates, the calculated yield due to the elastic break-up of $\alpha$ in $^3$He + n (Taken from Ref. 18).

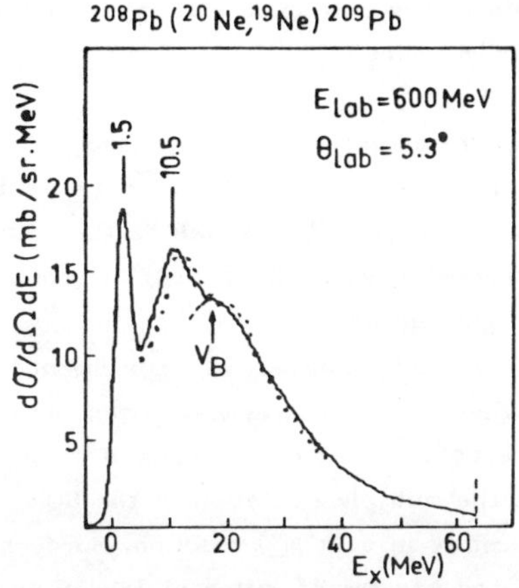


Fig.13-b) Excitation energy spectra from the reaction $^{208}Pb({}^{20}Ne,{}^{19}Ne)$ at 30 MeV/nucleon.
The dotted line is a calculation of the full break-up cross-section. (Taken from Ref. 19).

These spectra were obtained in single-arm experiments using magnetic spectrometers to detect at very forward angles the ejectile and reject the high-flux of incoming particles. A large continuum cross-section is observed under the "broad" bump, around 10 MeV. This part of the cross-section originates from fast direct processes where the projectile break into the detected ejectile and a light particle (here a neutron).

The distribution of the emitted neutrons in these processes are strongly focussed around the recoil and they travel at the beam velocity. On the opposite the neutron emitted by the unbound resonant SP states have much lower velocity.

The detection of a coïncidence between the ejectile and a neutron using EDEN time of flight spectrometer with an experimental set-up similar to the one shown in fig. 11 would help to disentangle the two processes and gives as a result a background free SP strength distribution.

The angular correlation between the ejectile and the neutron gives the spin of the sp state and the neutron branching ratios. After a subtraction of the statistical decay branch, the escape width can be obtained. Most of the microscopic structure of the decaying state is then accessible and shed light on the damping mechanisms.

Typically with the EDEN array and a excitation cross section of 10 mb, a coincidence rate of 1 c/s may be obtained. Decay branches at 1 % level could be determined.

- High-momentum giant resonances, multiphonon states and their decays

It is of interest to study single GR'S with a response that peaks at larger momentum transfer as well as the double phonon GR's. Both of these modes are expected at excitation energy of about 20–50 MeV depending on the mass and the GR considered.

In a large number of experiment, using heavy ions inelastic scattering, high-lying structures have been observed up to 60–70 MeV apparent excitation energy[21-22] which are claimed to be due to target excitations and interpreted as the multiple excitation of the GQR[23].

Recently in a $(\pi^+,\pi^-)$ reaction, Mordechai et al.[24] have observed a broad bump in the spectra of different target nuclei. It has been assigned as the double giant dipole resonance.

It is worth noting that in both cases the observed enhancements are sitting on a very large background and that the position in energy is near where one would expect the higher-momentum single-resonance states to occur. Again here in the case of collective excitations, experimental methods similar to the one mentionned in the previous section (SP state decay) are being investigated to establish on firm basis the location and the nature of these new modes.

An illustrative example is given by the work of J.A. Scarpacci[6] carried out at GANIL where inelastic spectra of $^{40}$Ca was measured up to 150 MeV excitation energy by means of the $^{40}$Ca + $^{40}$Ca reaction at 50 MeV/n. An array of CSI phoswitch detectors where placed around the target in coïncidence with the ejectile $^{40}$Ca detected at the focal plane of the spectrometer SPEG. Due to the different velocity and angular cone of the emitted protons corresponding to the different processes :

Transfer evaporation process $^{40}Ca + {}^{40}Ca \rightarrow {}^{40}Ca + {}^{41}Sc^{*}$
$\hookrightarrow p + {}^{40}Ca^{*}$

Proton knock-out $^{40}Ca + {}^{40}Ca \rightarrow {}^{40}Ca + {}^{39}K + p$ and

target inelastic excitation $^{40}Ca + {}^{40}Ca \rightarrow {}^{40}Ca + {}^{40}Ca^{*}$
$\hookrightarrow p + {}^{39}K$

the authors were able to unravel the different contributions to the inclusive cross-section as shown in fig. 14. The backward angle region where are concentrated the protons emitted by the excited target states shows clearly not only the population of the GQR around 18 MeV but a second bump at about the double of the energy of the GQR (see fig. 14 b).
The investigation of the particle decay of the two structures are undertaken to confirm the possible two-phonons nature of the second bump.
Almost identical experiments could be performed by using the EDEN detectors in heavy nuclei since the neutron decay of GQR is the main decay channel. Both the one phonon and two phonon collective states should have identical decay pattern.

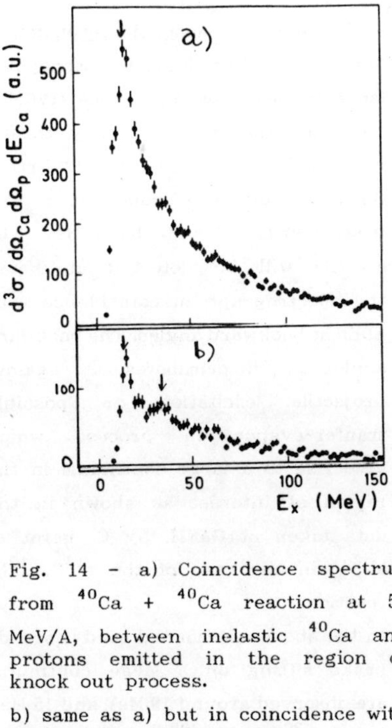


Fig. 14 - a) Coincidence spectrum from $^{40}Ca + {}^{40}Ca$ reaction at 50 MeV/A, between inelastic $^{40}Ca$ and protons emitted in the region of knock out process.
b) same as a) but in coincidence with backward at angles protons.
[Taken from Ref. 6].

Coupled channel calculations indicate that the cross section for multiple excitation of the GDR by Ar or Kr ions with incident energy between 60 and 80A MeV should be of the order of 100 mb/sr. It is then feasible to study for heavy nuclei the γ-decay by measuring triple coincidences between inelastically scattered Kr ions and two γ rays of about 13 MeV using the TAPS detection system.
Observing the double γ decay would unambiguously identity the observed

structure as a double GDR phonon selective stale.

- Decay of $M_1$ giant resonance

The giant dipole resonance, for the first time identified in mostly s-d and f-p shell nuclei via 200 MeV inelastic proton scattering[25] may be clearly observed in heavy nuclei by its $M_1$ $\gamma$-decay back to the ground state. The challenge will be to distinguish this decay from that of possible $1^-$ neighbouring states. The use of polarized beam and the measurement of the $\vec{p}$-$\gamma$ angular correlation will give a clear signature of $M_1$ states.

- The isovector monopole resonance

Contrary to the isoscalar monopole resonance (GMR), the properties of the isovector monopole resonance (IVGMR) are not well known. Althrough this mode has been discovered already three years ago in a pion change exchange reaction[26], there are still not firm conclusions on the resonance parameters. One can study the properties of the IVGMR using the ($^{13}$C, $^{13}$N) reaction at a beam energy of 30 to 80 A MeV. The ejectile will be detected at 0° in the spectrograph in coincidence with EDEN at backward angles. The backward angles discriminates against projectile excitation or possible tranfer-evaporation process which give rise to a large continuum in the region of interest as shown in the data taken at GANIL by C. Berat et al.[26] in the case of the ($^{13}$C, $^{13}$N) reaction.

In the above mentionned study, broads peaks sitting on a large continuum are observed around 19 MeV and 15 MeV for the $^{90}$Zr and $^{208}$Pb respectively. Since the log ft value for $\beta$-decay of $^{13}$C indicates that Fermi transition matrix element is dominant, thus the ($^{13}$C,$^{13}$N) reaction is selective for $\Delta T = 1$, $\Delta S = 0$ states including the IVGMR. For example the investigation of the coincidence experiment

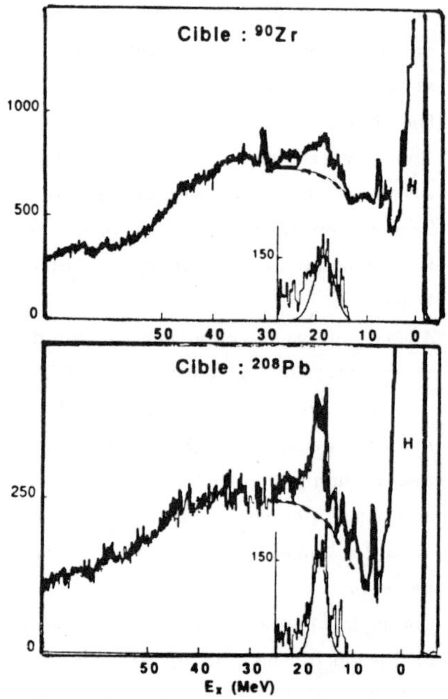


Fig. 15 - Residual energy spectra of $^{13}$N from the reaction ($^{13}$C,$^{13}$N) at 50 MeV/A energy incident energy on $^{90}$Zr and $^{208}$Pb targets (Taken from ref. 26).

$^{208}Pb \rightarrow (^{13}C,^{13}N) \leftarrow {}^{208}Tl^{*}$ is feasible. With a population cross-section

$\quad \hookrightarrow n + {}^{207}Tl$

of the IVGMR of 1mb/sr, one gets a coincidence neutron rate of 20 per hour. The ejectile neutron angular correlation should be isotropic, a signature for the decay of a $0^{+}$ state.

The spin flip-isospin flip channels have been studied using light and heavy ions reactions at intermediate energies (50 to 200 A MeV) to populate Gamow-Teller transitions. Charge exchange reactions such (p,n), (n,p) ($^{6}Li$, $^{6}He$), ($^{3}He$,t) ($^{12}C,^{12}N$) and ($^{12}C,^{12}B$) are very selective in exciting $L = 0$, $L = 1$, $\Delta S = 1$, $\Delta T = 1$ transitions. Almost no information exists of the decay of the Gamow-Teller resonances with the exception of a low energy ($^{3}He$,t p)[28] study on $^{208}Pb$.

Using the spectrograph to detect the ejectile and the EDEN array or charged particle detectors, a very large experimental program will be feasible.

## V - CONCLUSION

The construction of the superconducting cyclotron AGOR has already taken full advantage in the design stage of the pionnering work made at MSU on the first accelerator of this type. The range of energy and ions has been greatly enhanced, since its include now all ions from protons to very heavy nuclei. Polarized proton and deuteron beams at 200 MeV maximum energy will be available to study in a rather optimum energy window the many aspects of the largely unknown spin response of the nucleus.

The accelerator will serve experimental areas which are foreseen to be equiped by two different magnetics spectrometers, a very large general use scattering chamber and a number of multidetector systems for the detections of charged particles, neutrons, $\gamma$-rays, and electron-positron pairs. All these detectors systems are intended to operate in coincidence with magnetic spectrometer or other simple arm devices. The few examples of the physics experiment described in the previous section show that coming AGOR facility should be an very attractive center for nuclear structure and nuclear dynamics studies.

## REFERENCES


1) H. Schreuder, The AGOR cyclotron past the half-way mark, Proc. 12th conf. on Cyclotrons and their applications. (Berlin, 1989 to be published)
S. Galès, AGOR, a superconducting cyclotron for light and hevy ions. Proc. 11th Conf. on cyclotron and their applications.(Ionics, Tokyo, 1987), p. 184.

2) M. Morlet et al., Phys. Lett. B247 (1990) 228.

3) S. Galès, A.I. Vdovin, Ch. Stoyanov, Phys. Rep. 166 (1988) 127.
S. Galès, Proceedings of the International Symposium a Nuclear Structure (Dubna Press, June 1989, Dubna USSR) p.

4) A. Van der Woude, Prog. Part. Nucl. Phys. 18 (1987) 217.
W.T.A. Borghols et al, Nucl. Phys. A504 (1989) 231.

5) H. Laurent, Proceedings of the 2nd IN2P3-RIKEN Symposium on Heavy collisions (Obernai, France, 1980) ed. B. Heush et M. Ishihara (World, scientific) p. 1.

6) J.A. Scarpaci, Thesis Université Paris-Sud Orsay, Report IPNO-T-90-04.

7) H. Nifenecker and J.A. Pinston, Progress in Part. Nucl. Phys. 23 (1989) 273.

8) L.H. Harwood, J.A. Nolen and A.F. Zeller, ibid ref. 1, p. 315.

9) KVI, Groningen, Annual Report, p. 133.

10) ibid ref. 9) p. 132.

11) ibid ref. 9) p. 138.

12) ibid ref. 9) p. 143

13) ibid ref. 9) p. 143.

14) C. Glashausser et al., Phys. Rev. Lett. 58 (1987) 2404.
F.T. Baker et al., Phys. Lett. B237 (1990) 337.

15) W.G. Love and M.A. Franey, Phys. Rev. C24 (1981) 1073.

16) H. Sakai et al. Nucl. Phys. A441 (1985) 640.

17) F. Azaiez et al. Nucl. Phys. A444 (1985) 373.

18) C.P. Massolo et al. Phys. Rev. C34 (1986) 1256.

19) S. Fortier et al. Phys. Rev. C41 (1990) 2689.

20) N. Frascaria et al., IPN Orsay Annual Report (1987) p. 47.

21) N. Frascaria et al., Phys. Rev. Lett. 39 (1977) 918, Z. Phys. A294 (1980) 167, Nucl. Phys. A474 (1987) 253.

22) Ph. Chomaz et al., Phys. Lett. B139 (1981) 1, Nucl. Phys. A318 (1984) 167.

23) N. Frascaria et al. A482 (1988) 245C.

24) S. Mordechai et al., Phys. Rev. C41 (1990) 202.

25) N. Anantaraman et al. Phys. Rev. Lett. 46 (1981) 1318.
C. Djalali, Proc. Int. Sym. HESANS 83, ed. by N. Marty and N.V. Giai, J. Phys. 45 (1984) C4-375.

26) J.D. Bowman et al., ibid ref. 25, J. Phys. 45 (1984) C4-351.

27) C. Berat, Thèse Université de Grenoble, 1988, ISN 88-102,
C. Berat et al., Phys. Lett B218 (1989) 299.

28) C. Gaarde et al. Phys. Rev. Lett. 46 (1981) 902.

# EUROGAM : A NEW GAMMA SPECTROMETER FOR A NEW DIMENSION OF NUCLEAR STRUCTURE STUDIES

F.A. BECK

CRN - Division de Physique Nucléaire
F-67037 STRASBOURG CEDEX

*ABSTRACT :*

*A new gamma spectrometer, EUROGAM, is presently built in a France-UK collaboration. It should be operationnal with 45 Ge counters surrounded by BGO anti-Compton shields beginning 1992 at Daresbury. It will be transfered and completed to 70 counters beginning 1993 at Strasbourg with the aim to exploit the beams of the new accelerator Vivitron. The most advanced techniques in integrated electronics and in computing are used to control, monitor and store the high data acquisition rate expected from this large array. The lowering of the detection limits (factor* $10^2$ *in phase 1 and factor* $10^4$ *in phase 2 are expected) as compared to the present one (a few fractions of %) will able completly new developments in nuclear structure studies.*

## 1 INTRODUCTION

The combination of heavy ions accelerators and high resolution $\gamma$-ray multidetectors comprising germaniums counters surrounded by anti-Compton scintillators shields has permitted this last decade a quantitative gap in the way we approach the understanding of the microscopie properties of the nucleus. This 1980's generation of multidetector aways based on large national collaborations (TESSA3 in the UK, OSIRIS in Germany, CHATEAU DE CRISTAL in France, $8\pi$ in Canada, NORDBALL in Scandinavia) and on serveral North American detectors (HERA at Berkeley, SPIN SPECTROMETER at Oak Ridge, MULTISPECTROMETER at Argonne) have allowed the reduction of the detection intensity limit to fractions of a percent leading to an improved understanding both of the characteristics of the individual excitations of the nucleus and of the collective behaviour of the nucleus.

The most spectacular among the discoveries has been the experimental evidence for very regular cascade of $\gamma$-rays (some 15 to 20 transitions). This new type of $\gamma$ transitions "band" was discovered at Daresbury [1)] in the study of the $^{152}Dy$ nucleus. It was than more widely observed in the rare earth region (atomic number A = 130 to A = 150) as well in the mass range A = 190. Microscopic calculations permit an interpretation based on a second stable minimum for ("superdeformed") prolate ellipsoidal nucleus with ratio of the major to minor axis approaching 1.6 for A = 130 and A = 190 and approaching 2 in the region of A = 150.

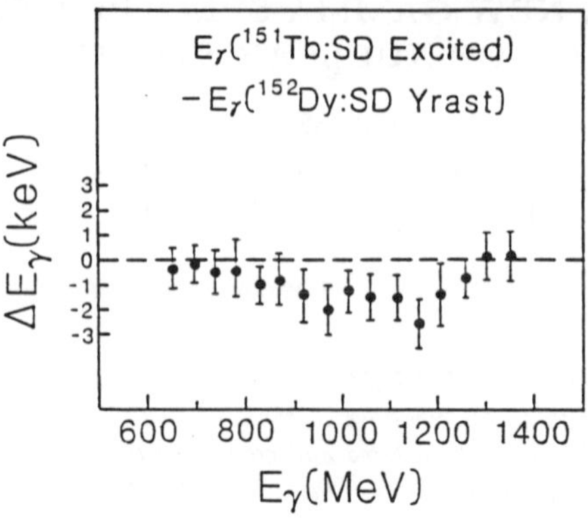


Fig. 1 : Differences in gamma-ray energies in superdeformed bands between the yrast band in $^{152}Dy$ and the excited band in $^{151}Tb$

Several very important properties of this SD bands are not yet experimentaly determined : the feeding mechanism which is done in a few transitions around spin $58\hbar$, the decay from the second well to the first well which seems to happen around spin $22\hbar$ also in a few transitions. The last linking transitions are not yet observed since the number of decay paths lead to very small intensities not measurable today. This effect makes unique spin assignments of SD bands impossible and so on do not able the determination of the static moment of inertia $\Im(1)$. Theoritical interpretation of this SD bands request also the knowledge of their parities and so on the determination of the multipolarity of the linking transitions. Other properties of the SD bands have raised up several questions with will need a quantative gap in intensity limit detection to be completly understood.

An other astonishing discovery has been made in a France-UK collaboration experiment performed at Daresbury and analysed at *Strasbourg*[2] : the data revealed SD bands whose $\gamma$ - ray transition energies are the same to within one part in a thousand for the $^{152}Dy$ (yrast band) and $^{151}Tb$ (excited band) nuclei (see Fig. 1) whereas difference larger by an order of magnitude are expected. Other examples of this phenomena involving the observation of "identical" bands in different nuclei have since been found in the rare earth region and the region A = 190. It poses many entirely open question relating to certain features of the mean nuclear field (the existence of pseudo symmetries for example).

Recently four superdeformed bands, with respectively 1.6 %, 2.1 %, 0.8 % and 1.1 % intensities, have been found in $^{193}Hg$ at *Daresbury*[3]. For the first time a crosstalk between two superdeformed bands has been demonstrated. The existence of such transitions (most probably E1 transitions) are interpreted as evidence for octupole shape correlations. At the same time two superdeformed bands found[4]) with CHATEAU DE CRISTAL in $^{147}Gd$ are interpreted together with the existing data in $^{146}Gd$ and $^{148}Gd$ obtained by other *authors*[5),6)] as a first hint concerning octupole distorsions (or at least octupole softness).

In general terms, one must reconcile the extraordinary "rigidity" of the superdeformed structure to complex structures present at the first stable minimum of the nuclear potential. The study of this feature will also permit a total new approach to clarify the properties of nuclei which have up to now been poorly investigated. For example, it was predicted many years ago by Mottelson and Valatin that nuclear superfluidity owing to a short range, two-body residual "pairing" interaction between nucleons in time-reversed orbits should break down ("quench") at high rotational frequencies. Information on this pairing which may eventually lead to a test of the theory is only just now becoming available. Certain other very interesting issues can also be probed (e.g. hyperdeformation, superdeformation at zero spin, complete spectroscopy and chaos, etc...).

The lowering of the detection limit will have also a great interest for other experimental studies like in particle - $\gamma$ coincidence measurements for nuclear mechanism studies or for observation of nuclei weakly produced far of the line of stability. In particular we expect to push our knowledge of N = Z nuclei towards A = 100 which corresponds to the double major stell closure N = Z = 50.

To reach this different physics targets the detection limit, which is today a few fractions of %, has to be improved by a factor $10^2$ to $10^4$ to achieve limits of $10^{-4}$ to $10^{-6}$. This goal needs a very important technical gap in the detector array as compared to the present existing tools : it has entrained a world wide grouping around two projects, GAMMASPHERE in USA and EUROBALL in Europe. The latter collaboration has got underway with the EUROGAM detector which is essentially financed by France and the UK.

EUROGAM construction has started in 1989 with the aim to have a first year of operation in 1992 at Daresbury with 45 Ge counters surrounded by BGO Compton suppressors (phase 1). At this time the possibility to use in coincidence the existing recoil mass spectrometer, especially for exotic nuclei studies, is also planed. The array will be transfered and completed to 70 Ge counters at Strasbourg beginning 1993.

The construction of EUROGAM has involved new and remarkable technical developments as well in detectors, as in electronics (integrated in VXI standard boards) and as in data acquisition techniques.

## 2 EUROGAM Design

### 2.1 General specifications

Gamma ray multidetector arrays are caracterised by two main quantities :

- the total photopic efficiency which indicated the data rate which can be reached.

- the resolving power which values approaches the quality of the data.

#### *2.1.1 Photopic efficiency*

The total photopic efficiency Q will depend on the individual photopic efficiency $\varepsilon_p$ and solid angle $\omega$ and the total number of counters N

$$Q = N\varepsilon_p\omega. \tag{1}$$

The data rate will depend on this parameters. So if the gamma decay path contains $M\gamma$ - rays (multiplicity) the coincidence rate of fold n will follow the relation

$$P_n^N(M) = C_n^N \sum_{x=0}^{n}(-1)^{n+x}C_x^n\{1-(N-x)\varepsilon_p\omega\}^M. \tag{2}$$

Fig. 2 shows the strong dependance of the possibility to detect high folds versus the increase of the total photopic efficiency Q.

The coincidence rate of order n depends on the same parameters given by the relation

$$\gamma^n = \sum_{x=n}^{M} C_x^n P_x^N(M) \sim Q^n. \tag{3}$$

This formula shows also that high fold data will produce high rate of low order coincidence (this will also help to reduce the problem of data rate flux).

#### *2.1.2 Resolving power*

If we consider a gamma cascade with an energy spacing $\Delta E$ (kev) the resolving power is defined as

$$RP = \left(\frac{\Delta E}{\sigma(E)} * \frac{1}{PT}\right). \tag{4}$$

where $\sigma$(E) is the energy resolution of the detector (intrinsic + Doppler broadening), PT the peak to total ratio measured with the detector and $\Delta E$ the energy spacing of the transitions of the gamma band. For the detection of fold n the resolving power will take the value $(RP)^n$.

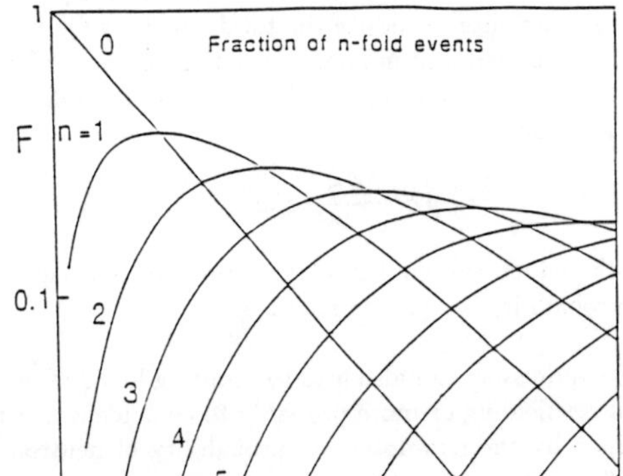


Fig. 2 : Fraction of n-fold events for a multiplicity M = 30 as a function of the total peak efficiency $N\varepsilon_p\omega$

The only detector having a reasonable energy resolution is the Ge detector. Unfortunately it can only be manufactured in a reduced size. This will lead to a small photopic efficiency as compared to the total efficiency and consequently to a reduced Peak to Total value. For the existing tools Ge detectors having a size around ∅ = 52 mm, L = 54 mm with intrinsic photopic coefficiency values $\varepsilon_p \approx 0.12$ and PT $\approx 0.17$ have been used. Only recently larger N-type have been developped (∅ = 70 mm, L = 70 mm) leading to enhanced $\varepsilon_p \sim 0.20$, PT $\approx 0.25$ values. To achieve better PT values, in absence of larger volumes, the only way is to try to suppress the events for which part of the energy escapes from the detector : this will be done by anti-Compton shields surrounding the Ge counter. The most efficient gamma detector, BGO, is used to have the best efficiency for the smallest amount of material : in fact the use of anti-Compton shields will reduce considerably the solid angle which can be covered by the Ge detectors and so contribute to reduce the total efficiency Q. The intrinsic PT values obtained after suppression will be reduced in nuclear reactions because of multiple hits, gamma or neutrons, in the detector. This probability is related to the multiplicity, for gamma rays it can be calculated by the approach relation

$$p \simeq \frac{M-1}{2}\varepsilon_u \omega. \tag{5}$$

The multiple hit will both reduce the quality of the spectrum (weaker PT values) and the efficiency (the single hit probability is equal to $i_s = 1 - p$).

The surrounding BGO suppressors can also lead to counting loss due to random suppression because of a non sufficient $\gamma$ shielding or more probably to coincident neutron detection. This reduction effect is quantified by the non-detection probability of neutron $i_n$ which has been evaluated to be around 0.85 in the previous arrays.

### 2.2 EUROGAM design phase 1

The objectives of EUROGAM are both to achieve the best photopic efficiency and to have a good resolving power with a minimized number of counters (costs). To determine the number of counters two factors have essentilally to be taken in account : the single hit probability and the Doppler broadening. Knowing that the maximum total efficiency which can be acchieved is around 0.4, due to the space occupied by the canning and the anti-Compton systems, and if a single hit probability larger than 0.94 is requested, the number N = 72 is deduced from equation (5) for M = 30. We will see later on that the Doppler broadening can be too large in some reaction cases, but this efect will only be taken in account for phase 2 since in phase 1 the recoil mass spectrometer can be a possibility used to correct the broadening.

Several type of Ge detectors have been studied with the code GEANT to find the best compromise between photopic efficiency and solid angle coverage. Tapered detectors (Fig. 3) have been found to give the best answer and with detectors of 70% efficiency, as compared to 7.6 x 7.6cm NaI scintillators for $E_\gamma$= 1.33 MeV, a photopic efficiency $\varepsilon_p \approx 0.23$ is achieved, each detector covering a solid angle $\omega \approx 5.1 * 10^{-3}$.

The 72 counters ball, see Figure 4, is based on a pentagon structure including 12 pentagons surrounded by 5 counters each. 60 detectors are all identical while the 12 central pentagons would need a special design and a little further distance. In phase 1 the aim is to let empty the 15 forward detectors space to can approach the recoil mass spectrometer and to fullfill the 45 regular holes.

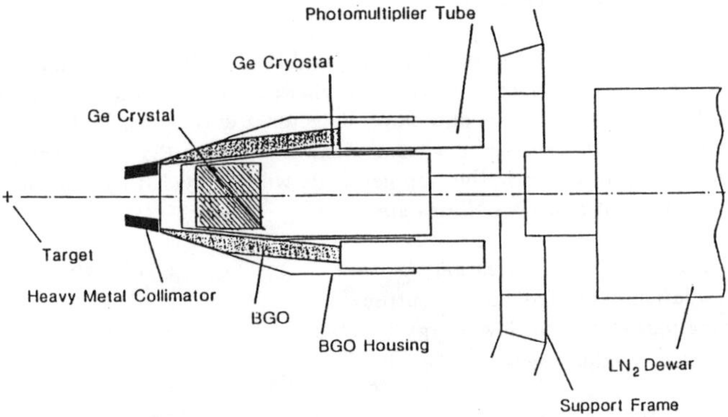


Fig. 3 : Schematic design of the EUROGAM tapered Ge detector surrounded by the BGO anti-Compton shield

Fig. 4 : EUROGAM general structure for phase 1

### 2.3 EUROGAM design phase 2

Different types of detectors have been imagined and studied (stacks of planars, segmented coaxial detector) to reduce Doppler broadening in phase 2 with the aim to increase also the total efficiency. Presently a prototype of a clover detector is developped by Intertechnique. It is based on a cluster of 4 detectors of $\emptyset = 50mm$ and L = 70mm (25 % efficiency each) in a geometry shown Fig. 5. In this case the Doppler angle will be divided by two and reduced to $4^{O}$. Additionnal important new possibilities are :

- possibility to measure linear polarization
- reduction in radiation dammage due to neutrons
- improved timing particularly for low energy $\gamma$ - rays
- no balistic deficit correction needed.

The aim is to use this new type of detectors around 90° where the Doppler broadening effect is maximum and to conserve a forward section and a backround section of each of 15 tapered detectors. The part of the ball between 60° and 120° will be covered by 48 clovers as shown in Fig. 6.

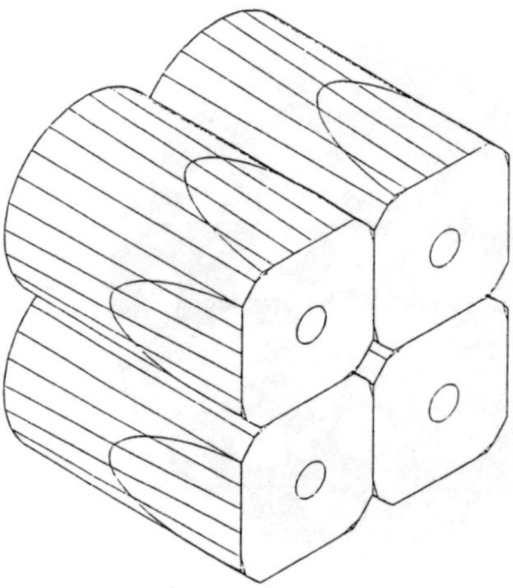

Fig. 5 : Schematic design of a clover Ge detector

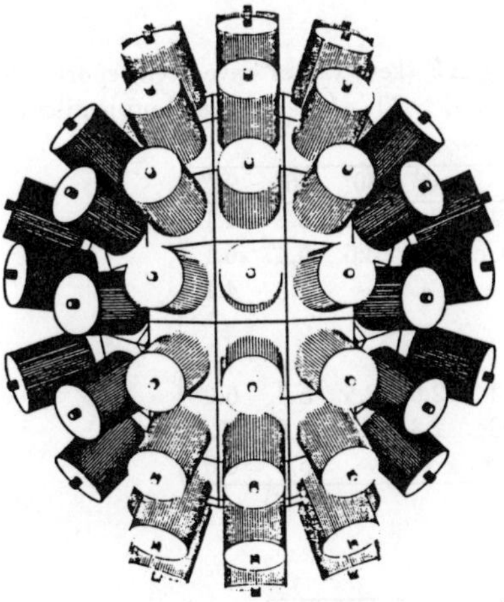

Fig. 6 : EUROGAM general structure for phase 2 ; the detectors between $60^0$ and $120^0$ are replaced by clovers detectors situated at 32cm form the target as compared to 20.5cm for the tapered Ge

## 2.4 Main caracteristics

In the following section we will compare the main caracteristics of phase 1 and phase 2 of EUROGAM to the multidetector TESSA3 where most of the new results have been found in the last years and which can be taken as representative of all other presently existing $\gamma$ - multidetectors arrays.

*Peak to total*

| detector | $(P/T)_{intrinsic}$ | $i_s$ | $i_n$ | $(P/T)_{realistic}$ |
|---|---|---|---|---|
| TESSA3 | 0.55 | 0.97 | $\simeq 0.85$ | 0.45 |
| tapered Ge | 0.62 | 0.94 | $\simeq 0.90$ | 0.52 |
| clover Ge | 0.60 | 0.94 | $\simeq 0.97$ | 0.55 |

The higher value obtained for $i_n$ with tapered and clover Ge is due to the proportionaly less surrounding BGO material as compared to TESSA3 detector. In the case of clovers both the larger distance (32cm compared to 20.5cm) and the better timing should allow neutron-gamma discrimination.

*Doppler broadening*

The Doppler broadening $\Delta E$ (kev) values have been reported for two typical kind of reactions with projectiles of mass $A \simeq 30$ (v/c $\sim$ 0.02) and projectiles of mass $A \simeq 80$ (v/c $\sim$ 0.04).

| | Angle | $0^0$ | $30^0$ | $60^0$ | $90^0$ |
|---|---|---|---|---|---|
| v/c = 0.02 | TESSA3 | 2.6 | 3.1 | 3.8 | 4.1 |
| | tapered Ge | 2.8 | 3.6 | 5.4 | 6.1 |
| | clover Ge | - | - | 3.8 | 4.1 |
| v/c = 0.04 | TESSA3 | 2.6 | 3.5 | 5.3 | 6.0 |
| | tapered Ge | 2.8 | 4.7 | 9.6 | 11.0 |
| | clover Ge | - | - | 5.3 | 6.0 |

*Individual efficiencies*

| | $\varepsilon_p$ | $\omega(10^{-3})$ | $\varepsilon_p\omega(10^{-3})$ |
|---|---|---|---|
| TESSA3 | 0.12 | 2.3 | 0.28 |
| tapered Ge | 0.23 | 5.1 | 1.17 |
| clover Ge | 0.27 | 4.6 | 1.24 |

*Arrays efficiencies and coincidence rates*

As seen above the array efficiency is characterized by the quantity $N\varepsilon_p\omega i_s$. From those value event rates can be calculated as a function of $\gamma$ - ray multiplicity M (see Fig. 2) as well as the $(\gamma)^n$ coincidence rates. The following table, calculated for M = 30 and taking as reference $(\gamma)^2$ in the TESSA3 case, shows clearly that $(\gamma)^4$ measurements will be possible in phase 1 and $(\gamma)^6$ in phase 2. This should enable $10^{-4}$ - $10^{-5}$ limits measurements in phase 1 and $10^{-6}$ values in phase 2.

| | $N\varepsilon_p\omega i_s$ | $(\gamma)^2$ | $(\gamma)^3$ | $(\gamma)^4$ | $(\gamma)^5$ | $(\gamma)^6$ |
|---|---|---|---|---|---|---|
| TESSA3 | 0.0043 | 1 | .01 | | | |
| Phase 1 | 0.0490 | 39 | 17 | 5 | 1 | |
| Phase 2 | 0.0890 | 130 | 105 | 60 | 26 | 9 |

## 3 ELECTRONICS AND DATA ACQUISITION

For this complex array a completly new and modern technology has been developped. The "guide lines" were :
- High degree of integration and minimizing of the cables number (the detector has to be reliable and also to can to be moved easely)
- Increased reability (large number of counters)
- Software control and monitoring of the electronic parameters
- Several buses for data handling
- Modularity of electronic and software allowing easy international collaborations in the construction and also upgrading in the future

Between different possibilities (Fastbus, Future bus, Multibus) the VXI standard has been choosed since he allows :
- electronic integration (logic and linear) in the VXI large boards
- it has both the VME bus and a local bus.

For the electronic circuits differents types of developments have been done : integrated circuits for the TAC's and the readout chips of the ADC's, SMD use for logical circuits and constant fraction discriminators, hybrids circuits for the linear amplifiers (with possibility of balistic correction) and for the fast amplifiers. Each Ge VXI board will include 6 complete Ge channels each having 4 ADC - 13 bits - 4.5 $\mu s$ conversion time (2 energy scales, 1 cross-over, 1 timing). Identically the BGO boards will contain 6 channels, each representing a complete BGO shield.

In the first phase the acquisition will be done with a fixed dead time of $15 \mu s$. A counting rate of $10^4$ ev/s with a mean value of 12 parameters is expected ; if necessary the data rate can be reduced to this value by requesting a higher fold minimum for the trigger without losing too much in the low order coincidents (see formula (3)).

In Figure 7 is represented the general architecture of the data acquisition system for EUROGAM. Beside the VXI crates it includes VME crates (histogrammers, event builder, on line sorting) all being connected throught a high data transfert bus DT32. A network of work stations will be used for electronic contral and monitoring, graphics, data files, etc... The system will be developped under UNIX using the VxWorks real time system for the on line tasks.

## CONCLUSIONS

Today 30 Ge detectors have been received and the delivery of the BGO Compton suppression shields has begun. The first integrated electronic cards in VXI standard should be achieved

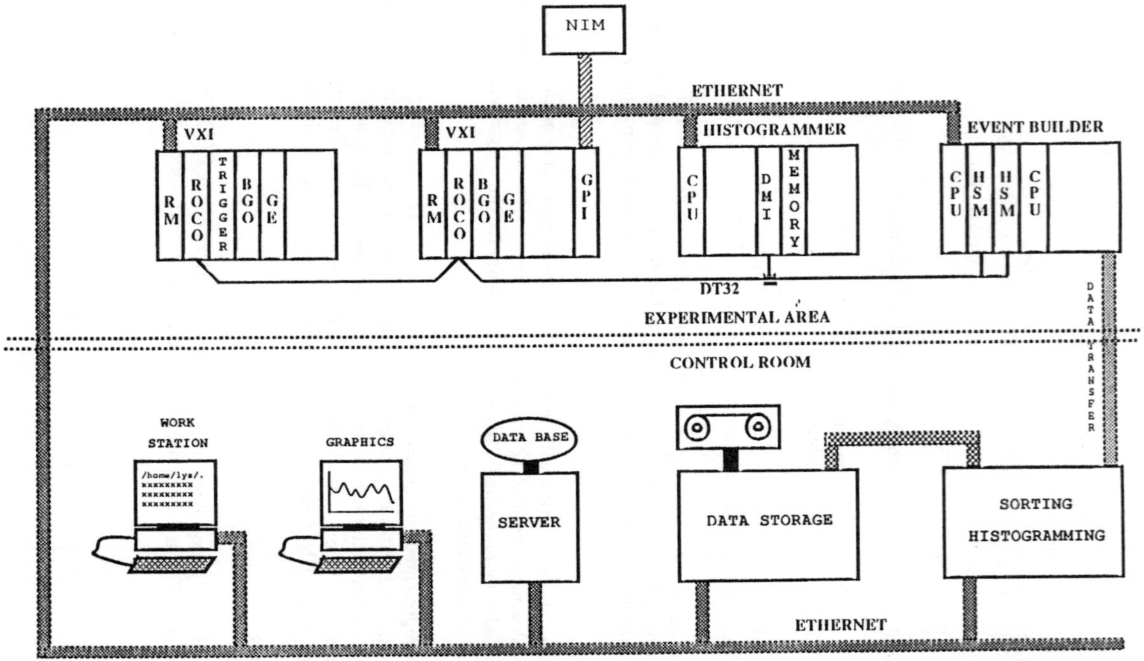


Fig. 7 : General scheme of the data acquisition system of EUROGAM based on integrated electronics in VXI boards and on work stations network for beginning May, so that the checking of the array should start in October 1991 at Daresbury. The first experiments are planed for January 1992 with 45 Ge detectors with the possibility to use in coincidence the existing recoil mass spectrometer.

The array will be moved one year later to France with the aim of explorating the Vivitron beam at Strasbourg. It will be completed there to the more efficient phase 2 with clover detectors. The first prototype of such a detector should be delivered by April 1991. Additional counters like a $4\pi$ particle array, a neutron wall, ... are also expected.

## ACKNOWLEDGMENTS

The EUROGAM collaboration benefits from numerous contributions from both countries France and UK. In France the collaboration implies 6 laboratories : CEN Bordeaux, ISN Grenoble, IPN Lyon, CSNSM Orsay, IPN Orsay and CRN Strasbourg. Many people work as well for the technical construction as in the discussion of the future experiments. My thanks go to all those involved.

## REFERENCES

1) P.J. Twin, B.M. Nyako, A.H. Nelson, J. Simpson, M.A. Bentley, H.W. Cramer-Gordon, P.D. Forsyth, D. Howe, A.R. Mokhtar, J.D. Morrison, J.F. Sharpey-Schafer and G. Sletten, Phys. Rev. Lett. 57 (1986) 811

2) T. Byrski, F.A. Beck, D. Curien, C. Schück, P. Fallon, A. Alderson, I. Ali, M.A. Bentley, A.M. Bruce, P.D. Forsyth, D. Howe, J.W. Roberts, J.F. Sharper-Schafer, G. Smith and P.J. Twin, Phys. Rev. Lett. 64 (1990) 1650

3) M.A. Riley, D.M. Cullen, J.F. Sharpey-Schafer, W. Nazarewicz, T. Bengtsson, R. Wyss, A. Alderson, I. Ali, C. Beausang, P. Fallon, P.D. Forsyth, F. Hanna, S.M. Mullins, J.W. Robert, P.J. Twin, R. Poyntes, P. Regan, R. Wadsworth, M.A. Bentley, A.M. Bruce, J. Simpson and G. Sletten, to be published

4) K. Zuber, D. Balouka, F.A. Beck, T. Byrski, G. de France, G. Duchêne, C. Gehringer, B. Haas, J.C. Merdinger, P. Romain, D. Santos, J. Styczen, J.P. Vivien, J. Dudek, Z. Szymanski and T.R. Werner, Phys. Lett. B, 254 (1991) 308

5) G. Hebbinghaus, K. Strähle, T. Rzaca-Urban, D. Balabanski, W. Gast, R.M. Lieder, H. Schnare, W. Urban, H. Wolters, E. Ott, J. Theuerkauf, K.O. Zell, J. Eberth, P. Von Brentano, D. Alber, K.H. Maier, W. Schmitz, E.M. Beck, H. Hübel, T. Bengtsson, I. Ragnarsson and S. Aberg, Phys. Lett. B240 (1990) 311

6) M.A. Delaplanque, C. Beausang, J. Burde, R.M. Diamond, J.E. Droper, C. Duyar, A.O. Machiavelli, R.J. Mc Donald and F.S. Stephens, Phys. Rev. Lett. 60 (1988) 1626

# INDRA : A new $4\pi$ Detector for Charged Particles and Nuclei at the GANIL Facility.

A.Chibhi, G.Auger, A.Benkiran, L.Bianchi, P.Bourgault, Y.Huguet, M.Le Guay, E.Plagnol, J.Pouthas, F.Saint Laurent., C.Spitaels, M.Tripon, J.P.Wieleczko
**GANIL, Caen, France.**

B.Berthier, Y.Cassagnou, J.L.Charvet, R.Dayras, R.Legrain, C.Mazur, J.Passerieux, E.Pollacco, C.Volant
**D.Ph.N, CENS, Saclay, France.**

C.O.Bacri, S.Barbey, D.Charlet, B.Borderie, A.Richard, M.F.Rivet, L.Tassan-got.
**IPN, Orsay, France.**

D.Cussol, J.M.Gautier, J.L.Laville, J.C.Steckmeyer, J.Tillier
**L.P.C., Caen, France.**

## 1. Introduction

The content of this conference as well as publications over the last few years [1] have shown very clearly the intense activity around the study of hot nuclei using heavy ion beams between 10 MeV/A and 1 GeV/A. These studies range from the search for a limiting excitation energy or temperature to searching for evidences of "sudden multifragmentation" associated or not with a phase transition.

This work is carried out both on the theoretical and on the experimental side. Theoretically a great deal of work has been done on sequential emissions of heavy fragments during the de-excitation process [2], on the characterisation of sudden multifragmentation [3], on the modification of phase transition signatures for system of small finite size [4] and on the study of non statistical fluctuations (intermittency analysis) at the transition energies [5].

The main conclusion of these studies, as seen by an experimentalist, is the fundamental requirement that event by event data acquisition and analysis be performed to extract the expected signals. An event by event analysis requires that for a given reaction the "total" multiplicity of light particles (n, p, d, t, $\alpha$) and heavier ions (Z>2) be measured, that all participating nuclei be identified and that their velocity or energy be measured.

This of course requires a $4\pi$ detector of remarkable performances which, at this date, is still beyond technical feasibility. However, rare experimental data and more numerous simulations have shown that phase transition signals were still measurable when only the charge and the energy of the emitted ions were measured [6]. With these requirements, the necessary $4\pi$ detector becomes feasible although still difficult to build.

# The INDRA Detector

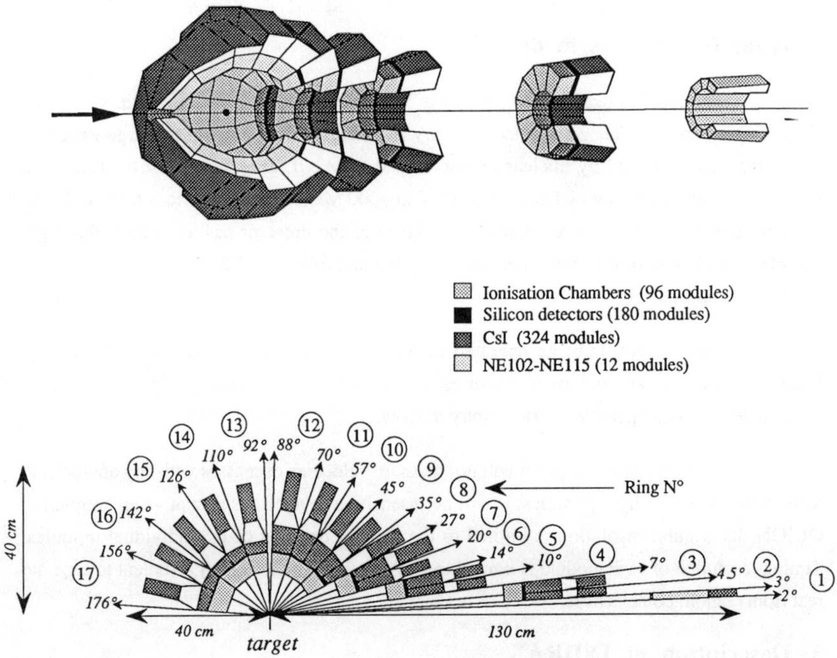


Figure 1.

For these reasons, the nuclear physics community working with the GANIL accelerator, has decided, in 1988, to build a new high performance $4\pi$ detector whose characteristics would satisfy the requirements given by their theoretical colleagues. This detector is named INDRA and the purpose of this presentation is to give a status report on the construction of this detector which is aimed to be operational by mid-1992.

## 2. What INDRA has to do.

The INDRA detector is therefore a "$4\pi$" detector covering ≈90% of the total solid angle. Beyond this, it is aimed to detect, identify and measure the energy of the large variety of nuclei that are produced by nuclear reactions at the GANIL energies. These include light particles with energies ranging from a few MeV to ≈200 MeV/A and heavy ions from 1MeV/A to more than 100 MeV/A. The angular modularity of the detector has to be such that light particle multiplicities of the order of 30 and heavy ion multiplicities of the order of 7 be detected with reasonable efficiency.

The detector is to be operated under vacuum conditions and has to be able to function, at least in part, with other detectors such as the neutron $4\pi$ ORION [7] detector, time of flight detectors and particle interferometry devices.

INDRA has its limits : it will not measure velocities or masses with any appreciable resolution except for light particles. It will be insensitive to neutrons except when coupled to ORION. Its angular resolution is limited to the angular opening of the individual modules. Timing resolution of each module is limited to 10nsec which is considered sufficient to separate real from random coincidences.

## 3. Description of INDRA

### 3.a The INDRA geometry.

INDRA is built with a mechanical design that has azimuthal revolution symmetry around the beam axis [8] (fig 1. and table 1). It is built as a series of rings, labelled 1 to 17 covering different and connected polar angular range. The "fragmentation" of the polar angular range is such as to balance the counting rate on each ring.

Each of these rings is azimuthally cut into a number of modules ranging from 8 to 24 depending on the ring considered. The type of detectors included in each module depends on the ring number.

# Description of the INDRA detector

| | Phoswich NE102 - NE115 | | | | | | | |
|---|---|---|---|---|---|---|---|---|
| Ring N° | $\theta_{min}$ (°) | $\theta_{max}$ (°) | N | $\Delta\phi$ (°) | e NE102 (cm) | e NE115 (cm) | $\Delta\Omega$ (msr) | d (cm) |
| 1 | 2 | 3 | 12 | 30 | 0.5 | 15 | 0.45 | 104 |

| | CsI | | | | | | Si | Ionisation Chamber | | | | |
|---|---|---|---|---|---|---|---|---|---|---|---|---|
| Ring N° | $\theta_{min}$ (°) | $\theta_{max}$ (°) | N | $\Delta\phi$ (°) | e (cm) | $\Delta\Omega$ (msr) | e (μm) | $\Delta\phi$ (°) | N | n CsI | d (cm) | $\Delta\Omega$ (msr) |
| 2 | 3 | 4.5 | 12 | 30 | 14 | 0.82 | 300 | 30 | 12 | 3 | 65.5 | 3.4 |
| 3 | 4.5 | 7 | 24 | 15 | 14 | 1.29 | 300 | | | | | |
| 4 | 7 | 10 | 24 | 15 | 14 | 2.10 | 300 | 30 | 12 | 4 | 31.5 | 11.4 |
| 5 | 10 | 14 | 24 | 15 | 14 | 3.59 | 300 | | | | | |
| 6 | 14 | 20 | 24 | 15 | 10 | 8.00 | 300 | 30 | 12 | 4 | 25 | 40.4 |
| 7 | 20 | 27 | 24 | 15 | 10 | 12.2 | 300 | | | | | |
| 8 | 27 | 35 | 24 | 15 | 10 | 18.7 | 300 | 30 | 12 | 4 | 12 | 96.0 |
| 9 | 35 | 45 | 24 | 15 | 10 | 29.3 | 300 | | | | | |
| 10 | 45 | 57 | 24 | 15 | 8 | 41.7 | No | 30 | 12 | 4 | 12 | 187 |
| 11 | 57 | 70 | 24 | 15 | 8 | 52.0 | No | | | | | |
| 12 | 70 | 88 | 24 | 15 | 6 | 79.8 | No | 30 | 12 | 2 | 12 | 160 |
| 13 | 92 | 110 | 24 | 15 | 5 | 79.8 | No | 45 | 8 | 3 | 12 | 239 |
| 14 | 110 | 126 | 16 | 22.5 | 5 | 93.5 | No | 45 | 8 | 4 | 12 | 340 |
| 15 | 126 | 142 | 16 | 22.5 | 5 | 76.6 | No | | | | | |
| 16 | 142 | 157 | 8 | 45 | 5 | 99.1 | No | 45 | 8 | 2 | 12 | 159 |
| 17 | 157 | 176 | 8 | 45 | 5 | 59.5 | No | | | | | |

Table 1.

Caption :

N : number of detectors per ring
e : Thickness of detector
$\Delta\Omega$ : Solid angle of detector
d : distance from target
n : number of CsI behind an ionisation chamber
$\theta$ : polar angle
$\phi$ : azimuthal angle

**Ring N°1 ($2° \leq \theta_{lab} < 3°$) :**

12 phoswich detectors (NE102+NE115) for fast counting rates at small angles. Ion identification has been checked to reach beyond Z=20. Identification thresholds are rather high due to the 0.5mm fast NE102 plastic thickness.

**Rings N°2 to 9 ($3° \leq \theta_{lab} < 45°$) :**

Each of these rings is divided into 24 sectors each made up of an axial ionisation chamber (depth=5cm, operated with CF4 or C3F8 at a pressure of 30torr) followed by a 300μm silicon detector backed by a Csi(Tl) detector of variable thickness (table 1). This stack of three different detector makes up the standard INDRA module. The combination of these detectors two by two will allow for the detection of products ranging from 200 MeV protons to Z≈30 ions of 1MeV/A. This very large dynamical range is usually obtained only by using stacks of a larger number of detectors. We have however paid special attention to the dynamical range of the associated electronics (≈1000 instead of a more conventional value of ≈100) and are now able satisfy the necessary requirements (see chapter 5). The coupling of the ionisation chamber and the silicon will detect and identify mostly low heavy ions. The coupling of the silicon and the CsI detectors will detect and identify slow light particles and fast heavier ions and the CsI detector alone will detect and identify the fast light particles (p,d,t,α,...)

Figure 2 gives a schematical view of a typical module. Mechanically, one may notice that the ionisation chamber is in fact common to 4 silicon detectors which are built on the same wafer.

On each of these rings, a special module is dedicated to calibration problems (essentially related to the non-linearity of the CsI(Tl) detectors) and a supplementary 2mm SiLi detector is introduced in front of the CsI(Tl) (fig 2). As each ring has azimuthal symmetry of revolution, a single calibration module should allow for the calibration of all the modules contained within the ring.

**Ring 10-17 ($45° \leq \theta_{lab} < 176°$) :**

For these rings, the modules are made up of an ionisation chamber followed by a Csi(Tl) detector. The reason for this configuration is that at these angles, fast heavy ions are not expected. However, if necessary for certain kinematical configurations, an intermediate 300μm Silicon detector could eventually be added as they are at more forward angles.

For the calibration modules, one per ring, a 50μm silicon and a 2mm SiLi detector are introduced in between the ionisation chamber and the CsI.

Altogether, INDRA is made up of :

# INDRA
# The detection Modules

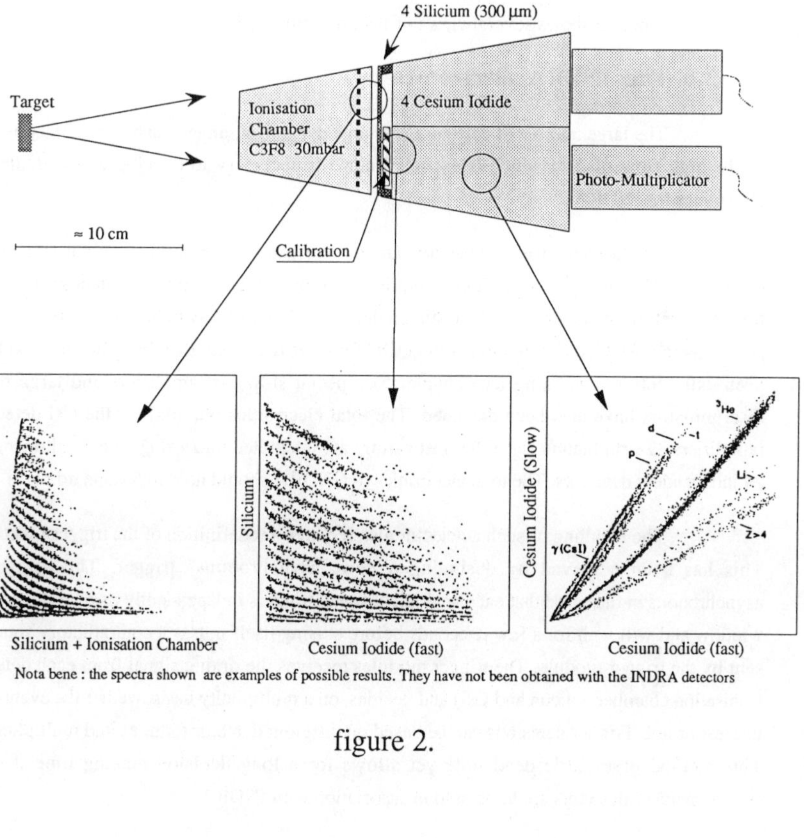


Nota bene : the spectra shown are examples of possible results. They have not been obtained with the INDRA detectors

figure 2.

12 Phoswichs detectors,
96 Ionisation chambers,
180 Silicon detectors,
324 CsI detectors,
24 Silicon calibration detectors

Figure 3 shows several views of the different rings.

## 3.b The INDRA electronics

The large number of detectors, the large dynamical range which is necessary as well as the high flow of data which is expected made it necessary to develop a new dedicated electronics for INDRA.

The large number of detectors and the intricate wiring implied a highly integrated electronics. The large dynamical range implied not only two gains, for each analogical signals form the ionisation chambers and the silicon detectors, but also a switching from conventional peak sensitive ADC to charge integration QDC for these detectors. Results of this method have been extremely encouraging (see chapter 5). Special slow-fast amplifiers and large range discriminators have also been designed. The total electronics required for the CsI detectors (amplifiers, discriminators and slow -fast gating) are integrated into a single board allowing for 24 independent detectors. The total electronics for INDRA should hold in 5 standard racks.

The handling of such a detector also requires a redefinition of the trigger functions. This has been achieved by designing a new "asynchronous" trigger. This trigger is asynchronous in the sense that each analogical signal triggers independently its own integration window and will wait for a few μseconds before erasing itself unless a contradictory signal is sent by the trigger module. The trigger modules receives the timing signal from each detector (ionisation chamber, silicon and CsI) and decides, on a multiplicity basis, wether the event is of interest or not. Trigger decisions can be based on different detector families and multiplicities. This method gives little dead time yet allows for a long decision making time if slow complementary detectors are to be used in association with INDRA.

Most of this electronics have been developed in the new VXI standard which is an extension of the VME bus adapted to precise analogical electronics.

This new standard has the advantage of allowing for large dimension boards. It has easy communication between modules and has a large supplementary BUS that can be dedicated for specific use. For example, this BUS is used for computer controled multiplexing of analogical signals present on the board. As it is an extension of the VME standard, it is fully compatible with this standard thus allowing fast readout times and is very easily integrated within the GANIL VME-VAX data acquisition system.

The entire electronics up to and including the VXI and VME crates will be positioned inside the experimental cave area. The different settings (high voltages, gas handling systems, scaler controls, gains, discriminator levels, trigger parameter...) will therefore be computer controled.

### 3.c Data acquisition and quality controls.

As has been stated above, the data acquisition will be controled by the standard GANIL system. However such a large detector with a total of 1550 independent channels requires some adjustments of the existing system. The number of possible identification matrices is in excess of 1500 and the amount of fast memory storage is of the order of 256 megabytes for spectrum storage. Even if such quantities are not out of the scope of present computers and large RAM disks for fast access, the permanent control of the quality of each detector is beyond the possibilities of the most dedicated physicist. We are therefore working on an automatic "quality control" software that should not only handle a survey of the different counting rates of the detectors (the ring design of INDRA largely facilitates this task since detectors on the same ring should count equally for inclusive data taking) but also the identification quality of each detector. Our present ideas on this topic concentrate on the automatic scanning of the matrices to determine the peak to valley ratio of the Z lines typically observed with $\Delta$E-E detectors (see chapter 5).

We will also be using an extensive test pulse system, analogical on the ionisation chamber and silicon preamplifiers and laser based directly on the CsI and plastic scintillators. This system will allow an automatic control of the electronic stability and linearity.

An independent software system is also designed to control the high voltages, the discriminator levels, the scalers, etc...

### 3.d Calibration procedures

The calibration of such a detector is a difficult task and the present outline of the methods that will be used is the following.

- On every ring, except ring N°1, a specific module is designed with special calibration Silicon detectors whose energy response is considered to be sufficiently well known to allow for self calibration using the "punch through" events. Having these modules should allow for the subsequent calibration of the other modules on the same ring by comparison of the spectra for each variety of charged ions detected.
- The pulser systems, triggered at regular intervals, will allow to monitor the stability and linearity of the electronics.

- The GANIL facility has the possibility of producing "cocktail" beams of p,d,t,α and heavier ions by stopping the incident beam in a thick carbon foil before the "α" spectrometer (fig 4). If sufficiently high intensity beams can be produced, secondary elastic and inelastic reactions on thick carbon targets, located at the target position of INDRA, will allow for the calibration of the modules both at forward and backward angles without dismounting the detector. Cross sections for protons of 30 MeV incident energy on $^{12}C$ (40mg/cm$^2$) show that a couple of hours of a $10^5$pps secondary beam should be sufficient for overall calibration.

## 4. Who does what for INDRA.

The INDRA detector is built by a joint effort of 4 laboratories.

The IPN laboratory of Orsay is taking care of all that is related to the CsI detectors : purchasing, factory quality controls, the VXI electronics and the laser pulser system. It is also in charge of part of the "calibration detectors" (50μm silicon and part of the 2mm SiLi) and is designing and constructing the VXI discriminator modules.

The D.Ph.N of Saclay is taking care of the equipment of the ionisation chambers which are mechanically constructed either by GANIL or by an independent company. It is also in charge of designing the preamplifiers, the amplifiers and the analogical pulser system.

The L.P.C. of Caen is in charge of design and construction of the reaction chamber which is already located in cave D5 at the GANIL site (fig 4.). It is also responsible for the phoswichs detectors and related VXI electronics as well as, in collaboration with the GANIL, of the trigger system.

The GANIL is responsible for the general co-ordination of the construction, the overall mechanical design, the mechanical construction of the ionisation chambers, the silicon detectors, the charge integration sensitive ADC, the acquisition system and the software control systems.

Figure 5. gives a schematical outline of these responsibilities.

## 5. Detector and electronic testing.

At this time, ring 4-5 has been almost completely constructed and tested during 2 beam periods.

The first beam period in June 1990 with a 60 MeV/A Kr beam in GANIL was used mostly to test the response of the silicon and CsI detectors. Figure 6.a shows a spectrum obtained at 4° in the laboratory and shows the ΔE-E matrix obtained using these detectors

# GANIL
## Accelerator and Experimental Areas

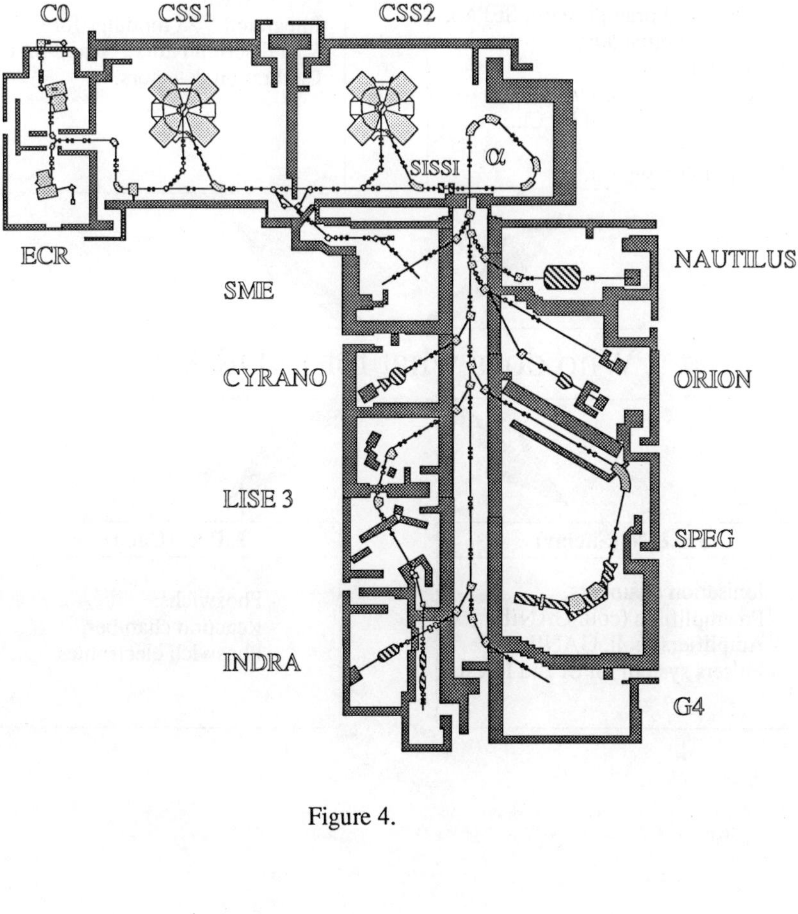


Figure 4.

**GANIL (Caen)**

- General coordination,
- Mecanical designs,
- Silicium and prampl. (coll. SEPN),
- D5 experimental area,
- Computer controls,
- Trigger (coll. LAL Orsay),
- VXI QDC for Si and Io. Ch.,
- Quality controls
- Simulation program.

**I.P.N. (Orsay)**

- CsI, Photo-mult., bases,
- Laser system,
- Integrated VXI modules for CsI,
- VXI discriminators : Si and Io. Ch.
- Calibration detectors.

## Who does what for INDRA ?

**S.E.P.N. (Saclay)**

- Ionisation chambers,
- Preamplifiers (coll. GANIL)
- Amplifiers (coll. GANIL)
- Pulsers system for Si and Io.Ch.

**L.P.C. (Caen)**

- Phoswichs,
- Reaction chamber,
- Phoswich electronics.

coupled to a standard peak sensitive ADC electronics for the silicon signal. The matrix shows that good Z identification has been achieved up to Z=37, essentially limited by the charge of the incident beam. In this case, the dynamical range obtained with the electronics associated with the Si detector is poor and lower Z values are clearly cut. Figure 6.b shows the same detectors associated with the charge integrating ADC. The lower Z thresholds has been eliminated. Figure 6.c shows an expansion on the lower part of this matrix showing that α particles, conditioned by their presence in the CsI detector, are clearly identified. Figure 6.d shows a non-expanded matrix obtained using a higher (x16) gain value. In this matrix, the α line is clearly separated. The proton line is not visible because the trigger threshold on the CsI electronics was set too high. Further improvement on this problem as well as on the amplifiers should allow for the detection of low energy protons and for better overall resolution.

The response of the CsI has also been found to be very linear as a function of light production and high quality light particle identification and energy measurements were obtained.

A later test, using a 220MeV (total) Si beam from the tandem-Booster facility in Saclay, was done to test the performances of the the Ionisation chamber-Silicon and Ionisation chamber-CsI modules. Figure 7.a shows the matrix obtained by the first module using the charge integration sensitive ADC. The inset shows an expansion of the lower part of the matrix, where the α line with a "punch trough" component is visible, and gives information on the threshold and resolution of the system.

The heavier ions correspond to Z≈30. Although not clearly visible on these figures, the high Z lines show, at low energies, the Bragg maximum indicating an energy threshold of less than 1MeV/A. Coupled to the double gain (x1 and x16) of the electronics, these results demonstrate that the detection dynamical range (≥1000) that was required is easily achieved and that the double or triple telescope modules of INDRA are sufficient to cover the necessary range.

Figure 7.b shows the results obtained with the Ionisation chamber-CsI telescope. Apart form deformation due to the non-linear response of the CsI to heavy ions, the quality of this matrix is sufficient to cover the energy and Z ranges expected at backward angles. figure 7.c shows the typical response of the CsI detectors where all light particles are identified and well separated from the heavier ions

## 5. Data analysis, simulation, and conclusions.

A large amount of work still remains to be done. Data analysis work has still not been started although the work that is performed on the calibration procedures and on the automatic quality controls should reduce the burden of this task. The "identical" response corresponding to detectors located on a given ring should simplify this problem.The goal for

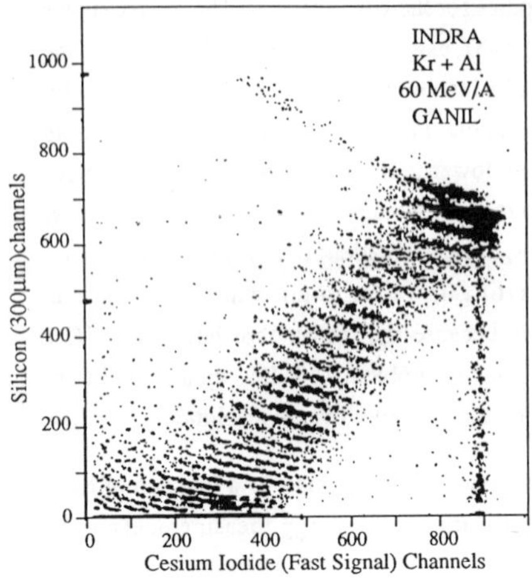


Fig. 6.a

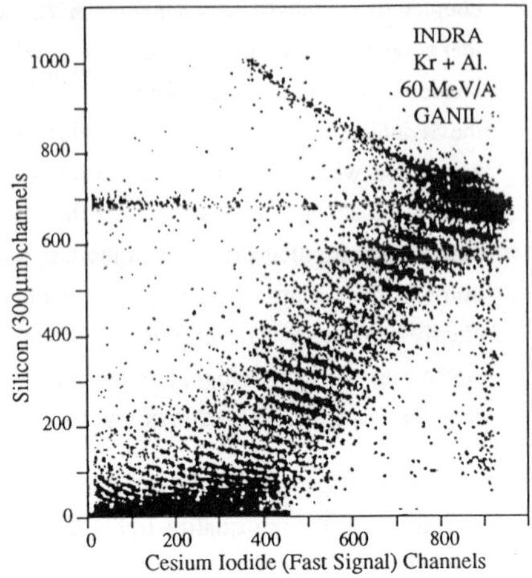


Fig. 6.b

Figure 6.a,b,c,d
Results of the tests performed
at the GANIL facility with
the INDRA modules

Fig. 6.c

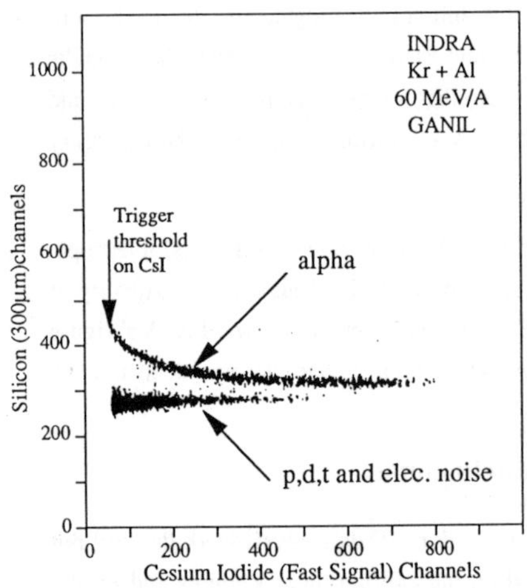


Fig. 6.d

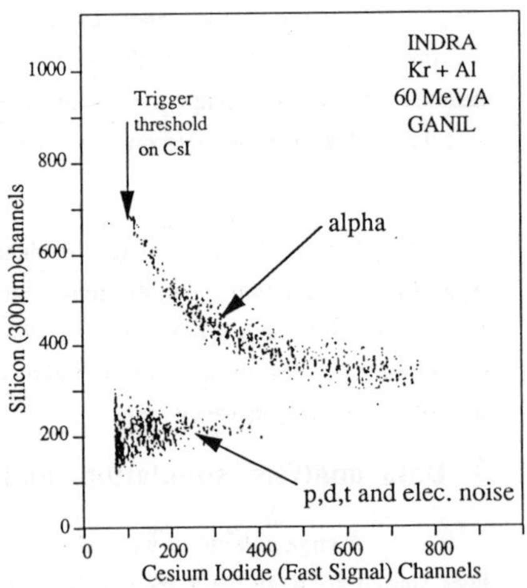

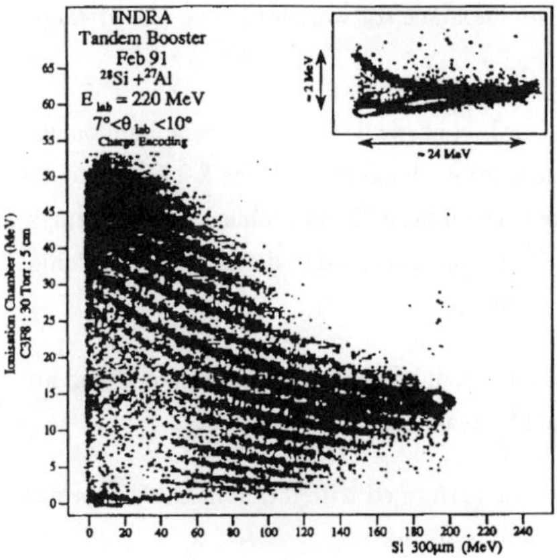


Fig. 7.a

Figure 7. a,b,c
Results of the tests performed
at the Tandem-Booster facility
in Saclay with the INDRA modules

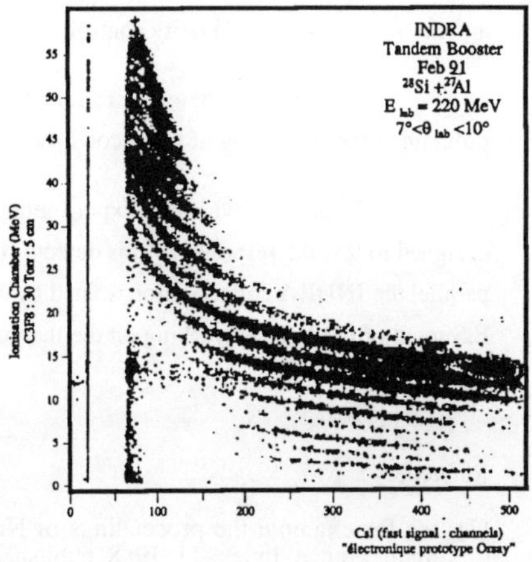


Fig 7.b

Fig 7.c

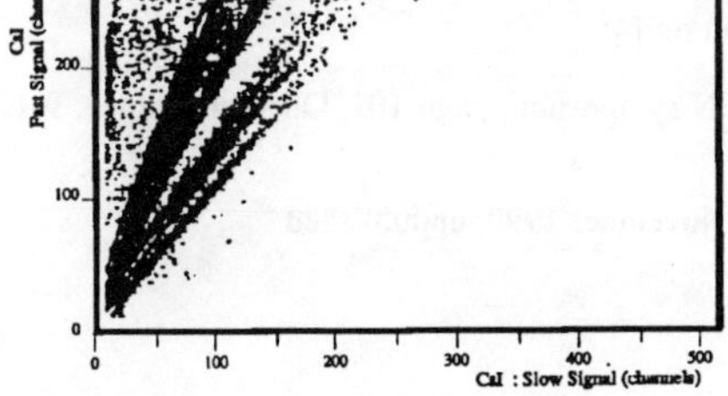

the near future is to deliver to INDRA users a semi-automatic software for Z-line identification and energy calibration.

As far as the construction of the detector is concerned, rings 1-7 are being mounted and should be ready for the end of 1991. The ionisation chambers of rings 8-17 represent a difficult task and prototypes are being constructed. The most difficult problem for these rings, is the glueing of the anode and cathode mylar foils that are also used to define the gas volume. Results on these problems should be available this summer.

Most of the prototypes of the electronic modules have been constructed and are being tested. The complete electronic set-up should be available early 1992.

Beginning of 1992, real-size tests will be performed with different GANIL beams and the acquisition and "quality control" software will be tested.

The INDRA detector is scheduled to be completely ready for the experimental program at the beginning of the second semester of 1992.

In parallel with this experimental work an INDRA simulation software has been designed to test the response of this detector to the predictions of various theoretical models. In parallel the INDRA collaboration, joined by X.Campi of the IPN Orsay and M.Ploszajczak and E.Suraud of GANIL, is working on the theoretical interpretation of future results.

References :


[1] see for example the proceedings of Nucleus-Nucleus Collisions III, page 217c, Saint Malo, France, June 6-11, 1988, published by North-Holland-Amsterdam

[2] L.G. Moretto, Nucl. Phys. A247(1975)211

[3] D.H.E GROSS in Nucleus-Nucleus Collisions III, page 217c, Saint Malo, France, June 6-11, 1988, published by North-Holland-Amsterdam

[4] X.Campi, Phys. Lett B352(1988)208

[5] M.Ploszajczak and A.Tucholski, GSI-90-40, Preprint, July 1990, accepted for publication in Nucl. Phys? A.

[6] W.Bauer, Phys. Rev C38(1988)1297 and ref [4]

[7] J.Galin et al. in the Second IN2P3-RIKEN symposium, page 102, Obernai, France, 9-12 April, 1990, published by World Scientific

[8] N.Copinet, University Thesis, Caen, 23 November 1990, unpublished.